I0057658

Searching for ORDER IN the COMPLEXITY of Evolving Worlds

ACKNOWLEDGMENTS

The SFI Press would not exist without the support of William H. Miller and the Miller Omega Program.

As part of a multi-member global initiative to reimagine political and economic theory, the Santa Fe Institute launched a new research theme on Emergent Political Economies (EPE) in 2022. Funded by the Omidyar Network, this research initiative has sought to develop new interdisciplinary frameworks and methods to better apprehend the core lynchpins that drive our political, economic, and social lives. At a 2023 EPE-sponsored workshop on "Complex-System Approaches to Twenty-First Century Challenges," scholars and practitioners gathered at the Santa Fe Institute to further extend our understanding of the economy as an evolving complex adaptive system. These latest volumes in the EECS series represent both the culmination of those discussions and an invitation to continue the conversation into the twenty-first century and beyond.

https://santafe.edu/EPE

THE ECONOMY AS AN EVOLVING COMPLEX SYSTEM IV

Volume One

R. MARIA DEL RIO-CHANONA
MARCO PANGALLO
JENNA BEDNAR
ERIC D. BEINHOCKER
JAGODA KASZOWSKA-MOJSA
FRANÇOIS LAFOND
PENNY MEALY
ANTON PICHLER
J. DOYNE FARMER

editors

© 2026 Santa Fe Institute
All rights reserved.

ᴘ꜀PR☙SS

THE SANTA FE INSTITUTE PRESS

1399 Hyde Park Road
Santa Fe, New Mexico 87501

The Economy as an Evolving Complex System IV, Vol. I

ISBN (HARDCOVER): 978-1-947864-66-5
Library of Congress Control Number: 2026930870

The SFI Press is made possible by the generous support
of the Miller Omega Program. These volumes were funded by a grant from the
Omidyar Network in support of the Emerging Political Economies theme.

Editorial Note

The order of editors in this book follows standard scientific publishing conventions. R. Maria del Rio-Chanona and Marco Pangallo are listed first, as they coordinated the editorial process. J. Doyne Farmer appears last, reflecting his guidance role. All other editor names are listed in alphabetical order.

IN ESSENCE OUR FAILURE was a vivid demonstration, which I have never forgotten, that theories, however plausible and "obviously" valid, can be destroyed totally by the obstinate facts of the real world.

HERBERT SIMON
Models of My Life (1996)

VOL. I
TABLE OF CONTENTS

Part I: Foundations of Complexity Economics

Part II: Methods & Concepts in Complexity Economics

INTRODUCTION

Penny Mealy, University of Oxford, Santa Fe Institute,
and Monash University;
Jenna Bednar, University of Michigan and Santa Fe Institute;
Eric D. Beinhocker, University of Oxford and Santa Fe Institute;
R. Maria del Rio-Chanona, University College London
and Complexity Science Hub;
J. Doyne Farmer, University of Oxford and Santa Fe Institute;
Jagoda Kaszowska-Mojsa, University of Oxford, Narodowy Bank
Polski,
and Institute of Economics, Polish Academy of Sciences;
François Lafond, University of Oxford;
Marco Pangallo, CENTAI Institute; and
Anton Pichler, Vienna University of Economics and Business
and Complexity Science Hub

In 1987, ten economists and ten natural scientists gathered at the Santa Fe Institute (SFI) for a ten-day workshop to explore a provocative idea: Could the global economy be understood as an evolving complex system? Few could have anticipated the significance of this meeting in igniting the development of a brand-new research field, now known as *complexity economics*. In reaching beyond equilibrium formulations to instead analyze the economy as an evolutionary system shaped by the interactions of diverse, adaptive and boundedly rational agents, the field sought to offer a more dynamic, realistic, and empirically grounded approach for understanding economic behavior. This shift in perspective would prove transformative, introducing new ways to understand everything from financial crises to technological change, to inequality and economic development.

What is Complexity Economics?

Complexity economics draws from the science of complex systems to understand the economy as an adaptive system where macro-level patterns emerge from the interaction of many diverse agents, rather than being imposed from above or derived from representative agents. The field shares with modern economics an increasingly empirical orientation but draws on interdisciplinary tools from physics, biology, computer science, and network theory to study economic phenomena across multiple spatial and temporal scales simultaneously. For instance, complexity economists pioneered the use of network analysis to map financial contagion, trace supply chain vulnerabilities, and understand how countries' productive capabilities shape their development paths. This empirical focus extends to uncovering universal patterns like power-law distributions in firm sizes, growth rates, and wealth: regularities that emerge across diverse economic contexts and can be explained through bottom-up mechanisms rather than top-down assumptions.

The emphasis on emergence naturally leads to a different approach to modeling economic behavior and dynamics. While agents in complexity economics may have goals or even utility functions, they are not assumed to achieve optimal outcomes. Instead, agents make decisions with limited information and cognitive capacity, attempting to pursue their goals but often falling short. This marks a conceptual shift: Agents are modeled as boundedly rational decision-makers who learn and adapt, whether through simple heuristics or more sophisticated learning algorithms—a perspective that aligns with behavioral economics research. This bottom-up approach reveals how realistic behavioral rules can generate sophisticated collective outcomes: technological innovation through combinatorial processes, the accumulation of productive capabilities that shape development trajectories (the economic complexity framework), and cascading effects that produce booms, busts, and structural transformations.

Agent-based modeling (ABM) is one of the most widely used tools for studying these dynamics (Axtell and Farmer 2025). These computer-based models simulate economies as constantly evolving processes where individual decisions aggregate into macro patterns that feed back to influence future behavior. Unlike traditional models, ABMs do not have to assume utility maximization. This allows them to simulate millions of interacting agents in complex environments that would be impossible to model with standard mathematical equations. ABMs can also capture economies out of equilibrium as they undergo transformations—whether adapting to technological disruption, navigating financial crises, restructuring in response to climate policies, or reorganizing during pandemics—generating endogenous cycles, tipping points, and regime shifts. Combined with advances in data science and machine learning, these methods are delivering falsifiable predictions and policy-relevant insights precisely when traditional models struggle most: during periods of disruption and transformation (Farmer 2024, 2025). The result is not just a set of different methodologies and techniques but a different way of seeing the economy: as a restless, evolving system whose patterns emerge from the interactions of millions of diverse, boundedly rational agents (Arthur 2021). It is this perspective that sets the stage for the field's contributions explored in this volume.

The History of This Series

The first volume of *The Economy as an Evolving Complex System*
(1988) marked a foundational shift in economic thinking. It
introduced a pioneering set of ideas and tools drawn from
nonlinear dynamics, evolutionary biology, neural networks, and
artificial life to analyze economic systems as decentralized,
adaptive, and out of equilibrium. Volume II (1997) deepened
this approach, expanding its empirical reach and emphasizing
features such as path dependence, bounded rationality, networked
interactions, and the co-evolution of institutions and behaviors.
By volume III (2005), complexity economics had grown into
a more sophisticated and multidisciplinary field, offering new
insights into finance, technological innovation, macroeconomics,
social interaction, and increasingly influencing conversations
beyond academia.[1]

Since then, the field has matured significantly (see Beinhocker
et al. 2026, ch. 2 in this volume, for a brief account of the
history of the field). Its intellectual foundations have deepened, its
methodological toolkit has become more refined and rigorous, and
its practical relevance to real-world policy has grown increasingly
evident (Arthur 2021). What began as an exploratory workshop at
the Santa Fe Institute has since evolved into an active community
of researchers and practitioners engaged in universities, think
tanks, central banks, and international institutions all around the
world.

Yet just as the field has evolved, so, too, has the complexity of
the global challenges it seeks to understand.

An Evolving Complex World: New Challenges,

---

[1]Beinhocker and Bednar (ch. 30 in this volume) provide a comprehensive
overview of the intellectual history and evolution of complexity economics,
from its founding origins at the Santa Fe Institute to its maturation as a
global, multidisciplinary field.

New Imperatives

Today's global economy is undergoing profound structural transformation. The race to reach net-zero emissions is well under way, as devastating climate impacts are becoming more frequent and severe around the world. The pace of technological change has never been faster or more consequential, with advances in areas like artificial intelligence poised to both revolutionize human productivity and upend jobs and sectors across the economy. Meanwhile, shifting balances of economic power are intensifying geopolitical tensions, disrupting global supply chains, and increasing the risk of conflict. At the same time, growing social divides between rich and poor, urban and nonurban dwellers, and those with differing levels of education continue to stretch the social fabric that holds our communities together, an effect amplified by polarizing discourse on social media platforms.

These changes are not marginal or linear processes. They are complex, adaptive, and path-dependent phenomena: precisely the kind of dynamics that complexity economics was designed to illuminate and understand.

This raises urgent questions: Can complexity economics rise to the challenge and meet the moment? To what extent can the field address such pressing global issues? Does viewing and analyzing the economy as an evolving complex system provide unique advantages over more traditional perspectives? And can the analytical approaches the field has heavily invested in— like agent-based models—provide more helpful insights than mainstream economic models?

These questions guided a recent workshop at the Santa Fe Institute, which provided the foundation of this book. Titled "Complex System Approaches to Twenty-First-Century Challenges: Inequality, Climate Change, and New Technologies," this workshop explored the applicability of the field's recent advances to real-world challenges and discussed key barriers in continuing to become more policy-relevant and impactful.

The workshop was one of the largest in the Santa Fe Institute's history, involving over sixty participants from a diverse range of disciplines and backgrounds. It involved ten sessions on various topics such as "Risk and Resilience in the Twenty-First Century: Are Our Analytical Approaches Up to the Challenge?"; "The Dynamics of Inequality: What Amplifies Unequal Outcomes and What Can Be Done About It?"; "How to Drive an Orderly Rather Than a Disorderly Green Transition"; "Fragmenting or Flourishing: How to Ensure Our Social Fabric Can Address Twenty-First-Century Challenges"; and "What Is the State of the Art in Complexity Economics and What Are the Most Pressing Methodological Challenges?"

This volume contains contributions from the workshop and beyond. The chapters present different perspectives on topics discussed in the workshop, as well as reflections on how the field of complexity economics has evolved and is progressing, and summaries of progress and future outlooks for key research strands within the discipline.

Methodological Advances: Machine Learning and the New Frontier of Agent-Based Models

Since volume III, one of the most significant and exciting methodological advances has been in agent-based modeling. While agent-based models have a long history (Richiardi, van de Ven, and Bronka, ch. 7), much of which has been advanced by researchers at the Santa Fe Institute, their adoption by mainstream economists and policymakers has remained somewhat limited. Early ABMs were often criticized for being overly complicated, containing too many free parameters and arbitrary assumptions, and for being difficult to calibrate or validate. These challenges made them difficult to compare and interpret, particularly against the backdrop of dominant equilibrium-based models (Pangallo, ch. 4).

However, a confluence of developments over the past decade has begun to shift this landscape. Increases in computational power, greater availability of high-resolution micro-level data (Borsos *et al.*, ch. 17), and methodological advances in estimation, calibration, and validation have opened the door to more empirically grounded, data-driven ABMs (Pangallo and del Rio-Chanona, ch. 9). These models increasingly exhibit greater rigor, transparency, and tractability, which is making them more attractive and user-friendly in applied policy settings.

And, indeed, many chapters in this volume reflect the growing influence and applicability of ABMs across a wide range of real-world contexts. In labor markets, ABMs are helping to model skill mismatches, unemployment, and the impact of technological change on workers (del Rio-Chanona *et al.*, ch. 23). In housing markets, they are offering insights into the emergence of bubbles, lending and borrowing behavior, and household responses to climate-related flood risks (Pangallo and del Rio-Chanona, ch. 9). In finance, ABMs are being used to study the emergence of stock-market crashes, heterogeneous expectations, and systemic risk (Borsos *et al.*, ch. 17). In climate and environmental policy, ABMs are increasingly used to explore the behavioral and network dynamics of risk propagation, insurance uptake, and adaptation strategies under uncertainty (Lamperti, Dosi, and Roventini, ch. 18; Filatova and Akkerman, ch. 21). And in macroeconomics, they are being employed to study production networks, consumption dynamics, inflation, and the coordination challenges inherent in fiscal and monetary policy (Dawid *et al.*, ch. 12; Hommes *et al.*, ch. 13).

Since volume III, these advances in agent-based modeling have been paralleled by dramatic improvements in data science, machine learning, and artificial intelligence. As Arthur Turrell (ch. 10) demonstrates, data-science techniques are transforming how policymakers measure complex economies, from web scraping to tracking price changes for vulnerable households to using computer vision for real-time crisis monitoring. These improvements in machine learning are also directly enhancing ABM capabilities, with algorithms increasingly used to empirically calibrate models (Dyer *et al.* 2022) and studies exploring whether large language models can be used to set realistic agent behaviors (del Rio-Chanona, Pangallo, and Hommes 2025).

Overall, these developments signal a shift in how ABMs are positioned within the field. What were once viewed as exploratory simulations are now becoming serious contenders: powerful tools capable of delivering falsifiable hypotheses, credible forecasts, and actionable policy insights.

Complexity at the Frontlines: Crisis, Contagion, and Criticality

In recent years, a series of global disruptions have thrust complexity economics into the forefront of policymaking. These events have not only tested the limits of traditional economic models but also showcased the unique strengths of complexity-based approaches in times of crisis.

The COVID-19 pandemic was a defining moment. As the virus spread across borders and sectors, policymakers from all nations were forced to grapple with nonlinear contagion dynamics, cascading supply shocks, and the challenge of managing economies through states of profound disequilibrium. In response, the complexity community mobilized rapidly: Within weeks, new out-of-equilibrium models were developed, designed specifically to predict how lockdowns would impact different industries (del Rio-Chanona *et al.* 2020), how these impacts were likely to propagate through economies (Pichler *et al.* 2020; Reissl *et al.* 2022), and how economies could be reopened in a way that reduced infection rates while increasing economic output (Pangallo *et al.* 2024). In addition to providing policymakers with strategic insights that were unattainable by conventional equilibrium-based models of the economy, a later ex-post analysis showed that the forecasts made by the modeling efforts were impressively accurate (Pichler *et al.* 2022).

In the pandemic's aftermath, inflation surged and central banks found themselves navigating unfamiliar terrain. Long reliant on dynamic, stochastic, general equilibrium (DSGE) models and econometric projections, several institutions began to explore alternatives better suited to uncertain, rapidly shifting environments. The Bank of Canada, for example, developed one of the first agent-based models specifically tailored for inflation targeting, macroeconomic forecasting, and policy analysis (Hommes *et al.*, ch. 13). This model, featuring boundedly rational households and firms interacting in a production network calibrated to the Canadian economy, not only outperformed DSGE models in forecasting gross domestic product (GDP) growth and consumption but also yielded unique insights into the key drivers of Canada's inflation surge.

Beyond inflation, complexity-based methods have proven increasingly useful for understanding systemic financial risk. For example, network models of financial contagion have shown how structural properties of interbank networks can amplify the risk of cascading defaults (Caccioli, ch. 15). And in an impressive review of ABM-relevant research and policy outputs of twenty-four central banks and seven related institutions, András Borsos, Adrian Carro, Aldo Glielmo, Marc Hinterschweiger, Jagoda Kaszowska-Mojsa, and Arzu Uluc (ch. 17) document the various contributions such models have made to help central banks perform stress testing, better understand the impacts of macroprudential policies, and monitor and analyze new risks arising from cybersecurity threats, cryptocurrencies, and the net-zero transition.

Complexity economics has also illuminated the vulnerability and criticality of global supply chains. While recent extreme weather events, geopolitical tensions, and disruptions from the COVID-19 pandemic and Russia's invasion of Ukraine have exposed the fragility of these networks, research has also shown that incentives within the system, like just-in-time inventory-management policies, can drive the system toward a highly critical state, such that even a small production delay can cascade through the network, generating major disruptions (Bouchaud, ch. 8). As has also been illustrated in financial systems (Caccioli, ch. 15), an uncomfortable reality facing policymakers is that many interventions aiming to drive greater efficiency may also weaken overall system resilience.

Economic crises may also occur without any external forcing, arising instead from endogenous dynamics of the economy, where each boom contains the seeds of a subsequent bust. Paul Beaudry, Dana Galizia, and Franck Portier (ch. 14) discuss research on endogenous business cycles and present a model that has strongly influenced the revival of the idea that exogenous shocks are not the sole drivers of economic fluctuations. Mathematically, such models exhibit forms of nonlinear dynamics such as limit cycles or chaos, which represent one of the key contributions of complexity-based methods to economics (Brock and Hommes, ch. 5).

Climate Change: Catalyzing the Case for Complexity

Several chapters in this book detail the contributions complexity economics is making to address climate change. Once perceived as a distant concern, climate change is now a defining force shaping lives, livelihoods, and the global economy. Its physical impacts, ranging from floods, fires, and storms to droughts, heat waves, and rising sea levels, are already impacting key sectors like agriculture, fishing, and tourism while placing mounting stress on infrastructure and public health systems.

Yet many traditional models in climate economics have struggled to capture the full economic severity of these risks. Standard integrated assessment models often assume gradual, marginal impacts, failing to reflect how shocks propagate through interconnected systems. In reality, damage to a single node, whether a household, firm, or region, can ripple outward through supply chains and financial networks, triggering broader effects such as migration, social unrest, or political destabilization (Filatova and Akkerman, ch. 21).

Moreover, many standard models poorly account for climate tipping points, such as the loss of Arctic sea ice or the thawing of permafrost, which can unleash nonlinear, self-reinforcing feedback loops that rapidly accelerate warming (Battiston and Monasterolo, ch. 19). As a result, these models have often underestimated potential damages. For instance, while prominent economists have suggested that 3.5°C of warming by the end of the century might be "optimal" from a cost–benefit perspective (Nordhaus 2018), most climate scientists argue that such a level would entail catastrophic consequences (Lamperti, Dosi, and Roventini, ch. 18).

New transition risks are also emerging with the introduction of climate policies, green technology investment, and shifting preferences away from emissions-intensive production. These developments can drive abrupt shifts in asset values, sectoral disruption, and major changes in global comparative advantage. Here, too, traditional models struggle to capture path-dependence, technological lock-in, investor uncertainty, asset stranding risks, and associated impacts on financial and labor markets—all of which shape the speed and smoothness of the green transition (Dumas and Andres, ch. 20). However, several new modeling approaches described in this volume better reflect the dynamics of cascading shocks, multiple equilibria, tipping points, and investor uncertainty, offering more realistic assessments of both physical and transition risks (Filatova and Akkerman, ch. 21; Battiston and Monasterolo, ch. 19; and Lamperti, Dosi, and Roventini, ch. 18).

Fortunately—and importantly—there are also solid empirical grounds for optimism. Recent contributions in complexity economics have demonstrated that climate action is not just a story of higher risks and costs but one of immense opportunity. Building on a rich body of work on technological forecasting (Lafond, ch. 25), Rupert Way *et al.* (2022) project that rapid transition to a net-zero energy system could generate $12 trillion in savings when compared to the cost of maintaining current fossil-fuel systems. These findings stand in sharp contrast to projections from many integrated assessment models used by governments and the Intergovernmental Panel on Climate Change, which have historically overestimated the costs of renewable energy and underestimated the pace of technological progress (Farmer *et al.* 2015). As this volume demonstrates, incorporating nonlinear dynamics, tipping points, and empirically grounded technological learning curves can fundamentally reshape how we assess the costs, benefits, and urgency of climate action.

Beyond the Black Box: The Process and Predictability of Technological Evolution

Beyond its role in addressing climate change, technological progress is a powerful driver of economic development and societal transformation. While mainstream economic models often treat it as an exogenous and largely unexplained source of productivity growth, key strands of work within complexity economics have sought to push conceptual, empirical, and modeling frontiers to better understand how technology evolves and how it might be better guided toward societal goals.

On the empirical front, François Lafond (ch. 25) outlines a recent wave of efforts to track cost-improvement trends across technologies and develop forecasting models grounded in observable patterns. Two statistical laws offer alternative lenses: Moore's law predicts costs will decline predictably over time, while Wright's law suggests costs fall with experience (i.e., learning-by-doing). The distinction has important implications for technological investment and innovation strategies. If Moore's law holds, we need to wait for progress; if Wright's law is more accurate, proactive investment and production subsidies can directly accelerate technological advancement. Leveraging World War II as a natural experiment, where production was driven by military necessity rather than market demand, Lafond, Diana Greenwald, and Farmer (2022) find that roughly 50% of cost improvements can be causally attributed to production-driven learning. This finding underscores the role that deliberate policy choices can play in shaping both the pace and the direction of technological progress.

Building on a lineage of important contributions on the nature of technological evolution (Arthur 2009), W. Brian Arthur (ch. 2) sheds new light on the process by which invention occurs. Transformative technologies, such as computers, X-ray machines, or the internet, don't emerge in a vacuum; they evolve through a process of combination and recombination of existing technologies. The global positioning system (GPS), for instance, delivers precise location data by integrating a suite of earlier technologies, including satellites, atomic clocks, radio transmitters, and mathematical algorithms, each of which built on prior technological breakthroughs. Novel technologies are then "encapsulated" and in turn become "modules" or building blocks for future generations of technologies, just as GPS itself has become a foundational building block for further innovations, powering technologies from smartphone navigation to drones.

Arthur describes technological innovation as a "generative system" that operates on two time scales: one fast, where existing elements are routinely combined for immediate purposes, and one slow, where novel elements occasionally enter the toolbox (sometimes from scientific discovery, sometimes from trial and error), driving long-run evolutionary change. This process is recursive, and the combinatorial space of possibility is continually unfolding (autopoietic), branching (or "splintering") into new domains, and effectively unbounded. As such, technological evolution has more in common with other combinatorial systems such as chemical synthesis, mathematical proofs, linguistic neologisms, legal systems, and software development (Valverde, Vidiella, and Duran-Nebreda, ch. 27) than an evolutionary perspective based on biological Darwinian descent with modification (though Arthur notes the combinatorial process is "complementary" with Darwinian evolution, and some evolutionary economists would likely view Arthur's account as providing a mechanism for variation within a broader, algorithmic view of "generalized Darwinism," e.g., Hodgson and Knudsen 2010; Beinhocker 2011).

Pathways to Progress: Capabilities, Complexity and the Dynamics of Growth

A central challenge in translating technological progress into broad-based growth and development outcomes across countries and regions is that innovation is often spatially bounded (Coyle, ch. 26). As a result, the ideas, industries, and infrastructure embedded in a particular place can shape—and often limit—the range of future possibilities. As shown by a rich and growing body of research on economic complexity and economic geography (Frenken and Neffke, ch. 28), places are more likely to generate new innovations, develop new products, or cultivate competitive sectors in areas that are closely related to their existing knowledge base.

~ 15 ~

This path-dependency arises in part because the knowledge required for innovation and industrial development is, as Friedrich Hayek (1945) famously noted, "beyond the span of the control of any one mind." It is instead distributed across networks of people, firms, and institutions. The depth of specialization is limited not just by the extent of the market (as Adam Smith [1775] argued) but also by the costs of coordinating complex, distributed knowledge. Even in an increasingly digital world, spatial proximity continues to play a key role in lowering these costs, enabling dense localized networks to combine and apply distributed expertise more effectively. While economists throughout the ages have, in various ways, articulated the importance of collective know-how embedded in places (Smith 1775; Veblen 1898; Hayek 1945; Lall 1992), only recently has the availability of granular data on patents, products, and skills enabled places' productive capabilities to be empirically analyzed (Neffke *et al.*, ch. 6; del Rio-Chanona *et al.*, ch. 23). Despite ongoing debates about the merits of different measurement approaches, studies within complexity economics consistently show that places with more complex or sophisticated productive activities tend to experience higher rates of economic growth (Neffke *et al.*, ch. 6).

In parallel, complexity economics has advanced new approaches to modeling economic growth and firm dynamics that better reflect observed empirical patterns. Traditional models, such as those following Gibrat's law (Gibrat 1931), assume that firm growth is random and proportional to size. Yet these assumptions often generate predictions, such as lognormal firm size distributions, that diverge from real-world data, which more closely follow heavy-tailed, Pareto-like distributions. To address such discrepancies, Robert Axtell and Omar Guerrero (ch. 11) propose a new stochastic theory of firm growth based not on multiplicative expansion but on labor reallocation dynamics. In their model, workers move between firms in search of better employment opportunities, making one firm's gain another's loss. This labor-flow mechanism departs from the independence assumptions of Gibrat-style models and successfully reproduces empirically observed firm-size distributions. In related work, José Moran and Massimo Riccaboni (ch. 24) explore compositional growth models, which explain aggregate growth by decomposing economies or sectors into smaller, independently evolving components—such as submarkets, firms, or products. These models help account for key empirical regularities, including the fat-tailed distributions of firm sizes and growth rates, that standard aggregate models struggle to replicate.

The Power of Feedback: How Complex Systems can Amplify Inequality and Undermine Democracy

An overarching theme that permeated all topics at the 2023 Santa Fe Institute workshop was the intertwined challenge of political economy, inequality, and democratic resilience.

Power dynamics run rife in many, if not most, twenty-first-century challenges, and even the most carefully designed policy solutions are often constrained or resisted by political realities. And with so many global developments—be it the rapid advance of artificial intelligence, the upheavals of climate change, or shifting geopolitical orders—carrying risks of further widening divides, there is an urgent need to better understand how to promote fairness, inclusion, and the long-term stability of democratic institutions.

Complexity economics offers a unique and powerful lens to understand inequality, not as a static distribution but as a dynamic, endogenous feature of complex adaptive systems. From the earliest agent-based models, such as Joshua Epstein and Robert Axtell's (1996) Sugarscape model, researchers have shown how inequality can emerge not just from differences in individual effort or ability but from the structure and dynamics of the system. A combination of asymmetries (including luck), path dependence, and the compounding processes common in economic systems (e.g., returns on capital, social network effects) can cause individual agent trajectories to diverge, creating emergent distributions with characteristics (e.g., power-law upper tails) that mirror empirical distributions of income and wealth (Yakovenko and Rosser, Jr. 2009; Li, Boghosian, and Li 2019; Palagi *et al.* 2023). Importantly, these disparities can emerge even under conditions of equal opportunity, revealing the limits of simple policy prescriptions that do not address systemic and structural drivers of inequality.

Moreover, complex adaptive systems can act as inequality amplifiers. Small initial differences, whether in wealth, education, or social connections, can be magnified through positive feedback loops and become self-reinforcing over time (Roithmayr 2014;

Trounstine 2018). Steven Durlauf, David McMillon, and Scott Page (ch. 22) show how one form of disadvantage can compound and reinforce others across different spheres, creating "system of system" effects. An unsafe environment may undermine health outcomes, leading to poorer school performance, and increased unemployment risk, which in turn can worsen health outcomes—a vicious cycle that compounds disadvantage. People suffering from poor health and economic prospects are also less likely to place trust in government and social services, further compounding exclusion and making effective policy interventions even harder to deliver.

~ 18 ~

Reducing inequality therefore requires more than redistribution. It demands interventions that disrupt feedback loops and reshape institutional structures that perpetuate disparities. As Durlauf, McMillon, and Page discuss, this could take several forms: weakening adverse features of a system (e.g., levying a wealth tax to reduce intergenerational income transfers to disrupt poverty traps in economic systems), dismantling structures that reinforce inequality (e.g., abolishing private prisons), or harnessing positive feedback loops to amplify equality-focused interventions (e.g., a policy that enhances access to capital markets for minorities). And since inequality is often produced by multiple interdependent systems, coordinating policy efforts across multiple domains is critical.

To inform such systemic approaches, policymakers require tools that reliably capture the structural characteristics and reinforcing mechanisms that propagate inequality in an economy. One example is described in Giovanni Dosi, Marcelo Pereira, Andrea Roventini, and Maria Enrica Virgillito (ch. 29), which outlines an extension of a multisector labor-augmented, general-disequilibrium stock-flow consistent ABM to investigate the effect of declining unionization on inequality in the United States. In addition to aligning with several empirical stylized facts in the US, the simulations suggest that declining unionization induces higher macro-level inequality, greater wage dispersion between firms, and stronger polarization in wage growth dynamics. Models like this serve as valuable policy laboratories, enabling governments to test reforms in silico, assess their capacity to disrupt entrenched inequality, and reduce risks before implementing policies in the real world.

Like economies, democracies are also complex adaptive systems, marked by feedback loops, network effects, and emergent dynamics. These systems perspectives help explain why identical constitutions can yield very different political outcomes (Putnam 1994; Elkins, Ginsburg, and Melton 2009; Bednar and Page 2018) and how seemingly stable democracies can be hollowed out and rendered fragile (Acemoglu and Robinson 2005; Ginsburg and Huq 2018). Polarization provides a vivid example: Models of elite opinion dynamics show how asymmetric feedback can drive runaway partisan divergence (Leonard *et al.* 2021), while agent-based simulations demonstrate how intolerance, exposure, and structural inequalities interact to push societies past tipping points into extreme polarization (Axelrod, Daymude, and Forrest 2021). ABMs have also been used to simulate protests and social movements, shedding light on the conditions that can give rise to uprising, civil unrest, and ultimately revolutions (Epstein 2002; Makowsky and Rubin 2013; Moro 2016; Thomas *et al.* 2025).

Democratic robustness depends on many of the same system properties that underpin resilience in other domains. Cross-cutting social ties and integrated information networks cultivate shared understanding and foster a sense of collective fate, while fragmented communication silos erode trust and weaken democratic norms (Horowitz 1985; Centola and Macy 2007; Axelrod, Daymude, and Forrest 2021; Bednar 2021). Diversity of representation expands the menu of solutions, redundancy in overlapping institutions provides important safeguards, and adaptability allows recalibration in response to shocks (Levin 2000; Bednar 2009). And democratic durability cannot rest on institutional design alone. Investing in social infrastructure such as libraries, parks, community centers, and street festivals is critical for sustaining cross-group connections, protecting the integrity of information systems, and building flexible arrangements that evolve as conditions change. Strengthening these elements is essential to ensuring that democracies remain resilient, legitimate, and capable of withstanding mounting pressures.

Shifting the Economic Paradigm: The Future of Complexity Economics

As this volume illustrates, complexity economics has evolved from a provocative workshop into a rich and empirically grounded field—one that is increasingly demonstrating its relevance for understanding and shaping economic issues in the twenty-first century. But the case for complexity economics is not just about better tools for addressing present-day challenges. At its core, it represents an opportunity to rethink the foundations of the economic paradigm.

Eric D. Beinhocker and Jenna Bednar (ch. 30) argue that complexity economics is well-placed to provide the scientific backbone for a new economic paradigm: one that better connects moral values, behavioral realism, systemic understanding, and institutional design. This involves more than refining the assumptions of economic models. In Beinhocker and Bednar's formulation, paradigms can be understood in terms of an "ontological stack"—a set of layered, mutually reinforcing ideas spanning from moral foundations to theories of behavior and economic systems to public narratives and practical policy applications. Complexity economics, they argue, has enormous potential to contribute to scientifically advancing the middle layers of this stack: the theories of behavior, economic systems, and processes of change that are critical for connecting normative goals to real-world outcomes. Without strong explanatory foundations, even the most well-intentioned paradigms falter. Indeed, as history has shown (from socialism to neoliberalism), flawed economic theories can lead to failed systems and unintended harm.

Of course, modern-day economics has changed significantly since the 1987 SFI meeting and is no longer the neoclassical, largely fact-free field that was critiqued in that meeting. Mainstream economics is now far more empirical and is increasingly grappling with the messy complexities of the economy that motivate complexity economics researchers. Yet, as Farmer (2024) argues, complexity economics departs not just in method but in fundamental theoretical foundations—seeking to replace core assumptions like utility maximization, rational expectations, and equilibrium with more realistic representations of economic behavior and dynamics. In this sense, complexity economics represents a revolutionary shift in how we theorize, simulate, and make sense of economic life.

This shift invites a more reflexive, adaptive, and dynamic economics—one capable of tractably grappling with systemic fragility and flexibility, innovation and inequality, and growth within environmental limits. It resonates with political economy movements that seek to move beyond both neoliberal market fundamentalism and rigid state control, and toward more pluralistic, democratic and ecologically responsible economic models. It aligns with contemporary calls for progress metrics beyond simplistic aggregate measures like gross domestic product, for institutions that foster collective intelligence and wisdom, and for public narratives that are grounded in cooperation, dignity, and resilience.

Such a major change won't happen overnight. Paradigm shifts can be generational projects. They require the demanding work of community building, methodological innovation, empirical validation, and interdisciplinary collaboration and synthesis. They also require imagination. The paradigm we build must not only better explain the world, it must help us envision and create a better one.

The chapters in this volume represent building blocks of such a shift. They do not offer a single blueprint. But together they sketch the contours of an emerging economic worldview: one that takes seriously the complexity of our challenges, the diversity of our societies, and the evolutionary nature of change. As we move forward, the task is not simply to refine the research agenda of complexity economics but to advance it as part of a broader effort to reshape the economic paradigm for a complex and uncertain century. ❧

Acknowledgments

The authors would like to kindly acknowledge the Santa Fe Institute's grant from the Omidyar Network on Emerging Political Economies, as well as funding from Baillie Gifford, the Open Society Foundation, Marie Skłodowska-Curie (H2020) grant no. 101023445, NAWA Bekker grant no. BPN/BEK/2024/1/00240, and the UKRI through ESRC grant PRINZ (ES/W010356/1).

REFERENCES

Acemoglu, D., and J. A. Robinson. 2005. *Economic Origins of Dictatorship and Democracy*. Cambridge, UK: Cambridge University Press.

Anderson, P. W., K. Arrow, and D. Pines, eds. 1988. *The Economy as an Evolving Complex System*. Boston, MA: Addison-Wesley.

Arthur, W. B. 2009. *The Nature of Technology: What it is and How it Evolves*. New York, NY: Free Press.

———. 2021. "Foundations of Complexity Economics." *Nature Reviews Physics* 3:136–145. https://doi.org/10.1038/s42254-020-00273-3.

Arthur, W. B., S. Durlauf, and D. Lane, eds. 1997. *The Economy as an Evolving Complex System II*. Reading, MA: Addison-Wesley.

Axelrod, R., J. J. Daymude, and S. Forrest. 2021. "Preventing Extreme Polarization Of Political Attitudes." *Proceedings of the National Academy of Sciences* 118 (50): e2102139118. https://doi.org/10.1073/pnas.2102139118.

Axtell, R. L., and J. D. Farmer. 2025. "Agent-Based Modeling in Economics and Finance: Past, Present, and Future." *Journal of Economic Literature* 63 (1): 197–287. https://doi.org/10.1257/jel.20221319.

Bednar, J. 2009. *The Robust Federation: Principles of Design*. Cambridge, UK: Cambridge University Press.

———. 2021. "Polarization, Diversity, and Democratic Robustness." *Proceedings of the National Academy of Sciences* 118 (50): e2113843118. https://doi.org/10.1073/pnas.2113843118.

Bednar, J., and S. E. Page. 2018. "When Order Affects Performance: Culture, Behavioral Spillovers, and Institutional Path Dependence." *American Political Science Review* 112 (1): 82–98. https://www.jstor.org/stable/26542118.

Beinhocker, E. D. 2011. "Evolution as Computation: Integrating Self-Organization with Generalized Darwinism." *Journal of Institutional Economics* 7 (3): 393–423. https://doi.org/10.1017/S1744137411000257.

Blume, L., and S. Durlauf, eds. 2005. *The Economy as an Evolving Complex System III*. Oxford, UK: Oxford University Press.

Centola, D., and M. Macy. 2007. "Complex Contagions and the Weakness of Long Ties." *American Journal of Sociology* 113 (3): 702–734. https://doi.org/10.1086/521848.

del Rio-Chanona, R. M., P. Mealy, A. Pichler, F. Lafond, and J. D. Farmer. 2020. "Supply and Demand Shocks in the COVID-19 Pandemic: An Industry and Occupation Perspective." *Oxford Review of Economic Policy* 36 (S1): S94–S137. https://doi.org/10.1093/oxrep/graa033.

del Rio-Chanona, R. M., M. Pangallo, and C. Hommes. 2025. *Can Generative AI Agents Behave Like Humans? Evidence from Laboratory Market Experiments.* arXiv preprint: 2505.07457. https://doi.org/10.48550/arXiv.2505.07457.

Dyer, J., P. Cannon, J. D. Farmer, and S. M. Schmon. 2022. "Calibrating Agent-Based Models to Microdata with Graph Neural Networks." In *ICML Workshop on AI for Agent-Based Modelling.* https://doi.org/10.48550/arXiv.2206.07570.

Elkins, Z., T. Ginsburg, and J. Melton. 2009. *The Endurance of National Constitutions.* Cambridge, UK: Cambridge University Press.

Epstein, J. M. 2002. "Modeling Civil Violence: An Agent-Based Computational Approach." *Proceedings of the National Academy of Sciences* 99 (S3): 7243–7250. https://doi.org/10.1073/pnas.092080199.

Epstein, J. M., and R. L. Axtell. 1996. *Growing Artificial Societies: Social Science from the Bottom Up.* Cambridge, MA: MIT Press.

Farmer, J. D. 2024. *Making Sense of Chaos: A Better Economics for a Better World.* New Haven, CT: Yale University Press.

———. 2025. "Quantitative Agent-Based Models: A Promising Alternative for Macroeconomics." *Oxford Review of Economic Policy,* https://doi.org/10.1093/oxrep/graf027.

Farmer, J. D., C. Hepburn, P. Mealy, and A. Teytelboym. 2015. "A Third Wave in the Economics of Climate Change." *Environmental and Resource Economics* 62 (2): 329–357. https://doi.org/10.1007/s10640-015-9965-2.

Gibrat, R. 1931. *Les Inégalités Économiques.* Paris, France: Sirely.

Ginsburg, T., and A. Z. Huq. 2018. *How to Save a Constitutional Democracy.* Chicago, IL: University of Chicago Press.

Hayek, F. A. 1945. "The Use of Knowledge in Society." *American Economic Review* 35 (4): 519–530. https://www.jstor.org/stable/1809376.

Hodgson, G. M., and T. Knudsen. 2010. *Darwin's Conjecture: The Search for General Principles of Social and Economic Evolution.* Chicago, IL: University of Chicago Press.

Horowitz, D. L. 1985. *Ethnic Groups in Conflict.* Berkeley, CA: University of California Press.

Lafond, F., D. Greenwald, and J. D. Farmer. 2022. "Can Stimulating Demand Drive Costs Down? World War II as a Natural Experiment." *The Journal of Economic History* 82 (3): 727–764. https://doi.org/10.1017/S0022050722000249.

Lall, S. 1992. "Technological Capabilities and Industrialization." *World Development* 20 (2): 165–186. https://doi.org/10.1016/0305-750X(92)90097-F.

Leonard, N. E., K. Lipsitz, A. Bizyaeva, A. Franci, and Y. Lelkes. 2021. "The Nonlinear Feedback Dynamics of Asymmetric Political Polarization." *Proceedings of the National Academy of Sciences* 118 (50): e2102149118. https://doi.org/10.1073/pnas.2102149118.

Levin, S. 2000. *Fragile Dominion: Complexity And The Commons.* Princeton, NJ: Princeton University Press.

Li, J., B. M. Boghosian, and C. Li. 2019. "The Affine Wealth Model: An Agent-Based Model of Asset Exchange That Allows for Negative-Wealth Agents and Its Empirical Validation." *Physica A* 516 (C): 423–442. https://doi.org/10.1016/j.physa.2018.10.042.

Makowsky, M. D., and J. Rubin. 2013. "An Agent-Based Model of Centralized Institutions, Social Network Technology, and Revolution." *PLoS One* 8 (11): e80380. https://doi.org/10.1371/journal.pone.0080380.

Moro, A. 2016. "Understanding the Dynamics of Violent Political Revolutions in an Agent-Based Framework." *PLoS One* 11 (4): e0154175. https://doi.org/10.1371/journal.pone.0154175.

Nordhaus, W. D. 2018. "Projections and Uncertainties About Climate Change in an Era of Minimal Climate Policies." *American Economic Journal: Economic Policy* 10 (3): 333–360. https://doi.org/10.1257/pol.20170046.

Palagi, E., M. Napoletano, A. Roventini, and J.-L. Gaffard. 2023. "An Agent-Based Model of Trickle-Up Growth and Income Inequality." *Economic Modelling* 129:106535. https://doi.org/10.1016/j.econmod.2023.106535.

Pangallo, M., A. Aleta, R. M. del Rio-Chanona, A. Pichler, D. Martín-Corral, M. Chinazzi, F. Lafond, *et al.* 2024. "The Unequal Effects Of The Health–Economy Trade-Off During The COVID-19 Pandemic." *Nature Human Behaviour* 8 (2): 264–275. https://doi.org/10.1038/s41562-023-01747-x.

Pichler, A., M. Pangallo, R. M. del Rio-Chanona, F. Lafond, and J. D. Farmer. 2020. *Production Networks and Epidemic Spreading: How to Restart the UK Economy?* arXiv preprint: 2005.10585. https://doi.org/10.48550/arXiv.2005.10585.

———. 2022. "Forecasting the Propagation of Pandemic Shocks with a Dynamic Input–Output Model." *Journal of Economic Dynamics and Control* 144:104527. https://doi.org/10.1016/j.jedc.2022.104527.

Putnam, R. D. 1994. *Making Democracy Work.* Princeton, NJ: Princeton University Press.

Reissl, S., A. Caiani, F. Lamperti, M. Guerini, F. Vanni, G. Fagiolo, T. Ferraresi, L. Ghezzi, M. Napoletano, and A. Roventini. 2022. "Assessing The Economic Impact Of Lockdowns In Italy: A Computational Input–Output Approach." *Industrial and Corporate Change* 31 (2): 358–409. https://doi.org/10.1093/icc/dtac003.

Roithmayr, D. 2014. *Reproducing Racism: How Everyday Choices Lock In White Advantage.* New York, NY: New York University Press.

Smith, A. 1775. *An Inquiry into the Nature and Causes of the Wealth of Nations.* W. Strahan and T. Cadell.

Thomas, E. F., M. Ye, S. D. Angus, T. J. Mathew, W. Louis, L. Walsh, S. Ellery, M. Lizzio-Wilson, and C. McGarty. 2025. "Repeated And Incontrovertible Collective Action Failure Leads to Protester Disengagement and Radicalisation." *arXiv preprint: 2408.12795,* https://doi.org/10.48550/arXiv.2408.12795.

Trounstine, J. 2018. *Segregation by Design: Local Politics and Inequality in American Cities.* Cambridge, UK: Cambridge University Press.

Veblen, T. 1898. "Why Is Economics Not an Evolutionary Science?" *The Quarterly Journal of Economics* 12 (4): 373–397. https://doi.org/10.2307/1882952.

Way, R., M. C. Ives, P. Mealy, and J. D. Farmer. 2022. "Empirically Grounded Technology Forecasts and the Energy Transition." *Joule* 6 (9): 2057–2082. https://doi.org/10.1016/j.joule.2022.08.009.

Yakovenko, V. M., and J. B. Rosser, Jr. 2009. "Colloquium: Statistical Mechanics of Money, Wealth, and Income." *Reviews of Modern Physics* 81 (4): 1703–1725. https://doi.org/10.1103/revmodphys.81.1703.

A BRIEF HISTORY OF THE EMERGENCE OF COMPLEXITY ECONOMICS

Eric D. Beinhocker, University of Oxford and Santa Fe Institute;
J. Doyne Farmer, University of Oxford and Santa Fe Institute;
Jenna Bednar, University of Michigan and Santa Fe Institute;
R. Maria del Rio-Chanona, University College London
and Complexity Science Hub;
Jagoda Kaszowska-Mojsa, University of Oxford, Narodowy Bank
Polski, and Institute of Economics, Polish Academy of Sciences;
François Lafond, University of Oxford;
Penny Mealy, University of Oxford, Santa Fe Institute,
and Monash University;
Marco Pangallo, CENTAI Institute;
Anton Pichler, Vienna University of Economics and Business
and Complexity Science Hub

For readers new to complexity economics, a brief history of how the field emerged may provide useful context for the chapters that follow in this volume. This is not intended to be a complete historical review (for a more in-depth account, see Beinhocker 2006) but instead its purpose is to give readers an overview of key strands of work that provide the foundations for modern complexity economics, as well as some entry points into the historical literature (up to the early 2010s). We have also chosen to focus on the social science foundations and only very briefly mention the immense contributions in ideas and methods that complexity economists have incorporated from the physical sciences. As such, we have inevitably omitted numerous important contributors and works, but the chapters that follow cite a much broader base of scholarship, as well as more recent research.

Complexity economics, a label introduced by W. Brian Arthur in *Science* in 1999, began to emerge as a distinct set of perspectives and methodologies in the mid-1980s, pioneered by a community of scholars associated with the Santa Fe Institute (SFI). As noted in the introduction to this volume, "Complexity economics draws from the science of complex systems to understand the economy as an adaptive system where macro-level patterns emerge from the interaction of many diverse agents, rather than being imposed from above or derived from representative agents (see Mealy *et al.* 2026, ch. 1 in this volume; and Arthur 2021 for a more extended definition). While SFI catalyzed the development of the field in the 1980s, the ideas underpinning complexity economics had deep historical roots in both the social and physical sciences.

Historical Antecedents in the Social Sciences

One could claim that the first complexity economist was the Islamic philosopher Al-Ghazālī (1058–1111) who viewed markets as necessary, self-organizing institutions arising from human interdependence and the division of labor. He observed that order in exchange does not require central design but emerges from decentralized interaction—provided moral and legal norms are in place. While couched in a religious context, his observations anticipate later complexity views (and Friedrich Hayek's) that macroeconomic order arises bottom-up from local interactions rather than from planning (Islahi and Ghazanfar 2011).

While unaware of Al-Ghazālī's thinking, seven centuries later, Western Enlightenment philosophers explored themes of bottom-up emergence. In the eighteenth century, Adam Ferguson (1767 [1979]) argued that social and economic institutions are "The result of human action, but not the execution of any human design." He further saw economies as evolutionary and path dependent, arising from habit, conflict, imitation, and adaptation, and neither optimal nor in equilibrium.

~29~

Ferguson's contemporary Adam Smith was also fascinated by such phenomena. His famous metaphor of the "invisible hand" (Smith [1977]1776) is often misinterpreted as a statement about the pursuit of individual self-interest leading to positive outcomes for society (or in more colloquial terms, greed is good). However, Smith scholars correct this interpretation (Rothschild 1994), noting that Smith was clear that he did not believe that greed is good but rather the metaphor was a comment on the phenomenon of emergence. As Gavin Kennedy (2009, 241) puts it, "Smith's identification of the processes associated with the unintended consequences of individual actions in such diverse phenomena as language, money, moral sentiments, exchange and markets... are usefully judged to be an early recognition of evolutionary 'emergent order.'"

Building on this classical liberal tradition, in the twentieth century, economists from the Austrian school such as Joseph Schumpeter ([1911]1934) and Friedrich Hayek (1945) explored ideas of spontaneous order creation, dynamic change, and disequilibrium. Hayek in particular can be seen as a precursor to complexity economics in that he understood the economy as a decentralized, adaptive system in which order emerges from local interactions and dispersed knowledge, rather than from equilibrium optimization or central design (Axtell 2016; Bowles, Kirman, and Sethi 2017).

In addition to ideas of self-organization and emergence, complexity economists have also built on a long history of evolutionary thinking in economics (Hodgson 1993), from Thorstein Veblen (1898) famously asking, "Why is economics not an evolutionary science?" to Richard Nelson and Sidney Winter's landmark *An Evolutionary Theory of Economic Change* (1982). Evolutionary perspectives have been particularly influential in management science, beginning with the behavioral foundations of institutional evolution articulated by Richard M. Cyert and James G. March (1963) and extending to models of organizational adaptation and search (Kauffman 1969; Cohen, March, and Olsen 1972; Beinhocker 1999).

A contemporary of Cyert and March, Herbert Simon is a particularly important inspiration for complexity thinkers. His works "The Architecture of Complexity" (1962) and "The Organization of Complex Systems" (1973) provide theoretical foundations for understanding the hierarchical and modular nature of complex systems and how such structures scale in ways that are stable and adaptable. Simon (1955; 1978) also presents a view of human decision-making as a process of "boundedly rational" heuristic search rather than optimization.

Following Simon's critique of neoclassical rationality, behavioral economists such as Daniel Kahneman and Amos Tversky (1979) and Gerd Gigerenzer (1991; 1999) began to build a substantial body of empirical evidence as to how real human beings make economic decisions, behavioral foundations used extensively by later complexity economists.

Two of the tools most frequently used by modern complexity economists—agent-based modeling and network analysis—also had important antecedents. Barbara Bergmann (1971; 1990) lay early foundations for the complexity perspective and agent-based modeling in her arguments against representative agents (see also Kirman 1992) and their inability to model phenomena such as agent heterogeneity, inequality, power, and discrimination. Thomas Schelling's (1971) model of housing discrimination provides a simple example of emergent phenomena and is one of the earliest and most influential agent-based models in economics (implemented using pennies on a paper grid). The use of network analysis in economics also has historical antecedents, notably Wassily Leontief's examination of the input–output structure of the US economy (1936).

Another historically important idea for complexity economists is the view of the economy as an order-creating, metabolic system operating far from equilibrium. This perspective can be traced back to Karl Marx, who in *Capital* (1867, vol. I, ch. 7) described the labor process as "purposeful activity aimed at the production of use-values" and as "the universal condition for the metabolic interaction between man and nature, common to all forms of society in which human beings live." In the 1970s, the economist Nicholas Georgescu-Roegen, while critical of Marx on many points, shared the view that economics had largely ignored the economy's relationship with nature, energy, and fundamental thermodynamic constraints. In *The Entropy Law and the Economic Process* (1971), he argued that the economy is a thermodynamically open, nonequilibrium system, transforming materials, energy, and information into more complex outputs while irreversibly generating waste and entropy. This work laid the foundations for ecological economics (e.g., Daly 1977) and for a complexity-based understanding of economic growth and development as a disequilibrium process embedded in biophysical systems (e.g., Ayres 1994).

Crossing Disciplines and the Emergence of Complexity Economics

While social scientists were exploring these various streams of thought, the twentieth century saw major developments in the physical sciences in understanding what later came to be called "complex systems." These advances cut across multiple disciplines, including work on nonlinear dynamical systems (e.g., Edward Lorenz), statistical physics (e.g., Kenneth Wilson), self-organizing systems (e.g., Lars Onsager and Ilya Prigogine), mathematical evolutionary theory (e.g., Sewall Wright and John Maynard Smith), computation (e.g., Alan Turing and John von Neumann), information theory (e.g., Claude Shannon), fractal geometry and power laws (e.g., Benoît Mandelbrot), and graph theory (e.g., Paul Erdős and Alfréd Rényi).

Thus, when the Nobel economist Kenneth Arrow and Nobel physicist Philip Anderson convened ten economists and ten natural scientists at the now-famous meeting at the Santa Fe Institute in 1987, there was fertile intellectual soil to explore (Waldrop 1992). At that meeting, the eyes of the economists were opened to the powerful concepts and methods employed by the natural scientists for analyzing complex systems, and the eyes of the natural scientists were opened to the challenges the economists faced in developing testable theories and models of a human social system as complex as the global economy.

The meeting yielded the first volume in this series, *The Economy as an Evolving Complex System* (1988), and launched a program of research at SFI, first directed by W. Brian Arthur, with funding support from then-CEO of Citigroup, John Reed.

The program was later directed through the 1990s by John Geanakoplos, Blake LeBaron, Lawrence Blume, and Steven Durlauf, with major contributions from J. Doyne Farmer, Samuel Bowles, and others.

Artificial Economies—the Development of Agent-Based Modeling

Foundational to this movement was the introduction of agent-based modeling (ABM) as a rigorous tool for economic inquiry (for an in-depth review, see Axtell and Farmer 2025). John H. Holland and John H. Miller (1991) offered an early manifesto for "Artificial Adaptive Agents," arguing that complex adaptive systems could model the evolution of economic behavior in ways traditional equilibrium analysis with representative agents could not (consistent with the other critiques mentioned, Bergmann 1971 and Kirman 1992). One of the early outputs of the ABM program was a collaboration between Brian Arthur, John Holland, Blake LeBaron, Richard Palmer, and Paul Tayler to build the Santa Fe Artificial Stock Market model (Arthur *et al.* 1997), demonstrating that ABMs could replicate key statistical emergent features of financial markets from bottom-up simulation of the interactions of heterogeneous agents.

~ 33 ~

Moving from financial markets to more general economic and social phenomena, Joshua M. Epstein and Robert Axtell's *Growing Artificial Societies: Social Science from the Bottom Up* (1996) introduced the influential Sugarscape model. In Sugarscape, simple agents follow local rules for movement, consumption, reproduction, and trade on a spatial landscape, yet generate complex macro phenomena such as wealth inequality, price formation, migration, and social stratification. The book demonstrated that macroeconomic and social regularities can be explained as emergent outcomes of micro-level interaction, without equilibrium assumptions or representative agents, helping motivate a significant body of ABM work in the decades that followed.

During the same period, economists William Brock and Cars Hommes (1998) provided an analytical counterpart to the SFI computational models, examining how the nonlinear dynamics of heterogeneous beliefs in a population of agents could lead to instabilities in financial markets (preceded by related work; see Chiarella 1992). These papers laid the foundation for further work by figures such as J. Doyne Farmer, Thomas Lux, and Jean-Philippe Bouchaud, showing that financial agent-based models could quantitatively replicate the statistical properties of real markets (e.g., Lux and Marchesi 1999; Farmer and Joshi 2002; Giardina and Bouchaud 2003).

In the 2000s, ABM work began to branch from financial markets into macroeconomics. Two economists helped lay important theoretical groundwork for that line of research. Axel Leijonhufvud (1981) reframed macroeconomies as coordination systems operating far from equilibrium, where instability arises endogenously from information failures, decentralized decision-making, and nonlinear adjustment processes rather than solely from exogenous shocks.

Alan Kirman (1992, 2010) systematically criticized the representative-agent paradigm that was foundational in macroeconomics and demonstrated that aggregate economic outcomes emerge from interacting heterogeneous agents, often in ways that cannot be inferred from individual behavior alone. He further showed how coordination failures, herd behavior, and nonlinear dynamics arise endogenously in markets.

Building on this theoretical work, economists including Stefano Cincotti and Herbert Dawid engaged in an ambitious large-scale model building project of the European economy called EURACE (Cincotti, Raberto, and Teglio 2010); Domenico Delli Gatti, Mauro Gallegati, and collaborators worked towards building a general macroeconomic ABM that could challenge widely used dynamic stochastic general equilibrium (DSGE) models (Delli Gatti *et al.* 2005; Delli Gatti *et al.* 2011). Around the same period, another Italian economist, Giovanni Dosi, and colleagues pioneered the use of ABMs to look at questions of economic growth integrating Keynesian demand fluctuations and Schumpeterian growth dynamics (Dosi, Fagiolo, and Roventini 2010).

Another early pioneer in agent-based modeling was Leigh Tesfatsion (2002), who articulated a methodology for what she called ACE (agent-based computational economics), set out best practices for model design and validation, and created infrastructure (tutorials, software resources, and edited collections) that made ABM more accessible to economists and students. Tesfatsion's work emphasized the use of ACE for both positive theory (explaining emergence, distributional outcomes, and disequilibrium dynamics) and policy experiments (what-if counterfactuals, regulation, and institutional design), creating a bridge with more orthodox economic methods (Tesfatsion and Judd 2006).

ABMs were also employed to study a variety of microeconomic phenomena. Kirman and Nicolas Vriend (2001) provided one of the earliest empirically grounded agent-based interpretations of market organization with their study of the Marseille fish market. That work showed how real-world markets operate through social learning and repeated interaction, rather than anonymous Walrasian price-taking. Other ABMs of individual markets included the liberalization of electricity markets (Nicolaisen, Petrov, and Tesfatsion 2001) and a detailed study of housing markets (Geanakoplos *et al.* 2012; Axtell *et al.* 2014) that analyzed the endogenous emergence of housing bubbles and tested policies for mitigating them (Baptista *et al.* 2016).

Agent-based models with many agents and realistic institutional complexity also provided motivation for further theoretical work. Game theory is a core tool in economics for exploring strategic interactions. Standard game theory assumes convergence to fixed points such as Nash equilibria, but Yuzuru Sato, Eizo Akiyama, and Farmer (2002) showed that, under bounded rationality, games may fail to converge to fixed-point equilibria and instead follow chaotic trajectories in the strategy space. Subsequently, Marco Pangallo, Torsten Heinrich, and Farmer (2019) revealed that such chaotic behavior is generic for competitive, complicated normal-form games. As most real-world economic phenomena can be characterized as competitive, complex, multiplayer games, this work provides important theoretical foundations for agent-based modeling, which doesn't assume equilibrium convergence and is arguably the only method that can faithfully model such real-world complexity.

Agent Ecologies, Scaling, and Networks

But ABMs were far from the only methodology employed by the growing community of complexity economists. W. Brian Arthur's highly influential "El Farol" paper (1994) demonstrated how a simple deterministic model of agents with heterogeneous expectations could produce complex endogenous fluctuations. The El Farol model was later formalized and generalized by Damien Challet and Yi-Cheng Zhang (1997) into the "minority game," helping launch an "econophysics" analytical literature studying economic fluctuations, phase transitions, and emergent order.

Other researchers demonstrated how "ecologies of expectations" (as Arthur put it) could evolve as agents updated decision rules, learned from their environment and each other, and
evolved. Physicist Kristian Lindgren (1991) provided a pioneering example demonstrating how evolving, interacting strategies
in a population of agents playing the prisoner's dilemma could
generate complex emergent dynamics that never settle to equilibrium. Another example is J. Doyne Farmer's work on market ecologies (Farmer 2002; Scholl, Calinescu, and Farmer 2021)
showing how the heterogeneous and evolving strategies of populations of agents in markets interact, and providing explanations
for key market dynamics, including market malfunctions.

In many of these systems of dynamically interacting,
heterogeneous agents, scaling laws emerged as a macroscopic
regularity. Vilfredo Pareto's 1890 characterization of income
distributions is usually credited as the first observation of a
power law in economics. Linguist George Kingsley Zipf in the
1930s observed that city sizes, firm sizes, and other phenomena
follow an inverse rank-size rule known as Zipf's law. The
complexity community built on these foundations; Xavier Gabaix
(1999, 2009) observed Zipf's law in city size distributions and
financial market statistical properties and linked these empirical
observations to stochastic growth processes.

Robert Axtell (2001) went on to show empirically that US
firm sizes follow a Zipf distribution, and later demonstrated that
this regularity holds robustly across countries, industries, and time
(2006). He argued that this scaling law is an emergent property of
simple, institutionally realistic microeconomic behaviors. It arises
from heterogeneous agents interacting through labor reallocation
under local increasing returns, generating persistent disequilibrium
dynamics rather than convergence to equilibrium.

Scaling laws are familiar territory to physicists; the physicist Geoffrey West, along with an interdisciplinary group of collaborators, showed empirically that cities and other social systems obey systematic scaling laws analogous to those found in biology, revealing deep regularities in how social, economic, and infrastructural quantities grow with system size (West, Brown, and Enquist 1997; Bettencourt *et al.* 2007; West 2017). The work demonstrates strong relationships between social interaction density and economic productivity, innovation, and inequality, with important implications for the role of urbanization in economic development.

The growing complexity-economics community has also embraced the adoption of network-based approaches in economics. This shift led both to detailed empirical characterizations of economic networks (Schweitzer *et al.* 2009) and to formal game-theoretic analyses of network formation and structure (Vega-Redondo 2007; Jackson 2008). An influential early line of research drew on insights from self-organized criticality (Bak, Tang, and Wiesenfeld 1988) and strategic complementarities (Durlauf 1993) to explain how microeconomic shocks can generate aggregate fluctuations through heavy-tailed distributions (Gabaix 2011) and network-mediated amplification mechanisms (Acemoglu *et al.* 2012). Other work (Jackson and Wolinsky 1996; Jackson 2008) formalized how economic networks form, how they shape diffusion and inequality, and how local interactions generate large aggregate effects.

Network-based complexity approaches had an important impact on conceptualizing and modeling financial stability and systemic risk. For example, Michael Boss *et al.* (2004) empirically characterized interbank lending networks as heterogeneous systems with power-law degree distributions, discussing how such structures generate systemic fragility. Network models of systemic risk gained particular prominence in the aftermath of the 2008 global financial crisis (e.g., Battiston *et al.* 2012; Thurner and Poledna 2013; Caccioli *et al.* 2014). For a comprehensive review, see Fabio Caccioli (2026; ch. 16 in this volume).

Another influential research program applies network analysis to the study of economic growth and development. Ricardo Hausmann and César Hidalgo introduced the Economic Complexity Index, which reveals how a country's long-run growth prospects are strongly shaped by its stock of productive knowledge. This knowledge is inferred from the structure of a bipartite network linking countries to the products they export, with greater product diversity and sophistication indicating greater underlying capabilities (Hidalgo *et al.* 2007; Hausmann *et al.* 2011). This framework is synthesized in *The Atlas of Economic Complexity* (Hausmann *et al.* 2014), which formalizes the concept of an economy's "product space"—the set of products a country is able to produce. The analysis emphasizes the path-dependent accumulation of capabilities: countries tend to diversify into products that are closely related to those they already produce, gradually expanding into more complex sectors. Over time, this process is associated with higher value added and rising standards of living, beyond what is explained by traditional factors such as capital accumulation or institutional
quality.

Building on this work, Luciano Pietronero and coauthors developed the fitness–complexity algorithm, offering an alternative network-based measure (Tacchella *et al.* 2012; Cristelli, Tacchella, and Pietronero 2015). Together, these contributions brought concepts of economic complexity and networks of capabilities (or industrial clusters or ecosystems) into discussions of economic growth, development, and policy.

The Meso Level: Technology and Institutions

While pioneering work was done to understand macro phenomena such as growth, instabilities, and distributions, along with micro-level work to understand market function, another group of scholars focused on the meso level, deepening our understanding of technology innovation and the evolution of institutions.

An early contribution is the volume edited by Giovanni Dosi *et al.* (1988), whose papers argue that innovation is an evolutionary, path-dependent, and institutionally embedded process, not a smooth response to prices or incentives, and explicitly challenges equilibrium and representative-agent approaches to modeling technological change.

W. Brian Arthur's *The Nature of Technology* (2009) offers a foundational theory, describing technology as an evolving, combinatorial system in which new technologies arise largely from the recombination of existing components rather than from *de novo* invention. Arthur shows how this recursive, modular process generates path dependence, increasing returns, and structural change, providing a micro-founded basis for theories of endogenous growth and economic evolution (see Arthur 2026, ch. 3 in this volume).

Of course, the other key factor in long-run economic growth is the development of institutions. Complexity researchers have played a central role in reframing cooperation and institutions as endogenous, evolving outcomes of strategic interaction, rather than simply assuming institutions as exogenous constraints or inevitable equilibrium outcomes.

Robert Axelrod (1984) lays important foundations with his work on the evolution of cooperation, most famously through his tournament of agents playing an iterated prisoner's dilemma. This work highlights the importance of reciprocity, reputation, and population structure, showing how adaptive strategies interacting over time can generate stable cooperative norms.

Samuel Bowles and Herbert Gintis have integrated evolutionary game theory, and collaborations with behavioral economists, anthropologists, and biologists, to show that cooperation can be sustained through social preferences, norms, and institutional enforcement, even when standard self-interest models predict defection (Bowles 1998; Bowles and Gintis 2011; Henrich *et al.* 2001; Henrich *et al.* 2004). Bowles combines methods from empirical population genetics and evolutionary game theory to show that a genetic disposition for individuals to behave altruistically towards in-group members could have coevolved with hostility towards "outsiders" and frequent group conflict (Bowles 2006; Choi and Bowles 2007).

Their work demonstrates that preferences themselves are shaped by evolutionary biological and cultural processes, and that markets, states, and communities co-evolve through feedback between incentives and norms (Bowles 2004). This perspective challenges the sharp separation between economics and sociology by treating institutions as adaptive mechanisms that stabilize cooperation in complex societies (Gintis 2016).

A complementary line of research on the evolution of institutions and norms was developed by H. Peyton Young, who provides rigorous game-theoretic foundations for how conventions, norms, and institutional rules emerge (Young 1993, 1998). Young demonstrated that when self-interested agents adjust their behavior through learning, imitation, and experimentation, collective outcomes depend on historical paths rather than converging to a unique equilibrium. This provides a formal explanation for why norms and institutions can be stable across diverse societies.

From 2001 to 2020, an annual SFI working group led by Bowles, Gintis, and Young, along with anthropologist Robert Boyd and economist Larry Blume, explored related themes under the title "The Coevolution of Individual Behaviors and Social Institutions."

Inequalities are an emergent result of institutional arrangements, technologies, and behaviors. Starting in 2006, Bowles and anthropologist Monique Borgerhoff Mulder led the cross-disciplinary "Dynamics of Wealth Inequality" project. This group has used a combination of dynamical-systems methods and network theory along with data from prehistory, recent small-scale societies, and other sources, to better understand the cultural, institutional, and technological forces driving economic disparities in the long run (Borgerhoff Mulder *et al.* 2009; Bowles and Fochesato 2024).

Finally, at the intersection of behavior, cooperation, and institutions, Scott Page has shown that groups of agents with diverse perspectives can outperform homogeneous groups of individually "optimal" decision-makers, challenging representative-agent logic (Page 2007). His work provides rigorous foundations for the role of cognitive diversity and decentralized problem-solving in enhancing institutional performance and system resilience (Page 2010).

From Critique to Innovation, Insight, and Relevance

The trajectory of complexity economics traces a shift from early conceptual critiques of equilibrium, rationality, and representative-agent models toward a mature, empirically grounded science of economic systems. Over several decades, researchers have developed new theoretical frameworks, computational tools, and empirical methods capable of capturing heterogeneity, behavioral and institutional realism, dynamics, path dependence, and emergence—first in stylized models and increasingly in data-rich, empirically validated analysis. What began as an exploratory synthesis of ideas from economics, physics, and other social and physical sciences has now grown into a coherent research program with demonstrated relevance for understanding markets, growth, systemic risk, inequality, institutions, technology, and climate change.

The chapters in this volume build on this intellectual history—and include contributions from a number of figures cited in this review—showcasing recent advances, and pointing toward the next phase of complexity economics: one focused not only on explanation, but also on better prediction, more effective interventions, and supporting the evolution of prosperous, just, and sustainable economic systems. ⅃

REFERENCES

Acemoglu, D., V. M. Carvalho, A. Ozdaglar, and A. Tahbaz-Salehi. 2012. "The Network Origins of Aggregate Fluctuations." *Econometrica* 80 (5): 1977–2016. https://doi.org/10.3982/ECTA9623.

Anderson, P. W., K. J. Arrow, and D. Pines, eds. 1988. *The Economy as an Evolving Complex System.* Reading, MA: Addison-Wesley.

Arthur, W. B. 1994. "Inductive Reasoning and Bounded Rationality." *American Economic Review* 84 (2): 406–411. https://www.jstor.org/stable/2117868.

———. 1999. "Complexity and the Economy." *Science* 284 (5411): 107–109. https://doi.org/10.1126/science.284.5411.107.

Arthur, W. B. 2009. *The Nature of Technology: What It Is and How It Evolves.* New York, NY: Free Press.

———. 2021. "Foundations of Complexity Economics." *Nature Reviews Physics* 3:136–145. https://doi.org/10.1038/s42254-020-00273-3.

———. 2026. "Combinatorial Evolution." In *The Economy as an Evolving Complex System IV,* edited by R. M. del Rio-Chanona, M. Pangallo, J. Bednar, E. D. Beinhocker, J. Kaszowska-Mojsa, F. Lafond, P. Mealy, A. Pichler, and J. D. Farmer. Santa Fe, NM: SFI Press.

Arthur, W. B., J. H. Holland, B. LeBaron, R. Palmer, and P. Tayler. 1997. "Asset Pricing Under Endogenous Expectations in an Artificial Stock Market." In *The Economy as an Evolving Complex System II.* Reading, MA: Addison-Wesley.

Axelrod, R. 1984. *The Evolution of Cooperation.* New York, NY: Basic Books.

Axtell, R. L. 2001. "Zipf Distribution of U.S. Firm Sizes." *Science* 293 (5536): 1818–1820. https://doi.org/10.1126/science.1062081.

———. 2006. *Firm Sizes: Facts, Formulae, Fables and Fantasies.* Technical report 44. Brookings Institution, Center on Social and Economic Dynamics. https://doi.org/10.2139/ssrn.1024813.

———. 2016. "Hayek Enriched by Complexity Enriched by Hayek." In *Revisiting Hayek's Political Economy,* edited by P. J. Boettke and V. H. Storr, vol. 21. Advances in Austrian Economics. Leeds, UK: Emerald Group Publishing Limited. https://doi.org/10.1108/S1529-213420160000021003.

Axtell, R. L., and J. D. Farmer. 2025. "Agent-Based Modeling in Economics and Finance: Past, Present, and Future." *Journal of Economic Literature* 63 (1): 197–287. https://doi.org/10.1257/jel.20221319.

Axtell, R. L., J. D. Farmer, J. Geanakoplos, P. Howitt, E. Carrella, B. Conlee, J. Goldstein, *et al.* 2014. "An Agent-Based Model of the Housing Market Bubble in Metropolitan Washington, DC." *SSRN Electronic Journal,* https://doi.org/10.2139/ssrn.4710928.

Ayres, R. U. 1994. *Information, Entropy, and Progress: A New Evolutionary Paradigm.* Woodbury, NY: American Institute of Physics Press.

Bak, P., C. Tang, and K. Wiesenfeld. 1988. "Self-Organized Criticality." *Physical Review A* 38 (1): 364–374. https://doi.org/10.1103/PhysRevA.38.364.

Baptista, R., J. D. Farmer, M. Hinterschweiger, K. Low, D. Tang, and A. Uluc. 2016. *Macroprudential Policy in an Agent-Based Model of the UK Housing Market.* Bank of England, Working Paper no. 619. https://doi.org/10.2139/ssrn.2850414.

Battiston, S., M. Puliga, R. Kaushik, P. Tasca, and G. Caldarelli. 2012. "DebtRank: Too Central to Fail? Financial Networks, the FED and Systemic Risk." *Scientific Reports* 2 (1): 541. https://doi.org/10.1038/srep00541.

Beinhocker, E. D. 1999. "Robust Adaptive Strategies." *Sloan Management Review* 40 (3): 95–106.

———. 2006. *The Origin of Wealth: Evolution, Complexity, and the Radical Remaking of Economics.* Boston, MA: Harvard Business School Press.

Bergmann, B. R. 1971. "The Effect on White Incomes of Discrimination in Employment." *Journal of Political Economy* 79 (2): 294–313. https://doi.org/10.1086/259744.

———. 1990. "Micro-to-Macro Simulation: A Primer with a Labor Market Example." *Journal of Economic Perspectives* 4 (1): 99–116. https://doi.org/10.1257/jep.4.1.99.

Bettencourt, L. M. A., J. Lobo, D. Helbing, C. Kühnert, and G. B. West. 2007. "Growth, Innovation, Scaling, and the Pace of Life in Cities." *Proceedings of the National Academy of Sciences* 104 (17): 7301–7306. https://doi.org/10.1073/pnas.0610172104.

Borgerhoff Mulder, M., S. Bowles, T. Hertz, A. Bell, J. Beise, G. Clark, I. Fazzio, *et al.* 2009. "Intergenerational Wealth Transmission and the Dynamics of Inequality in Small-Scale Societies." *Science* 326 (5953): 682–688. https://doi.org/10.1126/science.1178336.

Boss, M., H. Elsinger, M. Summer, and S. Thurner. 2004. "Network Topology of the Interbank Market." *Quantitative Finance* 4 (6): 677–684. https://doi.org/10.1080/14697680400020325.

Bowles, S. 1998. "Endogenous Preferences: The Cultural Consequences of Markets and other Economic Institutions." *Journal of Economic Literature* 36 (1): 75–111. https://www.jstor.org/stable/2564952.

———. 2004. *Microeconomics: Behavior, Institutions, and Evolution.* Princeton, NJ: Princeton University Press.

Bowles, S. 2006. "Group Competition, Reproductive Leveling, and the Evolution of Human Altruism." *Science* 314 (5805): 1569–1572. https://doi.org/10.1126/science.1134829.

Bowles, S., and M. Fochesato. 2024. "The Origins of Enduring Economic Inequality." *Journal of Economic Literature* 62 (4): 1475–1537. https://doi.org/10.1257/jel.20241718.

Bowles, S., and H. Gintis. 2011. *A Cooperative Species: Human Cooperation and Its Evolution.* Princeton, NJ: Princeton University Press.

Bowles, S., A. Kirman, and R. Sethi. 2017. "Retrospectives: Friedrich Hayek and the Market Algorithm." *Journal of Economic Perspectives* 31 (3): 215–230. https://doi.org/10.1257/jep.31.3.215.

Brock, W. A., and C. H. Hommes. 1998. "Heterogeneous Beliefs and Routes to Chaos in a Simple Asset Pricing Model." *Journal of Economic Dynamics and Control* 22 (8–9): 1235–1274. https://doi.org/10.1016/S0165-1889(98)00011-6.

Caccioli, F. 2026. "Understanding Financial Contagion: A Complexity-Modeling Perspective." In *The Economy as an Evolving Complex System IV,* edited by R. M. del Rio-Chanona, M. Pangallo, J. Bednar, E. D. Beinhocker, J. Kaszowska-Mojsa, F. Lafond, P. Mealy, A. Pichler, and J. D. Farmer. Santa Fe, NM: SFI Press.

Caccioli, F., M. Shrestha, C. Moore, and J. D. Farmer. 2014. "Stability Analysis of Financial Contagion Due to Overlapping Portfolios." *Journal of Banking & Finance* 46:233–245. https://doi.org/10.1016/j.jbankfin.2014.05.021.

Challet, D., and Y.-C. Zhang. 1997. "Emergence of Cooperation and Organization in an Evolutionary Game." *Physica A* 246 (3–4): 407–418. https://doi.org/10.1016/S0378-4371(97)00419-6.

Chiarella, C. 1992. "The Dynamics of Speculative Behavior." *Annals of Operations Research* 37:101–123. https://doi.org/10.1007/BF02071051.

Choi, J.-K., and S. Bowles. 2007. "The Coevolution of Parochial Altruism and War." *Science* 318 (5850): 636–640. https://doi.org/10.1126/science.1144237.

Cincotti, S., M. Raberto, and A. Teglio. 2010. "Credit Money and Macroeconomic Instability in the Agent-Based Model and Simulator EURACE." *Economics: The Open-Access, Open-Assessment E-Journal* 4 (1): 20100026. https://doi.org/10.5018/economics-ejournal.ja.2010-26.

Cohen, M. D., J. G. March, and J. P. Olsen. 1972. "A Garbage Can Model of Organizational Choice." *Administrative Science Quarterly* 17 (1): 1–25. https://doi.org/10.2307/2392088.

Cristelli, M., A. Tacchella, and L. Pietronero. 2015. "The Heterogeneous Dynamics of Economic Complexity." *PLoS ONE* 10 (2): e0117174. https://doi.org/10.1371/journal.pone.0117174.

Cyert, R. M., and J. G. March. 1963. *A Behavioral Theory of the Firm.* Englewood Cliffs, NJ: Prentice Hall.

Daly, H. E. 1977. *Steady-State Economics.* San Francisco, CA: W. H. Freeman & Co.

Delli Gatti, D., S. Desiderio, E. Gaffeo, P. Cirillo, and M. Gallegati. 2011. *Macroeconomics from the Bottom-Up.* Berlin, Germany: Springer.

Delli Gatti, D., C. Di Guilmi, E. Gaffeo, G. Giulioni, M. Gallegati, and A. Palestrini. 2005. "A New Approach to Business Fluctuations: Heterogeneous Interacting Agents, Scaling Laws and Financial Fragility." *Journal of Economic Behavior & Organization* 56 (4): 489–512. https://doi.org/10.1016/j.jebo.2003.10.012.

Dosi, G., G. Fagiolo, and A. Roventini. 2010. "Schumpeter Meeting Keynes: A Policy-Friendly Model of Endogenous Growth and Business Cycles." *Journal of Economic Dynamics and Control* 34 (9): 1748–1767. https://doi.org/10.1016/j.jedc.2010.06.018.

Dosi, G., C. Freeman, R. Nelson, G. Silverberg, and L. Soete, eds. 1988. *Technical Change and Economic Theory.* London, UK: Pinter.

Durlauf, S. N. 1993. "Nonergodic Economic Growth." *Review of Economic Studies* 60 (2): 349–366. https://doi.org/10.2307/2298061.

Epstein, J. M., and R. Axtell. 1996. *Growing Artificial Societies.* Cambridge, MA: MIT Press.

Farmer, J. D. 2002. "Market Force, Ecology and Evolution." *Industrial and Corporate Change* 11 (5): 895–953. https://doi.org/10.1093/icc/11.5.895.

Farmer, J. D., and S. Joshi. 2002. "Price Dynamics of Common Trading Strategies." *Journal of Economic Behavior & Organization* 49 (2): 149–171. https://doi.org/10.1016/S0167-2681(02)00065-3.

Ferguson, A. 1767 [1979]. *An Essay on the History of Civil Society.* Edinburgh, UK: Edinburgh University Press.

Gabaix, X. 1999. "Zipf's Law for Cities: An Explanation." *Quarterly Journal of Economics* 114 (3): 739–767. https://doi.org/10.1162/003355399556133.

———. 2009. "Power Laws in Economics and Finance." *Annual Review of Economics* 1 (1): 255–294. https://doi.org/10.1146/annurev.economics.050708.142940.

———. 2011. "The Granular Origins of Aggregate Fluctuations." *Econometrica* 79 (3): 733–772. https://doi.org/10.3982/ECTA8769.

Geanakoplos, J., R. Axtell, J. D. Farmer, P. Howitt, B. Conlee, J. Goldstein, M. Hendrey, N. M. Palmer, and C.-Y. Yang. 2012. "Getting at Systemic Risk via an Agent-Based Model of the Housing Market." *American Economic Review* 102 (3): 53–58. https://doi.org/10.1257/aer.102.3.53.

Georgescu-Roegen, N. 1971. *The Entropy Law and the Economic Process.* Cambridge, MA: Harvard University Press.

Giardina, I., and J.-P. Bouchaud. 2003. "Bubbles, Crashes and Intermittency in Agent-Based Market Models." *The European Physical Journal B* 31:421–437. https://doi.org/10.1140/epjb/e2003-00050-6.

Gigerenzer, G. 1991. "How to Make Cognitive Illusions Disappear." *European Review of Social Psychology* 2 (1): 83–115. https://doi.org/10.1080/14792779143000033.

Gigerenzer, G., P. M. Todd, and the ABC Research Group. 1999. *Simple Heuristics That Make Us Smart.* Oxford, UK: Oxford University Press.

Gintis, H. 2016. *Individuality and Entanglement: The Moral and Material Bases of Social Life.* Princeton, NJ: Princeton University Press.

Hausmann, R., C. A. Hidalgo, S. Bustos, M. Coscia, S. Chung, J. Jimenez, A. Simoes, and M. A. Yildirim. 2011. *The Atlas of Economic Complexity: Mapping Paths to Prosperity.* Hollis, NH: Puritan Press.

Hausmann, R., C. A. Hidalgo, S. Bustos, M. Coscia, A. Simoes, and M. A. Yildirim. 2014. *The Atlas of Economic Complexity: Mapping Paths to Prosperity.* Cambridge, MA: MIT Press.

Hayek, F. A. 1945. "The Use of Knowledge in Society." *American Economic Review* 35 (4): 519–530. https://www.jstor.org/stable/1809376.

Henrich, J., R. Boyd, S. Bowles, C. Camerer, E. Fehr, and H. Gintis. 2004. *Foundations of Human Sociality: Economic Experiments and Ethnographic Evidence from Fifteen Small-Scale Societies.* Oxford, UK: Oxford University Press.

Henrich, J., R. Boyd, S. Bowles, C. Camerer, E. Fehr, H. Gintis, and R. McElreath. 2001. "In Search of Homo Economicus: Behavioral Experiments in 15 Small-Scale Societies." *American Economic Review* 91 (2): 73–78. https://doi.org/10.1257/aer.91.2.73.

Hidalgo, C. A., B. Klinger, A.-L. Barabási, and R. Hausmann. 2007. "The Product Space Conditions the Development of Nations." *Science* 317 (5837): 482–487. https://doi.org/10.1126/science.1144581.

Hodgson, G. M. 1993. *Economics and Evolution: Bringing Life Back into Economics.* Ann Arbor, MI: University of Michigan Press.

Holland, J. H., and J. H. Miller. 1991. "Artificial Adaptive Agents in Economic Theory." *American Economic Review* 81 (2): 365–371. https://econpapers.repec.org/RePEc:aea:aecrev:v:81:y:1991:i:2:p:365-71.

Islahi, A. A., and S. M. Ghazanfar. 2011. *Economic Thought of al-Ghazālī.* MPRA Paper 53465, University Library of Munich. https://mpra.ub.uni-muenchen.de/53465/.

Jackson, M. O. 2008. *Social and Economic Networks.* Princeton, NJ: Princeton University Press.

Jackson, M. O., and A. Wolinsky. 1996. "A Strategic Model of Social and Economic Networks." *Journal of Economic Theory* 71 (1): 44–74. https://doi.org/10.1006/jeth.1996.0108.

Kahneman, D., and A. Tversky. 1979. "Prospect Theory: An Analysis of Decision Under Risk." *Econometrica* 47 (2): 263–291. https://doi.org/10.2307/1914185.

Kauffman, S. A. 1969. "Metabolic Stability and Epigenesis in Randomly Constructed Genetic Nets." *Journal of Theoretical Biology* 22 (3): 437–467. https://doi.org/10.1016/0022-5193(69)90015-0.

Kennedy, G. 2009. "Adam Smith and the Invisible Hand: From Metaphor to Myth." *Economic Journal Watch* 6 (2): 239–263.

Kirman, A. 1992. "Whom or What Does the Representative Individual Represent?" *Journal of Economic Perspectives* 6 (2): 117–136. https://doi.org/10.1257/jep. 6.2.117.

———. 2010. *Complex Economics: Individual and Collective Rationality*. London, UK: Routledge.

Kirman, A. P., and N. J. Vriend. 2001. "Evolving Market Structure: An ACE Model of Price Dispersion and Loyalty." *Journal of Economic Dynamics and Control* 25 (3–4): 459–502. https://doi.org/10.1016/S0165-1889(00)00033-6.

Leijonhufvud, A. 1981. *Information and Coordination*. New York, NY: Oxford University Press.

Leontief, W. 1936. "Quantitative Input and Output Relations in the Economic Systems of the United State." *Review of Economics and Statistics* 18 (3): 105–125. https://doi.org/10.2307/1927837.

Lindgren, K. 1991. "Evolutionary Phenomena in Simple Dynamics." In *Artificial Life II,* edited by C. G. Langton, C. Taylor, J. D. Farmer, and S. Rasmussen, 295–312. Reading, MA: Addison-Wesley.

Lux, T., and M. Marchesi. 1999. "Scaling and Criticality in a Stochastic Multi-Agent Model of a Financial Market." *Nature* 397:498–500. https://doi.org/10. 1038/17290.

Mealy, P., J. Bednar, E. D. Beinhocker, R. M. del Rio-Chanona, J. D. Farmer, J. Kaszowska-Mojsa, F. Lafond, M. Pangallo, and A. Pichler. 2026. "Introduction." In *The Economy as an Evolving Complex System IV,* edited by R. M. del Rio-Chanona, M. Pangallo, J. Bednar, E. D. Beinhocker, J. Kaszowska-Mojsa, F. Lafond, P. Mealy, A. Pichler, and J. D. Farmer. Santa Fe, NM: SFI Press.

Nelson, R. R., and S. G. Winter. 1982. *An Evolutionary Theory of Economic Change.* Cambridge, MA: Harvard University Press.

Nicolaisen, J., V. Petrov, and L. Tesfatsion. 2001. "Market Power and Efficiency in a Computational Electricity Market with Discriminatory Double-Auction Pricing." *IEEE Transactions on Evolutionary Computation* 5 (5): 504–523. https://doi.org/10.1109/4235.956714.

Page, S. E. 2007. *The Difference.* Princeton, NJ: Princeton University Press.

———. 2010. *Diversity and Complexity.* Princeton, NJ: Princeton University Press.

Pangallo, M., T. Heinrich, and J. D. Farmer. 2019. "Best Reply Structure and Equilibrium Convergence in Generic Games." *Science Advances* 5 (2): eaat1328. https://doi.org/10.1126/sciadv.aat1328.

Rothschild, E. 1994. "Adam Smith and the Invisible Hand." *American Economic Review* 84 (2): 319–322. https://www.jstor.org/stable/2117851.

Sato, Y., E. Akiyama, and J. D. Farmer. 2002. "Chaos in Learning a Simple Two-Person Game." *Proceedings of the National Academy of Sciences* 99 (7): 4748–4751. https://doi.org/10.1073/pnas.032086299.

Schelling, T. C. 1971. "Dynamic Models of Segregation." *Journal of Mathematical Sociology* 1 (2): 143–186. https://doi.org/10.1080/0022250X.1971.9989794.

Scholl, A., A. Calinescu, and J. D. Farmer. 2021. "How Market Ecology Explains Market Malfunction." *Proceedings of the National Academy of Sciences* 118 (39). https://ideas.repec.org/a/nas/journl/v118y2021pe2015574118.html.

Schumpeter, J. A. [1911]1934. *The Theory of Economic Development.* Cambridge, MA: Harvard University Press.

Schweitzer, F., G. Fagiolo, D. Sornette, F. Vega-Redondo, A. Vespignani, and D. R. White. 2009. "Economic Networks: The New Challenges." *Science* 325 (5939): 422–425. https://doi.org/10.1126/science.1173644.

Simon, H. A. 1955. "A Behavioral Model of Rational Choice." *Quarterly Journal of Economics* 69 (1): 99–118. https://doi.org/10.2307/1884852.

———. 1962. "The Architecture of Complexity." *Proceedings of the American Philosophical Society* 106 (6): 467–482. https://www.jstor.org/stable/985254.

———. 1973. "The Organization of Complex Systems." In *Hierarchy Theory,* edited by H. H. Pattee, 1–27. New York: George Braziller.

———. 1978. "Rationality as Process and as Product of Thought." *American Economic Review* 68 (2): 1–16. https://www.jstor.org/stable/1816653.

Smith, A. [1977]1776. *An Inquiry into the Nature and Causes of the Wealth of Nations.* Chicago, IL: University of Chicago Press.

Tacchella, A., M. Cristelli, G. Caldarelli, A. Gabrielli, and L. Pietronero. 2012. "A New Metrics for Countries' Fitness and Products' Complexity." *Scientific Reports* 2:723. https://doi.org/10.1038/srep00723.

Tesfatsion, L. 2002. "Agent-Based Computational Economics: Growing Economies from the Bottom Up." *Artificial Life* 8 (1): 55–82. https://doi.org/10.1162/106454602753694765.

Tesfatsion, L., and K. L. Judd, eds. 2006. *Handbook of Computational Economics, Volume 2.* Amsterdam, Netherlands: Elsevier.

Thurner, S., and S. Poledna. 2013. "DebtRank-Transparency: Controlling Systemic Risk in Financial Networks." *Scientific Reports* 3 (1): 1888. https://doi.org/10.1038/srep01888.

Veblen, T. 1898. "Why Is Economics Not an Evolutionary Science?" *Quarterly Journal of Economics* 12 (4): 373–397. https://doi.org/10.2307/1882952.

Vega-Redondo, F. 2007. *Complex Social Networks.* Cambridge, UK: Cambridge University Press.

Waldrop, M. 1992. *Complexity: The Emerging Science at the Edge of Order and Chaos.* New York, NY: Simon & Schuster.

West, G. B. 2017. *Scale: The Universal Laws of Growth, Innovation, Sustainability, and the Pace of Life in Organisms, Cities, Economies, and Companies.* New York: Penguin Press.

West, G. B., J. H. Brown, and B. J. Enquist. 1997. "A General Model for the Origin of Allometric Scaling Laws in Biology." *Science* 276 (5309): 122–126. https://doi.org/10.1126/science.276.5309.122.

Young, H. P. 1993. "The Evolution of Conventions." *Econometrica* 61 (1): 57–84. https://doi.org/10.2307/2951778.

———. 1998. *Individual Strategy and Social Structure.* Princeton, NJ: Princeton University Press.

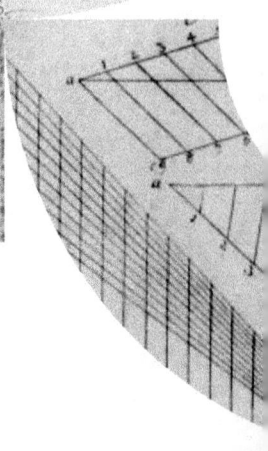

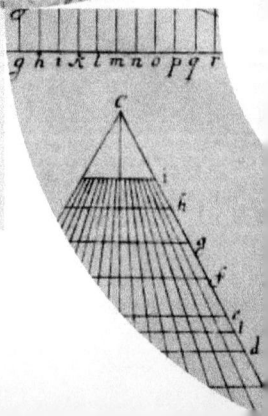

PART I

Foundations of
Complexity Economics

COMBINATORIAL EVOLUTION

W. Brian Arthur, Santa Fe Institute and SRI International

Abstract

This chapter discusses a mechanism that lies behind evolution in a number of systems: combinatorial evolution. In this mechanism, novel elements are created by combining earlier existing elements, and they become available as building blocks for creating yet further novel elements. This defines a mechanism of heredity: The system evolves. Combinatorial evolution was introduced as a concept in Arthur (2009) and it since has received widespread attention. This chapter examines how combinatorial evolution works and shows that it occurs naturally in a class of systems that includes technology, chemistry, mathematics, computation, language, and parts of biology.

In this chapter I want to talk about a mechanism that lies behind evolution[1] in a number of systems: direct combination. Novel elements can emerge from the combination of pre-existing elements, and become available as potential building blocks for creating yet further elements. The system has a means of heredity—it evolves. Chloroplasts, the small but essential organelles in plant cells responsible for photosynthesis, are believed to have originated from cyanobacteria—photosynthetic bacteria that were once free-living—combining with primitive

---

[1]Evolution, according to Stephen J. Gould (1989), has two main meanings: that "lineages alter their form and diversity through time by a natural process of change—'descent with modification'" or that "all organisms are related by ties of genealogy or descent from common ancestry along the branching patterns of life's tree." I use evolution in its second, stronger sense of "ties of genealogy" via direct heredity or direct parentage.

eukaryotic cells. The early cells engulfed (swallowed without digestion) the bacteria and the novel combination allowed the host cells to harness sunlight for energy and led eventually to the development of modern plant forms.[2]

Evolution via combination is by no means rare in biology. And it is the main mechanism, I believe, in another large system that evolves: technology. Novel technologies combine or are assembled from other, previously existing technologies and often become building blocks for creating yet further novel technologies. A global positioning system (GPS) provides location information by combining satellites, atomic clocks, radio transmission, and mathematical algorithms, all of which existed as earlier technologies. Atomic clocks aboard the satellites provide extremely precise time, and by measuring the time delays among signals received from multiple satellites, navigators can determine their position with high accuracy. We can say that GPS has its origins in—has its "parentage" in—satellites, atomic clocks, and radio receivers, and these in turn have parentage in the technologies that created *them*. We can also say that GPS becomes a fundamental building block of technologies such as modern ship and aircraft navigation. Notice there is a mechanism of heredity here, but it isn't variation and selection. GPS does not derive from successive variations of previous navigation systems; rather, it derives from combination.

In my book *The Nature of Technology: What It Is and How It Evolves*[3] I described this mechanism and called it *combinatorial evolution*. I wrote about it largely as a process within technology, while mentioning that it occurs in other fields as well—mathematics, for example. On deeper investigation I have

---

[2] This idea was first proposed by Margulis (1970) and is now widely supported (see Keeling 2010).

[3] This chapter draws heavily from this book (Arthur 2009).

discovered there is a small but significant literature on evolution by combination: mostly in biology, in the history of chemical metabolism, and in complex systems (e.g., Simon 1962; Margulis 1970, 1981; Smith and Szathmáry 1995; Holland 1975; Kauffman 1993; Wagner 2014).[4] I find this literature fascinating, but I also find it scattered and more than a little restricted. For the most part, writings are specific to the particular domain they are describing, and they talk about new entities—novel structures within cells, new metabolic reactions, new pattern formations—being created via combination and therefore pushing evolution forward by one step or at most two. Missing is a description of a general, continuing process of evolution being driven by combination.

In this short overview, I want to provide such a description. In particular I want to introduce a certain interesting class of systems in which this form of evolution is present. I will use technology to illustrate how the mechanism works and will briefly indicate how it shows up in other fields.

A Class of Systems That Show Combinatorial Evolution

To set a context for combinatorial evolution, I want to talk about a class of systems that are at base collections of elements. Language is such a system; it contains a collection of elements—vocabulary words and punctuation symbols—that can be combined to form expressions—sentences or speech utterances for immediate purposes—without limit. Wilhelm von Humboldt ([1836]1988, 91) famously observed that language allows for the "infinite use of finite means." A quite different system, a coinage system,

---

[4]In particular, Herbert Simon's paper of 1962 is a classic and it anticipates some of the ideas here. It shows that complex systems may often evolve in a hierarchical way by integrating or combining simpler constructions (intermediate forms) into higher-level ones. This promotes modularity, enables complex systems to evolve more rapidly by reducing the overall complexity of changes needed, and allows such systems to survive the transition providing the intermediate forms are stable.

similarly contains a collection of elements: coins of different denominations, which in combination can produce arbitrary cash amounts for payment or for tendering change. This system also allows for infinite use of finite means. Languages and coinage systems are examples of *combinatorial systems*: They have elements—building blocks if you like—that can be combined into *expressions* or *utterances* for immediate purposes, much as LEGO brick elements can be used to form toy-model constructions for immediate purposes. Combination follows a set of allowable rules or *grammar*, and we can think of the system as a *collection* or *toolbox* of reusable element types that allows new and purposed combinations to arise. If the system is one in nature, combination might happen haphazardly with not all results having a useful purpose or conforming to the grammar; if it is directed by human design, likely the combination will be properly formed and comply with the grammar.

⚏

Some combinatorial systems contain a fixed set of unchanging element types that can be used repeatedly for different immediate purposes, but do not evolve. I want now to go beyond this and look at combinatorial systems where *novel* element types can be generated and added to the toolbox from time to time. These, too, are formed from existing elements via combination. This may happen randomly, or it may happen because a particular combination is found to be useful repeatedly, or it may happen because a new element answers some ongoing need. Whatever the case, new elements may be generated from time to time and added to the toolbox, where they become available as new building blocks to form expressions and create further novel elements. I will call such systems *generative combinatorial* systems, or simply *generative systems*.

As an example of a generative system, think of an (imaginary) computer language.[5] It has both a basic vocabulary of instructions and a library of often-used functions that carry out recurring tasks like calculating logarithms or plotting data. These library functions are prewritten code, included so that programmers don't have to write them each time they use them, and together with the basic instructions they form the toolbox of the language. Programming applications—"expressions" if you like— are constructed from the toolbox, and from time to time new library functions are added to it, also constructed from the basic instructions and existing library functions. These become building blocks available for immediate coding and for generating yet further library functions. In this way the language evolves. Of course, the evolution in this case would not be over eons, it would be over months or years. If we looked at the elements in a programming library we could construct a family tree of their ancestry back to basic instructions, but it wouldn't be a neatly branching tree; it would be a crisscrossing network because most library functions have multiple ancestors.[6]

We now have a class of systems to work with, but how exactly does *evolution* happen within such systems?

What really evolves in a generative system is its collection of element types. The steps are not particularly complicated, so let me delineate them here:

1. Some recurring or ongoing need (or opportunity) arises.
2. A combination is found or emerges that fulfills this need.
3. The combination becomes an element in its own right, and it becomes *encapsulated*, existing as its own entity. It then

---

[5] I say *imaginary* because actual computer languages have several libraries, and computer language design is by no means simple (Stroustrup 1994).
[6] For empirical evidence of this see Valverde, Vidiella, and Duran-Nebreda (2026).

becomes a member of the library or toolbox, available for further combination.

Step 1 means that some need or purpose arises that poses an open problem or creates an opportunity that demands to be fulfilled. In Step 2, a solution to this problem or opportunity is found, either deliberately or by showing up randomly. In Step 3, this solution or successful combination, if used often enough, becomes a module. It becomes "selected." It gets its own label and becomes encapsulated in a device, or method, or element available for further combination. Encapsulation gives us fresh building blocks, and these may or may not replace old elements. In this way the toolbox changes in composition as its elements change and provides for new and fresh immediate expressions. It adapts, in other words, by providing novel elements that reflect new needs in its changing environment. The overall system evolves slowly over time as its elements change and provide for new combinations.

~ 61 ~

The details of specific systems that conform to this overall description differ in each field where we find combinatorial evolution. Some may offer a toolbox that can contain only a limited number of elements. Some may evolve novel elements via variation and selection in addition to the mechanism above—in fact, occasionally significant new combinations arise via variation and selection. Some may generate novel elements that render previous ones obsolete and we can say then they have both active and inactive elements.

Notice that the systems I am describing operate on two timescales: a fast one where they create expressions or utterances for immediate use; and a slow one where they occasionally create novel element-types by combining previous ones. It is not hard to show that generative systems exhibit in common other interesting—and rich—features. I list these separately in Box 1.

Combinatorial Evolution in Technology

I have been talking generally so far, and in what follows I want to discuss actual examples of systems that show combinatorial evolution.[7] Let me begin with technology. By *technology* (singular) I mean simply a means to a human purpose: the steam engine, the railway locomotive, a statistical method, a sorting algorithm. By *technology* (plural) I mean the overall collection of such means. The context will make clear which is meant.

How does combination arise with technologies? Actually, it is something that arises naturally. Technologies are assembled from parts or subassemblies, and so by nature they are combinations.[8] There is a reason for this. To do its work, a technology uses a phenomenon, usually several, and the phenomenon must be set up, properly prepared, and properly controlled. This requires many parts.

A hydroelectric power plant, for example, uses water falling under gravity to turn huge turbines, which move generators whose rotating magnets move electricity in neighboring stationary coils of wire—this latter effect is its base phenomenon. But the water must be stored, channeled into intake pipes (called penstocks), and directed into the turbines that move generators so that electricity flows, and the electricity must be stepped up in voltage and fed into transmission lines. So the system is an amalgam of reservoirs, penstocks, turbines, stators, transformers, and high-voltage transmission lines. Hydroelectric power generation is

---

[7]For details on the ideas presented here, see Arthur (2007, 2009). On technology see also Koppl *et al.* (2023).

[8]This notion that technologies are combinations of other technologies goes back at least to Thurston ([1878]1971) and has been proposed by several authors since, such as Usher ([1929]1988) and Gilfillan (1935).

Box 1. Generative combinatorial systems exhibit certain common properties.

No upper limit. Expressions and building blocks that can be created are in principle infinite in number.

Perpetual novelty. Such systems can continually create expressions and building blocks that have not been used before; they are open and never complete.

Two timescales. On a fast timescale, such systems create expressions—utterances—by routine combination of their toolbox elements for immediate purposes. On a slow timescale they generate new toolbox elements by occasionally combining ones that already exist.

Recursiveness. Elements are formed from subparts that are combinations of elements formed from subparts . . . . Each element forms a hierarchy of combinations down to individual basic elements.

Changing internal composition. Individual elements may alter their internal composition over time. An electric car contains a battery subsystem, which may change and improve over the life of the vehicle. Generative systems are fluid, highly configurable, and changeable over time.

Autopoiesis. Such systems create novel elements from existing elements in a self-sustaining way. Technically we can say that they are *autopoietic*[a]—they create themselves out of themselves.

Splintering. Such systems or toolbox collections easily splinter into special subsystems with their own methods or grammar.[b] Chemistry splits into inorganic chemistry, organic chemistry, biochemistry, nanochemistry, and so on. Each of these may become its own specialty.

An ecology. Elements within systems often form an ecology: They may support each other, compete with each other, negate each other. And often they interact with their outside environment.

Adaptation. Such systems adapt by generating novel building blocks that better conform to the needs of their current environment.

---

[a] The term *autopoietic* was introduced in another context by Varela, Maturana, and Uribe (1974).
[b] For evidence of splintering (or splitting) in technology, see Lafond and Kim (2019).

an individual technology put together from a large toolbox of technologies.[9]

On a day-to-day basis technology or its toolbox libraries make possible individual constructions, expressions or utterances such as the Airbus A350, or the Hutong Yangtze River cable-stayed bridge. Technology is a language in everyday use. Its projects may be mundane or imaginative and elaborate.

So much for technologies and the toolbox they create. We still need to ask how novel toolbox elements in technology are generated. This is equivalent to Darwin's question of how species form. Some may indeed arise from variation and selection, and this is by no means infrequent (Basalla 1988). But if the novel technology—radar, GPS, the polymerase chain reaction—is radically new, it does not come from successive modifications of earlier technologies. It arrives independently in its own right. We are really asking here how *invention* happens. There is a literature on this from the 1930s,[10] but it isn't very satisfactory. Explanations of invention here inevitably invoke at the heart of the process some "act of insight" or "fortuitous configuration . . . in thought," but the fortuitous act is never quite spelled out.

My own explanation (Arthur 2007) is that invention is really a form of problem-solving, and as such is not particularly mysterious. It follows a distinct process. When some recurring or ongoing need arises, certain people—I call them *originators*—begin to look for a means to solve it.[11] They consider an overall

---

[9]There are many such toolboxes, because technology tends to splinter. Individual technologies are based on particular families of phenomena— optical ones, chemical ones, electrical ones, electronic ones, digital ones— so we have optical technology, chemical technology, electrical technology, electronic technology, digital technology, and so on.

[10]See for example Usher ([1929]1988), Kaempffert (1930), and Gilfillan (1935).

[11]Invention may also proceed via harnessing new phenomena such as X-rays; here a similar process works (Arthur 2007).

principle—the idea of some phenomenon in use—in fact usually several principles, keeping in mind how each could be set up with the pieces or functionalities they have at their disposal. As they explore this, difficulties or hindrances arise and become subproblems that need to be resolved. These, too, need solutions, and so the process goes back and forth between problem and subproblem until a satisfactory solution has been reached. In 1972, when Gary Starkweather at Xerox Parc was looking for a means to print images from computers, he mulled over several candidate principles. One was to use a cathode ray tube to "burn" text onto paper surrounding it. A different, promising one was to use the computer to direct a laser beam to "paint" images on a copier drum; highly focused images could then be rolled off the copy machine. In the course of pursuing this principle several subproblems arose. One was that the lasers of the time were heavy and cumbersome and couldn't be moved rapidly enough to scan back and forth across the drum. Starkweather resolved this by keeping the laser fixed and having it direct its light at a revolving set of mirrors so that the beam could scan across the drum much as a lighthouse's beam scans across the horizon. After lengthy effort getting the individual components to work and resolving subproblems, the laser printer emerged (Starkweather 1980). It was the direct offspring of the computer, the laser, several optical technologies, and xerography.

Invention, then, at base is problem-solving using resources at hand. And of course it requires human agency (at least for now). It requires imagination, the ability to envisage some principle in use and the parts that might be needed for it, and it requires deep engineering skills, but it does not require genius or mysterious insight.

Once a novel technology or invention comes into being, it may become encapsulated and added to the toolbox. It then becomes a new building block, a new "species" if you like, available for further combination. We are now firmly in evolutionary territory. The collective of technology evolves or builds out via its "parentage"[12] in the previous technologies it builds from, and this way the overall collection of technologies bootstraps itself upward from the few to the many and from the simple to the complex. In its early days, the transistor plus a few other electronic components produced simple logic circuits; these served as building blocks for more complex circuits like adders, registers, and control units; and these in further rounds of combination produced arithmetic, graphic, and neural-network processors. Each stage contributed new building blocks for future technologies, and in this way over several decades the complex emerged from the simple.

The reader might check whether technology fulfills the properties listed in Box 1.

An obvious question at this stage arises: What use is it to know that technology evolves—and that it evolves via combination? My answer is this. In both our day-to-day thinking and in economic thinking we tend to see "technology" either as some vague macro thing that just exists, or as a set of independent objects—railroads, the steam engine, radar, mass production— that arise independently and sporadically. If, however we see technologies arising out of previous technologies and making possible future technologies, we see where technologies have come from and how they relate and interrelate over time. We also become conscious that our economy is not something given but something constantly forming and re-forming at all levels. Since the nineteenth century, economists have been aware that technology viewed as a "means of production" makes possible—

---

[12]"Parentage" need not include all components of a new technology. Practically speaking, I think of it as the base technologies that made the technology possible.

creates the skeleton of—the economy (Arthur 2021). We therefore begin to see where our economy has come from, and that it is never fully predictable and never at rest.

Change is ongoing in all subparts of technology, at all scales, and at all times; and this means that change is ongoing in all parts of the economy, at all scales, and at all times. The economy is ever creating itself anew.

Other Fields That Show Combinatorial Evolution

I want to show now that combinatorial evolution occurs in other fields besides technology, though it may not be the main driver in these. I would caution the reader that I am not a specialist in these areas, so my treatments here will be necessarily brief and simplified. My purpose is to indicate how combination applies across a number of fields without delving into full complexity. Specialists may find detailed analyses in domain-specific literature.

> **Chemistry.** In multistep chemical synthesis, a chemical product may typically combine with another product to create a novel product, which combines with another product to create a further product. For example, ethene reacts with oxygen to form ethylene oxide, which can further react with water to produce ethylene glycol. Ethylene glycol can in turn react with terephthalic acid to form polyester. In this way more complicated molecular structures arise from simpler ones and create new molecular structures for addition to the chemical "library" (Carey and Sundberg 2007). It was by such a process—a very long, drawn-out one—that in ancient times metabolic chemistry bootstrapped its way from the early formation of simple organic molecules, such as amino acids and nucleotides, to the development of more complex molecules capable of catalyzing reactions. These molecules facilitated

the breakdown of nutrients for energy and the synthesis of new cellular components. This gradual increase in complexity led to the formation of protocells, which acted to encapsulate and concentrate metabolic reactions, and in due course these simpler molecules combined and organized into the complex metabolic systems we see in modern organisms (Wagner 2014; Luisi 2016). A contemporary version of combinatorial evolution shows up in the new field of combinatorial synthesis, where diverse building blocks are deliberately, experimentally combined to produce a high multiplicity of outcomes. Specific combinations are then selected based on their alignment with a target outcome (Beck-Sickinger and Weber 2002).

Biological Evolution. In biology, combination is by no means rare; in fact, it turns out to be surprisingly frequent. It arises in several mechanisms, of which I will highlight four: symbiosis, horizontal gene transfer, the evolution of genetic regulatory networks, and recombination.[13] With symbiosis, biological entities may coexist for mutual benefit and occasionally merge or combine to form new structures (Margulis 1970, 1981), and this may be the cause of certain major transitions in evolution (Smith and Szathmáry 1995). Mitochondria and multicellular organisms likely originated this way. With horizontal gene transfer, organisms— especially bacteria and archaea—acquire genetic material directly from different species or organisms with different

---

[13] Combination also "underlies the evolution of proteins, in that many proteins are constructed of structural motifs, which evolution has recombined in different ways to construct new proteins, just as [happens] with technologies" (Erwin 2024). Modern evolutionary theory of course includes other processes such as genetic drift, developmental bias, niche construction, and multi-level selection.

lineages; this foreign DNA integrates into the recipient's genome and, if stable, can be inherited by subsequent generations. Over time, horizontal gene transfer can lead to the emergence of new traits or new phenotypes in populations (Ochman, Lawrence, and Groisman 2000). With gene regulatory networks, we can say that these are built out of circuits that form a toolbox from which elements are combined for different development purposes (Erwin 2024). And with recombination, large pieces of genetic material are reshuffled and freshly combined, and the variation that results may or may not be selected for. In this case our mechanism and the basic one of variation and selection overlap. This occurrence of the two mechanisms I've been talking about working together is frequent, especially in natural systems. Combination may involve a long series of intermediate steps and these steps can be regarded as variations on which selection acts (e.g., Szostak, Bartel, and Luisi 2001). Not all these mechanisms fit the generative model I described above, but at least roughly, symbiosis and gene regulation do.

These examples of combination in chemistry and biology are well known. In three other fields—conceptual ones— this is less the case. Here, combinatorial evolution also takes place but it does so more in an abstract, nonphysical way: Existing concepts are combined—possibly implicitly—into new, encapsulated concepts, and in turn these become available for further combination. This sounds more than a touch ethereal, but as a mechanism of heredity in our human world it is nonetheless real.

Consider these fields as examples:

Mathematics. If we take the main objects that form the bones of mathematics—concepts, structures, axioms, methods, theorems and their proofs—a few of these are elemental, but the vast majority are constructions made up from simpler objects. Arithmetic is built from an accepted number system, axiom system, and from methods for addition, subtraction and division. Algebra is built from arithmetic, inherits its methods, and adds its own. Linear algebra is constructed from elementary algebra, and so on. In mathematics, construction—combination—is central. Mathematics once in place is used for mundane purposes or expression especially in commerce, engineering, and science. Novel elements arise if they prove useful and become encapsulated as concepts in their own right. Encapsulation is central in mathematics. Theorems are logical arguments encapsulated, methods are procedures for manipulation encapsulated, concepts are definitions in terms of simpler objects encapsulated. Encapsulation makes the discipline tractable,[14] much as abstract terms make specialized discourse tractable.

Evolution, the process of novel objects arising from combination of existing ones, is most clearly seen in the construction of proofs. Describing Andrew Wiles's proof of Fermat's Last Theorem, mathematician Kenneth Ribet remarks, "You turn a page and there's a brief appearance of some fundamental theorem by Deligne, and then you turn to another page and in some incidental way there's a theorem by Hellegouarch—all of these things are just

---

[14]Bertrand Russell (1919) commented dryly, "The process of deducing logic from first principles is an arduous one."

called into play and used for a moment before going on to the next idea" (quoted in Singh 1997). Mathematics architects its structures from existing structures which are themselves architected from previous structures. It builds itself in response to the external world and from elements it already possesses.

Language. In language, evolution mostly shows up in its more gradual form of descent with modification: Vocabulary changes slowly in meaning, and usage changes too (Crowley 1997). But evolution by combination also shows up: From time to time language reduces (combines, if you like) a complicated set of concepts into a single, usable word or neologism.[15] The fairly new word "Brexit" denotes Britain's decision to leave the European Union; and the suffix *-gate*, as in Watergate, stands for minor government malfeasance. Of course Brexit isn't a fixed combination of other words; it is at best a loose conceptual assemblage of simpler unstated actions and concepts. But it is an encapsulation that saves words and therefore earns its place in the language.

~71~

Legal System. At its core the legal system is a set of concepts and precedents that derive from combinations of earlier concepts and precedents. Such legal synthesis means analyzing existing legal principles, rules, and doctrines, and integrating them to create new legal concepts or to develop deeper understanding of existing ones. Of course the law changes by other processes, and in this one "parentage" would be hard to establish, so here evolution does become ethereal. Still, I like to keep this mechanism in mind because it forms a key generator for the evolution of the law, and it

---

[15] The process is called *lexicalization* (Brinton and Traugott 2005).

gives a sense of historical linkage and provenance to legal ideas.

The reader might think of other fields where combinatorial evolution applies.[16] In the above systems, earlier elements can combine to add to a library of elements available for creating additional elements. We can see each system simultaneously as a language, a chemistry, or a technology.

Discussion

If we define our combinatorial evolution mechanism as the creation of new "toolbox" elements in a system via combination, we can see from the examples I've given that this mechanism often goes along with the Darwin's basic variation and selection one. Sometimes, as with technology, combination is followed by variation and selection—once a primitive aircraft exists, variations on it follow. Sometimes, as in biology and chemistry, a fresh combination follows from many steps of variation and selection. And sometimes, as with horizontal gene transfer, the two mechanisms overlap. The exact relation between the two mechanisms doesn't matter that much. What matters in combinatorial evolution is that new "toolbox" or "library" elements for further use are created via combination.

Before we close, let me make a few more general remarks about combination.

Combination is rarely if ever some new thing or event accomplished in single stroke. In the natural realm, it nearly always requires exploration of the space of possibilities along with a series of intermediate steps that lead to the novel combination. In the human realm, it may involve random exploration—tinkering— but more likely it calls for systematic, tested choices among

---

[16]For applications in archaeology see Lehner (2023) and Laursen (2023).

alternatives. In both cases combination is not what is put into a system; it is what comes out. Of course, standard variation and selection also involve "exploration" and many experimental intermediate steps before a final result—a new species, say— emerges. The overall outcome, whether evolution by combination or evolution by variation and selection, is nonetheless real.

I said earlier that combination can be ordinary and mundane, but sometimes—especially in nature—it can be remarkable. New combinations, Philip Anderson pointed out in 1972, may differ in quality and possibilities from their constituent components: Water has a "wetness" that neither oxygen nor hydrogen atoms possess—wetness we can say "emerges" from combination. On reflection, I believe what is remarkable about combinatorial systems is not simply the emergent qualities they generate, it is the *world* they generate. Combination—call it creative construction—in the human realm makes possible the world of music and of literature; in nature it makes possible the world of biochemistry and of life.

One of the puzzles of Darwin's original mechanism is how incremental modifications can bring into being elaborately constructed forms. How does the mammalian eye emerge from "numerous, successive, slight modifications" (Darwin 1859, 189) of more elementary structures? This question has been greatly argued over, and largely resolved in Darwin's favor—a chain of small modifications can indeed bring about novel forms. Still, as an evolutionary mechanism, it is easier to see how combination can quickly generate elaborate structures. It has a large adjacent possible, meaning it can make large leaps, which can directly allow the emergence of complicated new forms. If combination has this advantage, why don't we see more of it in nature? The reason is that novel combinations are unlikely to be viable precisely because they are *not* slight modifications; they may be haphazard in form and unlike any object that preceded them, and therefore not well adapted to survive, whereas an incremental Darwinian change will be more likely to survive simply because it is close to something that already survives. In the human domain, by contrast, we can create combinations that do survive, so that here evolution is relatively swift.

One last point. The combinatorial mechanism I have described is a shorthand version of how evolution works. In a given system, both expression and generation of novel element types require an apparatus to make routine combinations and to generate novel building blocks, and this apparatus normally lies outside the system we are looking at. A programming language requires not just a set of allowable instructions and a library that adds to these; it requires human (or artificial) minds that write code for these. The code requires computational machinery for its execution, and additions to the library require judgment from experienced practitioners. Expression and generation do not happen within the language itself. The same can be said for even the basic variation and selection mechanism. It is easy to talk about, but its realization in a biological world of organisms and their ecosystems requires a panoply of molecular machinery for copying and correcting DNA, for transcription, gene expression, and the construction of organisms themselves; such machinery is very much the subject of modern biological research. Variation and selection are simple, but the apparatus to realize these in the natural world is not.[17]

---

[17] This panoply of machinery must of course itself evolve, which implies that some evolutionary systems may evolve "symbiotically with"—in tandem with—other evolutionary systems.

Conclusion

In this chapter I have been describing a mechanism that lies behind evolution in a number of systems. Novel elements are created by combining earlier existing elements, and they become available as building blocks for creating yet further novel elements. This defines a mechanism of heredity, so we can say the system evolves. We can also say in general that from original simplicity complexity grows—it too evolves.

Combinatorial evolution occurs naturally in a class of systems that routinely combine elements for certain purposes. This class includes chemistry, mathematics, computation, technology, language, and several parts of biology. Such systems operate on a fast time-scale: They create expressions—utterances if you like—by routine combination of their member elements for immediate purposes. And they operate on a slow timescale: They generate new member elements by occasionally combining ones that already exist. It is this set of changes on the slower timescale that we call *evolution*. Systems in this class show several common features: recursion, openness to novelty, adaptation, and autopoiesis. Their evolution is *organic* in that they build new elements on top of previous ones; *indeterminate* in that we cannot predict what novel elements will arise next; and *history dependent* because the occurrence of novel elements is partly random and the sequence of their arrival determines what happens next in time.

Our combinatorial mechanism in no way negates Darwin's original, beautiful perception of successive selected modifications leading to novel structures—novel species. Combination and natural selection operate alongside each other in many systems. Combinatorial evolution is not rare, and where we see it we become aware of a connected, deep, ancestral history that we did not see before. ❦

Acknowledgments

I am grateful to Ronan Arthur, Doug Erwin, Doyne Farmer, François Lafond, Rika Preiser, David Simpson, and Andreas Wagner for useful comments. All views expressed in this paper are strictly my own.

REFERENCES

Anderson, P. 1972. "More Is Different." *Science* 177 (4047): 393–396. https://doi.org/10.1126/science.177.4047.393.

Arthur, W. B. 2007. "The Structure of Invention." *Research Policy* 36 (2): 274–287. https://doi.org/10.1016/j.respol.2006.11.005.

———. 2009. *The Nature of Technology: What it is and How it Evolves.* New York, NY: The Free Press.

———. 2021. "Foundations of Complexity Economics." *Nature Reviews Physics* 3:136–145. https://doi.org/10.1038/s42254-020-00273-3.

Basalla, G. 1988. *The Evolution of Technology.* Cambridge, UK: Cambridge University Press.

Beck-Sickinger, A., and P. Weber. 2002. *Combinatorial Strategies in Biology and Chemistry.* Chichester, UK: Wiley.

Brinton, L. J., and E. C. Traugott. 2005. *Lexicalization and Language Change.* Cambridge, UK: Cambridge University Press.

Carey, F. A., and R. J. Sundberg. 2007. *Advanced Organic Chemistry Part B: Reaction and Synthesis.* Cham, Switzerland: Springer.

Crowley, T. 1997. *An Introduction to Historical Linguistics.* Oxford, UK: Oxford University Press.

Darwin, C. 1859. *On the Origin of Species by Means of Natural Selection.* London, UK: John Murray.

Erwin, D. 2024. Personal communication, October 15, 2024.

Gilfillan, S. C. 1935. *The Sociology of Invention*. Chicago, IL: Follett Publishing Company.

Gould, S. J. 1989. *Wonderful Life: The Burgess Shale and the Nature of History*. New York, NY: W. W. Norton & Co.

Holland, J. H. 1975. *Adaptation in Natural and Artificial Systems*. Ann Arbor, MI: University of Michigan Press.

Kaempffert, W. 1930. *Invention and Society*. Chicago, IL: American Library Association.

Kauffman, S. 1993. *The Origins of Order*. New York, NY: Oxford University Press.

Keeling, P. 2010. "The Endosymbiotic Origin, Diversification, and Fate of Plastids." *Philosophical Transactions of the Royal Society B: Biological Sciences* 365 (1541): 729–748. https://doi.org/10.1098/rstb.2009.0103.

Koppl, R., R. C. Gatti, A. Devereaux, B. D. Fath, J. Herriot, W. Hordijk, S. Kauffman, R. E. Ulanowicz, and S. Valverde. 2023. *Explaining Technology*. Cambridge, UK: Cambridge University Press. https://doi.org/10.1017/9781009386289.

Lafond, F., and D. Kim. 2019. "Long-Run Dynamics of the U.S. Patent Classification System." *Journal of Evolutionary Economics* 29:631–664. https://doi.org/10.1007/s00191-018-0603-3.

Laursen, S. T. 2023. "The City of Dilmun and Evolving Technology, c. 2000 BC: Combinatorial Evolution and the Emergence Urban Heterogeneity." *Journal of Urban Archaeology* 8:65–84. https://doi.org/10.1484/J.JUA.5.135659.

Lehner, M. 2023. "Combinatorial Evolution and Heterogeneous Cohabitation at the Giant Pyramids." *Journal of Urban Archaeology* 8:21–46. https://doi.org/10.1484/J.JUA.5.135657.

Luisi, P. L. 2016. *The Emergence of Life: From Chemical Origins to Synthetic Biology*. Cambridge, UK: Cambridge University Press.

Margulis, L. 1970. *Origin of Eukaryotic Cells*. New Haven, CT: Yale University Press.

———. 1981. *Symbiosis in Cell Evolution*. New York, NY: Freeman and Co.

Ochman, H., J. G. Lawrence, and E. A. Groisman. 2000. "Lateral Gene Transfer and the Nature of Bacterial Innovation." *Nature* 405 (6784): 299–304. https://doi.org/10.1038/35012500.

Russell, B. 1919. *Introduction to Mathematical Philosophy*. London, UK: Allen & Unwin.

Simon, H. 1962. "The Architecture of Complexity." *Proceedings of the American Philosophical Society* 106 (6): 467–482. https://www.jstor.org/stable/985254.

Singh, S. 1997. *Fermat's Last Theorem: The Story of a Riddle that Confounded the World's Greatest Minds for 358 Years*. London, UK: Fourth Estate.

Smith, J. M., and E. Szathmáry. 1995. *The Major Transitions in Evolution*. New York, NY: W. H. Freeman.

Starkweather, G. 1980. "High-Speed Laser Printing Systems." In *Laser Applications*, edited by M. Ross and F. Aronowitz, vol. 4. New York, NY: Academic Press.

Stroustrup, B. 1994. *The Design and Evolution of C++*. Boston, MA: Addison-Wesley.

Szostak, J. W., D. P. Bartel, and P. L. Luisi. 2001. "Synthesizing Life." *Nature* 409 (6818): 387–390. https://doi.org/10.1038/35053176.

Thurston, R. H. [1878]1971. *A History of the Growth of the Steam-Engine*. Port Washington, NY: Associated Faculty Press.

Usher, A. P. [1929]1988. *A History of Mechanical Inventions*. New York, NY: Dover.

Valverde, S., B. Vidiella, and S. Duran-Nebreda. 2026. "Coevolution of Software and Innovation: Constraints, Tinkering, and Symbiosis." In *The Economy as an Evolving Complex System IV*, edited by R. M. del Rio-Chanona, M. Pangallo, J. Bednar, E. D. Beinhocker, J. Kaszowska-Mojsa, F. Lafond, P. Mealy, A. Pichler, and J. D. Farmer. Santa Fe, NM: SFI Press.

Varela, F., H. Maturana, and R. Uribe. 1974. "Autopoiesis: The Organization of Living Systems, its Characterization and a Model." *BioSystems* 5:187–196. https://doi.org/10.1016/0303-2647(74)90031-8.

von Humboldt, W. [1836]1988. *On Language: The Diversity of Human Language-Structure and its Influence on the Mental Development of Mankind*. Translated by P. Heath. Cambridge, UK: Cambridge University Press.

Wagner, A. 2014. *The Arrival of the Fittest: How Nature Innovates*. New York, NY: Current.

COMPLEXITY ECONOMICS AND GENERAL EQUILIBRIUM

John Geanakoplos, Yale University and Santa Fe Institute

Introduction

In the following pages I give my understanding of general-equilibrium economics and complexity economics (i.e., agent-based modeling). The two are spiritual siblings, because they are based on ground-up modeling via many heterogeneous agents. I argue that the difference between them is that general equilibrium (GE) involves supply equaling demand across many interdependent markets, or, in other words, a multidimensional fixed point of doing and expecting. This framework is a prerequisite for complete rationality. By contrast, agent-based modeling (ABM) drops simultaneous equations and replaces rational utility maximization with behavioral rules. The advantage is that, freed from the computational burden of solving simultaneous equations, ABM can operate in much higher dimensions and include realistic levels of detail that are unimaginable in equilibrium models. The downside is that one can never be sure that, as the economy evolves into new situations, the assumed behavioral rules won't become unrealistically irrational.

My suggestion is to treat GE and ABM as complementary. Economists should solve stylized GE models to find out what rational behavior is. Once rational behavior is understood, it can be mimicked or approximated by behavioral rules in a much richer ABM model. Conversely, one can check that an ABM model is compatible with full rationality by building a simpler GE model and seeing if similar features appear. Of course, some ABM models are designed to display features that emerge only in a world of boundedly rational agents.

I describe four ongoing research projects in which I began with a GE-style model and then tried to enrich it (with others) into an ABM model. These are mortgage prepayment models, models of monies and inflation, models of booms and busts with collateral, and models of housing.

General Equilibrium

EQUILIBRIUM IN TWO WORDS

Inspired by the popularity of the 2001 Hollywood movie *A Beautiful Mind*, about the brilliant mathematician and Nobel Prize–winning game theorist John Nash, the Indian Game Theory Society was launched in 2003 with a conference in Mumbai. Thousands of people came to hear Nash, though fewer than a hundred stuck around for the rest of us. Five of the speakers and our families, including Nash and his wife, left immediately afterward for a tour of Rajasthan and other Indian sites. Mrs. Nash had divorced John soon after the events of the movie, as his mental condition spiraled downward. But she continued to look after him, and when in old age his hormones changed and his mental balance returned, she married him again.[1]

---

[1] Though he was close to "normal" by the time of our conference and trip, he enjoyed teasing us by seeming to lapse into the maladies portrayed in the movie. On one of the first days we were taken to see a famous sitar player. After the performance we went backstage and the sitar player, who had been warned he would be meeting the great John Nash, said to him, "You look like a man who has talked to God." Nash said, "I used to, but I find it is getting more difficult."

Another day Nash came running in to the room, where we were relaxing, waving the *Herald Tribune*. "I see that Jennifer Connelly [the beautiful actress

Each visit to a new city began with a press conference and pictures, which the next morning would appear on the front page of the city newspapers with all our pictures excised except for Nash's. At the first conference a young woman reporter rose and said, "Of course we have all seen the movie, and supposedly Nash equilibrium is explained there, but it wasn't really so clear. Could each of the speakers say, in a word, what is equilibrium?"

Four of us, including Nash, gave somewhat long-winded answers. I talked about thinking about what other people are thinking about what you are thinking. Then it was Robert Aumann's turn to answer.

"This reminds me," Aumann said, "of the first press conference that Nikita Khruschev, the premier of the Soviet Union, gave to western reporters. A reporter rose and asked Khrushchev to say a word about the health of the Russian economy. Khrushchev replied 'Good.' The reporter laughed and said he didn't mean literally one word, it would be OK to say two words about the health of the Russian economy. Khrushchev replied, 'Not good.'"

Aumann said, "In one word, game theory means interaction. In two words, rational interaction."

The same two words describe general equilibrium.

GENERAL EQUILIBRIUM IN A FEW MORE WORDS

Thales and other ancient thinkers seemed to have the glimmerings of supply and demand 2500 years ago. By the 1870s the marginalists William Stanley Jevons, Carl Menger, and

---

who played his wife] is marrying the actor who played my imaginary best friend in the movie. You know what that means? That means the whole time they were supposed to be thinking about me they were just getting it on."

We visited a famous Jain temple (I forgot where) and there was mysterious writing on the walls. Nash said "It is not Sanskrit or any other language I know. Could someone bring a mirror; perhaps it makes sense backwards."

especially Léon Walras had introduced many of the important elements of equilibrium theory into economics. It was not really until the 1950s and the birth of general equilibrium with Ken Arrow and Gérard Debreu that equilibrium became axiomatically formalized.

The name *general equilibrium* denotes three features of the theory. First, general is meant to apply to a situation with many *heterogeneous* agents trading many heterogeneous goods, in contrast to the representative agent models, in which the economy behaves as if it is directed by a single desired goal, and to the partial equilibrium models with only two goods. The theory is also general in that the agents are described not concretely with functional forms, but by *axioms*, and the qualitative nature of their interactions is described by provable *theorems* derived from the axioms. Finally, equilibrium to Arrow and Debreu meant that each agent pursues their goals *rationally*.

Léon Walras (1834–1910) described in his book *Elements of Pure Economics*, first published in 1874, a framework or model for understanding the behavior of economic agents interacting in the marketplace. His work was extended and refined by many authors, most notably Irving Fisher (1867–1947), Sir John Hicks (1904–1991), Paul Samuelson (1914–2009), Ken Arrow (1921–2017) and Gérard Debreu (1921–2004), into what is now called modern general-equilibrium theory. For over one hundred years, the Walrasian model has been *the* central paradigm in economic theory, especially if it is regarded as subsuming, as a special case, the partial equilibrium framework deriving from William Stanley Jevons (1835–1882), Carl Menger (1840–1921) and Alfred Marshall (1842–1925), and the many extensions and applications it has found in macroeconomics, finance, international trade, and public finance.

The general-equilibrium methodology rests on four cornerstones: (1) agent optimization; (2) market clearing; (3) perfect competition; and (4) rational expectations. Agents are supposed not to act mechanically, for example, by repeating actions from the past no matter how circumstances change, or following simple rules of thumb. Instead, it is supposed that they hold their objectives constantly in mind, adapting their actions when the environment changes or when their anticipation of the future changes, in order to advance as far as possible toward their goals. Social welfare is measured in the general-equilibrium tradition by how completely individuals believe the system permits them to achieve these *a priori* goals. The Walrasian tradition thus subscribes wholeheartedly to an individualistic political philosophy by reducing aggregate action to the sum of individual actions motivated by individual objectives, and then evaluating social success by aggregating individual opinions of personal success.

~ 83 ~

The Walrasian tradition builds an extremely strong hypothesis of rationality into its definition of equilibrium. Even though their trades determine the prices, agents are presumed to anticipate correctly all the prices at which they will trade before they commit themselves to a single market transaction. This paradox is at the heart of general equilibrium, and a similar paradox is at the heart of Nash equilibrium. For general equilibrium, it means that equilibrium requires solving the demand equals supply equations, one for each market, before any agent has acted, and thus before any agent could reveal their own demand. Moreover, the point of general equilibrium, as opposed to partial equilibrium, is that each market-clearing equation depends on all the prices, not just its own price, so the equations are interdependent, and thus difficult to solve. Similarly, Nash equilibrium requires each agent effectively to anticipate what the other agents are doing, recognizing that what they are doing reflects their anticipations of what she is doing, which depends on what they think she thinks they will do *ad infinitum*. That can be demystified only through a fixed point, defined on the potentially high-dimensional space of every agent's potential actions, as Nash showed.[2]

The actions of the many heterogeneous agents affect the prices, which affect each of them. Thus, the agents interact. The premise of general equilibrium is that the macroscopic price is the result of individual actions, and understanding economic aggregates is best done from the bottom up. Second, the general-equilibrium tradition maintains that to understand how the economy evolves, it is crucial to take into account how each individual herself evolves as they rationally change anticipations of what lies ahead. Thus, interaction and rationality describe general equilibrium, as Aumann said.

Since their beginnings, the social sciences have paid lip service to the importance of mathematics. Recall the famous inscription on Plato's Academy: *Let no one ignorant of geometry pass through these doors.* Of all the social sciences, economics has been the most profoundly affected by mathematics. Indeed, while Plato believed that a mind trained in the logic of mathematics would be more philosophical, he could hardly have anticipated the modern discovery that to do economics is to do mathematics (albeit of an elementary nature). Every important economics department has several professors specializing in mathematical economics, in addition to a number of others who occasionally use mathematical tools in their theorizing or their econometrics.

Solving large systems of simultaneous interdependent equations, or finding a fixed point on a high-dimensional space, is one

---

[2]Bounded rationality models of general equilibrium might also require solving simultaneous equations or using fixed points; full rationality absolutely requires it.

reason for the importance of mathematics in economics. A simpler explanation is that the fundamental economic primitives are price and commodity, and both of these naturally lend themselves to representation by numbers. (Think of how much more difficult it is to quantify a neurotic disorder.) Not only that, but the price and quantity of a commodity are related to each other in a mathematically interesting way; they can be multiplied, yielding revenue. A third reason is that the axiom–theorem approach to economics gives understanding, while examples or simulations leave doubt and occasionally misunderstanding.

As an example of the axiomatic approach, I mention the three most famous and basic theorems about Walrasian general equilibrium, proved in 1951, 1954, and 1971, respectively. The first is that under quite general assumptions, mainly diminishing marginal utility and increasing marginal cost, equilibrium must exist, no matter how many interdependent commodity markets and no matter what the specific utilities of the consumers or the technologies of the firms. Second, equilibrium with complete markets is always Pareto efficient, meaning that there is no feasible way to make everybody better off. Finally, for almost every choice of utilities and endowments, there is only a finite number of equilibria, and each moves differentiably as the environment is perturbed differentiably.

APPLIED GENERAL EQUILIBRIUM

Economists were quick to build general-equilibrium models with utilities and endowments and production technologies that were calibrated to reality, in order to test the effects of policy interventions. One of the workhorse models was the famous $2 \times 2 \times 2$ model of international trade with two inputs and two outputs and two countries. In that model it was possible to prove theorems about the qualitative effects of changes, and to compute quantitative effects in the calibrated model. Moving to larger and more realistic models was impossible because computing equilibrium was so difficult.

In 1967 Herb Scarf, a mathematician on the economics faculty at Yale University, changed all that when he invented the Scarf algorithm for computing fixed points and general equilibrium. Scarf coined the phrase *applied general equilibrium*, whose purpose was to build general-equilibrium models with many commodities and many heterogeneous agents based on data about real agents, using his algorithm to find equilibrium. My first job, as an undergraduate, was to program the Scarf algorithm for an international trade model and then present it with him to policymakers in Washington, DC. Some years later, faster algorithms were derived, sometimes specialized to the particular general-equilibrium model. Quite large classical GE models are now easily computed. In its modern form, applied general equilibrium is now usually called *computable general equilibrium*.

There is a limit, however, to the size of a model algorithms can solve. As Nimrod Megiddo and Christos Papadimitriou (1989) made clear some years later, the computational complexity of fixed points is TFNP, (P < TFNP < NP), which is harder than the complexity of linear programming or concave programming. So as the dimensions rise, the computation time grows quickly. Short-horizon models (with just a few periods), even with lots of agents and goods, are manageable. Daily trading with hundreds of commodities over several years is unmanageable.

Macroeconomics is now most often couched in long-horizon, even infinite-horizon, economies where computational limitations bind. This led to a return to representative (single) agent economies, or economies with two agents. Often the modeler will look for a steady-state equilibrium, and then linearize around the steady state (even when the steady state is not stable). Some progress is being made in heterogeneous agent economies by treating a first period in full generality, then assuming a steady state with no uncertainty from period 2 onward.

Agent-Based Models and Complexity Economics

There is an absolutely fundamental spiritual similarity between general equilibrium and agent-based modeling (sometimes called complexity economics) because both are bottom-up methodologies that start with heterogeneous agents. The difference is that agent-based modeling drops the simultaneous equations and fixed points of general equilibrium, and, *a fortiori*, of full rationality, replacing utility maximization and rational expectations about the equilibrium with behavioral rules.[3] In Robert Aumann's language, agent-based modeling is interaction without full rationality.[4]

Abandoning equilibrium and fixed points eliminates the need to solve simultaneous and interdependent equations. Agent-based modeling is therefore vastly less computationally burdened than general equilibrium. It becomes possible to include every transaction of every agent, to show where every agent stands at every moment, to really see what is physically going on, mirroring the finest level of data. The model can become more realistic, and it can be validated by checking its predictions against many details in the real world that must be glossed over in any general-equilibrium model. Computable general-equilibrium models in practice usually do not even come close to including the rich detail

~ 87 ~

---

[3] The fundamental difference is not that general equilibrium posits utility maximization, and agent-based modeling does not. It is that GE requires simultaneous and interdependent equations, and ABM does not. The implication is that GE can always accommodate full rationality, and ABM cannot always do so.

[4] So far it also seems that ABM relies for results on simulations and not theorems, although the theorems may yet come.

they could. And if they did, they would still end up with a tiny fraction of the details that agent-based modeling can encompass.

There are three dimensions along which agent-based modeling can surpass general equilibrium. The first is the much greater detail it affords by being freed from the computationally expensive need to compute fixed points.

The second is that, even if real-world agents were utility maximizers, except in very simple environments, one really does not know what their utilities are. One sees their past behavior in reality and has to infer what their utilities might be. It is much simpler to fit a behavioral rule to their past decisions than to rationalize them with a utility function.[5] Agents in agent-based models often mimic the behavior of real agents more exactly than do the general-equilibrium agents.

Finally, agents are not fully rational in reality, and phenomena that emerge because of their irrationality can be brought to light in agent-based models.

Agent-based dynamics come from iterating behavioral rules for each agent that depend on how the other agents have behaved in the past, but not on *fully rational* anticipations of how they will behave in the future. The conceit of agent-based modeling is that the complex adaptive system created by interacting agents is driven primarily by the physics of their interactions and some limited sense of expectations that can be be expressed as a function of the past. The general-equilibrium approach might be necessary if the evolution of the system is crucially driven by the evolution of truly rational adaptive thinking of the agents.

---

[5]However, it is critical that the behavioral rule be connected somehow to incentives, if not explicitly derivable from a known utility.

COMBINING AGENT-BASED MODELING WITH GENERAL EQUILIBRIUM

Of course, one can build behavioral rules depending on the past that imitate behavior stemming from anticipations about the future. For example, suppose that a contrarian C wants to do the opposite of what a matcher M is going to do. C's behavioral rule might be to go right next time if M went left last time. One might interpret that as if C anticipated M will go left again, so C will do the opposite of what C expects M will do. But rationalizing behavior as anticipation-based is far from the full rationality of Nash equilibrium models. In rational models, C would have to think about what M was thinking about C. M would understand C's logic, and, anticipating that C would go right, M might go right instead of repeating left. But C should also understand this longer logic, and go left after all. But M should understand this . . . The fully rational thing for each agent to do is flip a coin. Of course, flipping a coin is a behavioral rule that an agent-based model can posit. The trouble is that this rule cannot be discovered without solving for rational behavior. The example of the contrarian and the matcher illustrates that, in general, any kth order thinking model, or learning algorithm (that can be represented by a behavioral rule), can be put into a context in which its behavior is very far from rational.

~89~

In my opinion, agent-based modeling and general equilibrium should be thought of as complementary brothers in arms. One can start with a general-equilibrium model that identifies the broad strokes of a picture. An agent-based model can then take it into much greater detail. Moreover, once one can see in the equilibrium model where reasoned anticipation plays a vital role, the agent-based modeler can proxy for that by positing a behavioral rule that mimics rational behavior, for example, by going 50–50 left–right in the above example, or by identifying some signature in past outcomes in the equilibrium model that can trigger the action in the equilibrium model derived from reasoned behavior.[6]

I give four examples of models that I have conceived of as equilibrium models and then gradually transformed (with the help of others) into agent-based models. In three of the models, rational behavior discovered in general-equilibrium optimization can be mimicked in the agent-based model. In the fourth inflation model, one aspect of the general-equilibrium model is that a liquidity trap emerges, which might be hard to replicate in an agent-based model. If the central bank injects a large quantity of money (as in quantitative easing) into the economy, then prices will rise if agents think the money will stay there permanently. If agents are convinced that the central bank will eventually withdraw the money, the injections will not create inflation, as if there is a liquidity trap. Agent-based modeling is continuing on all four models, but just beginning in the monetary model. It is not yet so clear that agent-based models will naturally capture the liquidity trap.

Mortgage Prepayments

The most convincing evidence for the (financial) value of agent-based models is that Wall Street has used them for over three decades to forecast prepayment rates for tens of millions of individual mortgages.

I describe the prepayment model I formulated in the early 1990s as head of fixed income research at Kidder Peabody and then as head of research at Ellington Capital Management starting in

---

[6]In section 5, I explain how anticipations of *future* volatility in the equilibrium model leads lenders to ask for more collateral. In the agent-based model one can use past volatility to trigger tougher lending, which is normally quite close to the rational lending behavior since past volatility is usually a good predictor of future volatility.

1995, which, to the best of my knowledge, was the first agent-based prepayment model. We shall see how the Kidder model began as a conventional aggregate model and evolved into an agent-based model, even before detailed loan-level information became available. Some of the criticisms of agent-based modeling can be seen to have merit; the model that worked in the 1990s needed substantial changes after 2000. But the approach was still generally superior to aggregate modeling.

~91~

THE MORTGAGE PREPAYMENT PROBLEM

Aggregate Predictions versus Agent-Based Predictions

There are approximately 50 million residential mortgages extant in the United States today (and 85 million mortgages of all kinds). In each mortgage, the borrower is obliged to make a monthly coupon payment, calculated either as a fixed percentage of the original balance, or as a variable percentage of the original balance, for a period of years, often thirty, though occasionally fifteen or even shorter. Nearly every single mortgage gives the borrower an option each month to pay off the remaining balance (or prepay) in lieu of all future payments. Prepayments radically change the cash flows and valuation of the mortgages to the holder. Partly to reduce this risk, starting in the 1970s, mortgages were aggregated into *pools*; shares of the pools were then sold as securities in a process called *securitization*. Typically, pools consisted of a large number of individual mortgages issued at approximately the same time, with approximately the same payment conditions (e.g., fixed-rate mortgages of about 8% issued in the first half of 1986).

Conventional mortgage prepayment modeling was pioneered by academics, and then taken up in a series of working papers by various Wall Street investment banks throughout the early 1990s. Typical of those models was one at Kidder Peabody.

Since the purpose of pooling was to diversify the risks inherent in any homeowner's decision, and since buyers in the secondary market bought pools of mortgages and not individual mortgages, it seemed perfectly natural to the pioneers of prepayment modeling to try to predict the *aggregate* prepayment of each pool directly, even though those prepayments are the sum of individual homeowner prepayments. An apparently decisive argument for this approach was that it is much simpler to predict one number than 10,000 numbers, and that the individual mortgage outcomes were not available in the data anyway.

Rational Behavior?

In figure 1, the solid line is an example of a monthly prepayment history (expressed at an annualized rate) up to 1999 on the pool of 1986 Fannie Mae 8% coupons. The figure shows that *homeowners do not prepay optimally.* Historical aggregate prepayments are never exactly 0% or 100%; in fact, they are rarely more than 7% in any month (62% annualized). So the conventional approach looked for common-sense patterns in the data, and tried to refine the common sense by using statistical methods to estimate parameters *a priori* given functional forms for aggregate prepayment behavior.

The first pattern that becomes evident from the data is that prepayments go up rapidly when the new mortgage rates (into which homeowners can refinance) become sufficiently lower than the mortgage rate of the existing loan. This conforms to common-sense incentives, without exhibiting the bang-bang behavior of optimal option exercise.

The second pattern that was observed in the data is that prepayment rates are not a function only of current conditions. An 8% coupon pool of mortgages prepays at a slower rate the

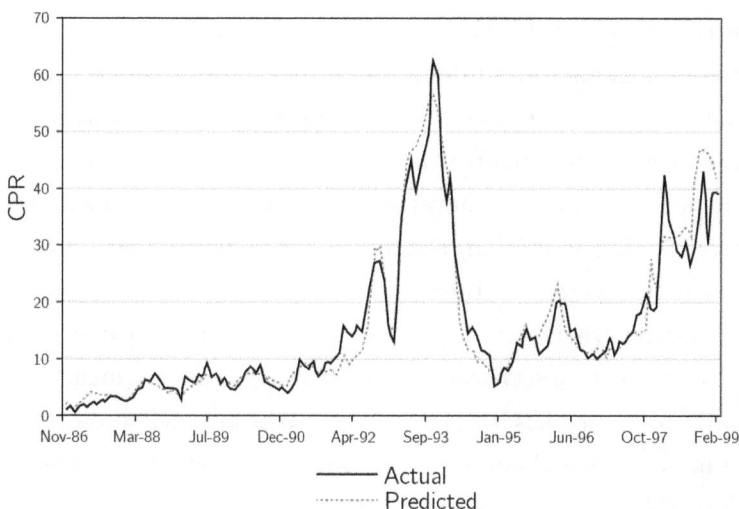

Figure 1. Conditional payment projections can be very reliable.

second time rates drop to 6.5% (in 1998) than it does the first time rates drop to 6.5% (in 1993). This phenomenon is called burnout.

Modeling the Future

Macroeconomists on Wall Street in the early 1980s would forecast prepayments one year ahead according to how they thought the macroeconomy was going and whether they thought interest rates were going up or down. These were unconditional forecasts, just like the weather forecasts of the time (showers tomorrow clearing up by Tuesday). Starting in the late 1980s, it became clear that, for thirty-year mortgages, one had to make ten-year predictions or more, and that these had to be conditional predictions: If interest rates and housing prices follow such and such a path or scenario, then prepayments will do such and such. Unconditional predictions are hopelessly naive; one cannot know the ten-year future path of interest rates or housing prices, yet prepayments will surely differ depending on those scenarios. Conditional predictions have the possibility of being correct; it is at least imaginable that one could predict the future path of prepayment rates, conditional on the future path of interest rates and housing prices. The new benchmark became forecasting in 1986 what prepayments would be in January 1993, conditional on the interest rates and housing prices through 1992, but without conditioning on anything else that wasn't known in 1986, like the actual prepayments in 1992.

Homeowners who are choosing whether to prepay do not know the future scenario of interest rates; at any moment they only know what has happened so far. To be rational, in the sense of modern general-equilibrium theory, about whether to prepay or to wait for an even better moment to prepay, the homeowners need to know the precise probability distribution over all future scenarios, including the probability of different continuation scenarios conditional on what will happen up until any future time t. In short, homeowners would need to hold a picture of a tree of future events and assign a conditional probability to each of the many branches leaving each of the many nodes. The optimal prepayment strategy that minimizes the expected discounted payments would depend on the tree and also on the conditional probabilities assigned to all future branches.

The reader should notice that the number of paths grows exponentially with time, so even knowing the tree, it might seem impossible at first glance to model rational agents.

The Conventional Model

The conventional prepayment model of the early 1990s moved beyond the noncontingent forecasts of the 1980s, but without any agents. It made *aggregate* predictions at the pool level, based solely on past macro variables without taking into account homeowner expectations about the future. There were no interactions and there was no rational optimization.

The model essentially reduced to estimating an equation with an assumed highly nonlinear functional form for prepayment rate (the fraction of existing homeowners who prepay)

$$Prepay(t) = f(age(t), seasonality(t),$$
$$originalrate - newrate(t), burnout(t),$$
$$parameters) + error(t)$$

where $originalrate - newrate(t)$ is meant to capture the incentive to refinance at a given time t, and burnout

$$burnout(t) = \sum_{\tau=1}^{t} \max(originalrate - newrate(\tau), 0)$$

is the summation of this incentive over past periods. Mortgage pools with large burnout tended to prepay more slowly. However, the model provides no explanation for burnout.

A last important point is that the function f looked highly nonlinear in the data. To fit its twists and turns, the regression included many parameters.

Once the parameters are estimated, the conventional model makes prepayment predictions along any hypothesized path of future interest rates, as described in the last section. At any point, the prepayment rate depends on the contemporaneous interest rate or mortgage rate, and also on past interest rates and mortgage rates along the hypothesized path because of burnout. The value of owning a security is then obtained by running a Monte Carlo simulation. A thousand paths are generated according to some distribution. The prepayment function spits out the cash flows at each node along each path. All these cash flows are discounted along each path and then averaged over the thousand paths.

The model did a pretty good but not great job of predicting prepayments in the near term and in valuing mortgage pools. For predictions beyond five years, the model failed badly out of sample. The conventional explanation was that homeowner behavior was evolving and over five years would change. An implication was that the parameters had to be frequently reestimated, and that only the most recent five years of observed behavior was useful in estimation; the previous twenty years of data had to be discarded. There were also serious divergences between the model valuations and the trading prices of various mortgage securities called IOs (valued too high) and POs (valued too low).

THE OPTIMIZING AGENT-BASED MODEL

The conventional prepayment model used an exogenously specified functional form to describe aggregate pool behavior. Though the functional forms were motivated by common-sense considerations of homeowner incentives, there was no explicit optimization from which prepayments were derived. In particular, there was no mention of homeowner expectations in their behavior. Finally, the parameters needed to be reestimated frequently to keep up with evolving behavior.

By contrast, the agent-based prepayment model my team and I developed at Kidder Peabody in the early 1990s, and then refined many times at Ellington beginning in 1995, starts from hypothetical individual homeowners and follows every single one of their individual mortgages. Though we had absolutely no data on individuals in the pool, we imagined individuals and produced aggregate prepayment forecasts by simply adding up over all the imagined individual agents.

We hypothesized that each homeowner prepays optimally subject to frictions, with rational probability assessments of future paths of interest rates. The frictions decayed exponentially over time at a rate we called the smart factor. Finally, and most importantly, the homeowners were taken to be heterogeneous. Far fewer parameters needed to be estimated because the optimization itself forces some of the behavior, which then does not need to be estimated. The exponential decay in the frictions enabled predictions that are reliable for more than five years and permitted regressions on the entire history of data.

In the agent-based model, each homeowner faces a friction cost c of prepaying, which includes some quantifiable costs such as closing costs, as well as less tangible costs like time, inconvenience, and psychological costs.[7] The homeowner is also subject to an alertness parameter a, which represents the probability the agent is paying attention each month, and also a turnover rate T denoting the probability of selling the house (and thus prepaying the mortgage) for reasons unrelated to desires to refinance, such as divorce or moving.[8] The agent is assumed aware of her cost and alertness and turnover, and subject to those limitations chooses her prepayment optimally to minimize the expected present value of her mortgage payments. The prepayment frictions c and especially a explain why prepayments don't jump to 100% as soon as the option is in the money.

A critical part of the model stems from the observation that homeowner behavior is evolving over time; it is getting more efficient or rational. Quantifying the rationality frictions as cost c and alertness a makes it easy to quantify their reduction. Define the smart factor $s < 1$ and set $c_t = c_0 s^t$ and $a_t = 1 - (1 - a_0)s^t$, where t is calendar time. Over calendar time agents become more rational. The world does not stand still; technology improves and people become more sophisticated. Transactions can happen over the internet, reducing the cost of prepayment. And homeowners can be alerted by email or the internet that they

---

[7] The cost is paid in the month the agent prepays, and not otherwise; this cost is a disincentive to prepay.

[8] Turnover rate T might be allowed to depend on observables like seasonality, as in the conventional model.

should prepay (by refinancing into the sender's new mortgage), improving the alertness of homeowners. The smart factor s quantifies the rate at which costs decline and alertness increases over time as prepayment behavior becomes more rational. By choosing the right smart factor parameter s it becomes possible to reconcile the whole history of prepayment evolution and to forecast what will happen going forward.[9]

To compute the optimal moment to prepay, the homeowner is given the tree of future interest rates and the conditional probabilities that are implied by the derivatives market about future interest rates. The Ellington traders themselves compute these market-implied probabilities in order to value the securities they trade. It is consistent to impute these same contingent probabilities to the homeowners. Thus, every homeowner imagines the same tree of future interest rates, with probabilities assigned for each branch at each node so as to match all the derivative prices in the markets. The homeowner optimally solves an option exercise problem by backward induction, given c, a, T, s.

The optimal prepayment calculation is made tractable by supposing that the interest rate tree is recombining. That means the probability of interest rates at calendar time $t + 1$ depend on what interest rates were at calendar time t, not on where they were in any previous time period. The number of nodes then grows quadratically in the number of modeled time periods (say, months in the life of the mortgage), not exponentially.

The backward induction prepayment calculation not only specifies prepayments, it also calculates the value of the mortgage, eliminating the need for the Monte Carlo simulation. It also provides additional information, like the gain in value to the issuer

---

[9]Of course, s might not be constant, and could differ between cost and alertness.

of the mortgage due to the rationality frictions $c_t > 0$ and $a_t < 1$.

The most important part of the optimizing agent-based model is that agents are taken to be heterogeneous. Agent heterogeneity is a fact of nature. It shows up in the model as a distribution of costs, alertness, and turnover rates. Each agent is characterized by a triple (c_0, a_0, T) of cost, alertness, and turnover, which (together with s) specifies through the optimization precisely when and under what circumstances the agent will prepay. In particular, it gives a forecast of whether or not the agent will prepay at each point along any future scenario of interest rates (that is in the tree). Nothing needs to be assumed or estimated about the functional form of the agent's behavior.

~ 99 ~

Different agents (c_0, a_0, T) will of course prepay differently. The distribution of these characteristics throughout a pool at inception determines the aggregate pool prepayment behavior along any interest rate scenario. Assuming that every large pool has the same distribution of types (c_0, a_0, T), this distribution of *individual* types determines the aggregate pool prepayment behavior along any interest rate scenario for every pool. We then chose the distribution of types that minimized the difference between the model and historical *aggregate* pool prepayments, across the different pools and their different interest rate histories and different prepayment histories.

The agent-based model tells us a story rather than just fitting curves. Prepayment speeds increase continuously as interest rates go down because the benefit of prepaying overcomes the cost c of more and more heterogeneous homeowners. Burnout is also a natural consequence of heterogeneous agents. The agents with low costs and high alertness prepay faster, leaving the remaining pool with slower-paying homeowners, automatically causing burnout. The same heterogeneity that explains why only part of the pool prepays in any month also explains why the rate of prepayment burns out over time. The agent-based approach disciplines the functional forms and shrinks the number of parameters that need estimating.

It is also important that the agent-based approach captures behavior that the modeler does not need to have understood in advance. To capture burnout with a parameterized curve, one must have already identified the phenomenon and built it into the curve. In contrast, once there is a distribution of costs and alertness to explain the continuous increase in prepayment speeds as interest rates drop, burnout is automatically created. A similar effect comes with the increase in interest rate volatility. The homeowners in the *optimizing* agent-based model rationally wait longer to exercise their prepayment option when volatility increases, so prepayment speeds decline. The prepayment-volatility effect is captured even if the modeler was unaware of it. The conventional model does not show that behavior. It could only be included by putting in more parameters and an explicit dependence of prepayment rate on interest rate volatility.

Once the behavior of the optimizing agents with rational expectations is understood, their individual behavior can be pretty accurately mimicked by a behavioral rule that depends on a few inputs (which we called option costs) that need to be computed from the tree of interest rates but cover all types. Although the backward induction calculation is quite fast, once the individual behaviors are boiled down to behavioral rules, the computations become lightning-fast.

The agent-based model had one more interesting wrinkle I called *contagion*. The conventional model identified a media effect in which prepayment speeds seemed to jump up when interest rates hit record lows over the past five years. We found it fit the data better to make alertness a_t rise when there were more households prepaying at times just before t: Presumably this is because people are more likely to be alerted to the advantages of prepaying from friends who have just done so. This again reflects the agent-based approach where agents respond not just to exogenous factors like the interest rates but also to what other agents are doing.

The optimizing agent-based model only involved one endogenous market, for mortgages. Interest rates and turnover, which of course involve interactions in many markets were taken as exogenous to the model. There were no simultaneous equations. Thus there was no problem in transforming the optimizing agent-based model into a behavioral agent-based model.

Triumph of the Agent-Based Model

With a relatively small number of parameters we closely fit thousands of data points, including prepayments from 1986 to 1995 for all coupons issued either by Fannie Mae or Freddie Mac each year since 1986. As an example of how the model fits, in figure 1 the grey dotted curve shows the prepayments the estimated parameters give on the FNMA 1986 8% coupon. The in-sample fit from 1986 to 1995 is very good, for this coupon and the others, but so is the out-of-sample fit over the three years 1996–1999. It is hard to imagine another macroeconomic series with so many ups and downs that can be fit as well with so few parameters in and out of sample. We see that an agent-based approach to prepayments can retrospectively fit a long history and even make reasonable (conditional) predictions several years into the future.

The agent-based model fit substantially better than the conventional model. Another thing the smart factor did is bring the model valuations of interest-only (IO) and principal-only (PO) securities much closer to their market values. The increase in future speeds implied by the smart factor naturally lowered the price of IOs and raised the price of POs, without artificially altering near-term predictions of speeds which could not reasonably be expected to differ much from current speeds (assuming constant interest rates).

Even if the agent-based model had fit only equally well, it is surely a better model. It tells a story. And the agents respond to changing environments in sensible ways because they are optimizing. There is always a grave danger with *a priori* functional forms fitted to past data that the future will be very different from the past, moving into regions for which there is no prior data and forcing the prediction to rely entirely on an extrapolation of some arbitrary functional form.

Limitations of the Agent-Based Model and the Conventional Model

The agent-based model is not immune to the criticisms I just levied against the conventional model. Several new patterns of behavior emerged that showed that the agent optimization goal posited above for the agent-based model, and the functional form for prepayments assumed in the conventional model, no longer captured what homeowners were thinking.

In the first decade of 2000, a new feature appeared: the cash-out refinance. Homeowners began to refinance their mortgages even when interest rates barely dropped, or even rose, in order to get bigger loans. Second, Fannie and Freddie began to extend loans to a larger group of homeowners whose behavior was significantly different, invalidating the hypothesis that all pools started with the same distribution of homeowner types. Homeowners actually began to default, which was not a concern in the model described above. Third, after 2007 some homeowners could not prepay when interest rates went down.

The agent-based model needed a substantial new factor. The loan-to-value (LTV), which depends crucially on house price appreciation, which had not been foreseen in the model of the early 1990s. This illustrates an important limitation of agent-based models, or any model: Behavioral rules, functional forms, or even utility maximization can become inappropriate (or in need of revision) as the world changes.

On the other hand, starting in the 2000s, vastly more *individual* data became available, making the whole world realize that the heterogeneous agent-based approach to prepayments is unavoidable.[10]

Monies and Inflation

Economists and central bankers (and their models) were surprised by the extraordinary inflation of 2020–22, just like they were by the worse inflation of the 1970s, for which there is still no definitive explanation. I argue that a robust predictive model of price levels must be agent-based, monitoring disaggregated transactions between heterogeneous agents, including the methods of payment.

Most modern monetary models have very few agents, and have banished the quantity of money and even money itself (in favor of interest rates). Curiously, just as in reality we are getting more new monies, the models exiled all money, blinding us to the inexorably emerging new monies and the repercussions their growth will have for inflation.

In my opinion the COVID-19 inflation was due in large part to the unprecedented deficit expenditures by the Trump and Biden administrations of 25% of GDP in two years, *nearly all with new printed money*. This printing money aspect has gone virtually unnoticed in the popular press and is underappreciated in policy circles. Similarly, in my opinion the 100% inflation over the latter half of the 1970s and the early 1980s was due in part to the rapid

---

[10] The effects of observable individual borrower characteristics (for example, age or occupation) can be incorporated in the model (when they become available) by allowing them to modify the cost, alertness, and turnover.

growth of a *new money* called credit cards. This explanation for the 1970s inflation has never been seriously considered by economists, even though it was given by Alfred Kahn, the inflation czar that Jimmy Carter appointed to deal with it. I believe money market cards, Bitcoin, stable coins, and a host of potential future monies will cause another big inflation if the Federal Reserve ignores them.

The orthodox models of inflation use too little data, have almost no general theorems, and are bad at making predictions except when inflation doesn't change. Most economists and the Fed failed to predict the 2020–22 inflation. For two years from June 2020, when month-over-month inflation hit 6% annualized, until March 2022, by which time annualized month-over-month inflation passed 10%, the Fed stood idle, convinced first that there would be no inflation, and then that it would be transitory. After the Fed finally started raising rates, there was wide disagreement in predicting how fast inflation would come down. Larry Summers, one of the few economists who predicted that the Trump–Biden spending spree would cause inflation, destroyed his short-lived reputation as a seer by predicting that it would take two years of 7.5% unemployment or one year of 10% unemployment to get inflation under control.

One rationale for these failures of the orthodox models is that COVID supply-side disturbances were so unprecedented that failure to predict their consequences is unsurprising. However, inflation predictions weren't so good before. The average fifteen-month inflation predictions by prominent economists and policymakers from 2004 to 2023 was off from the actual inflation on average by 1.8%. (Recall the disagreement about whether 2008–2014 monetary injections [QE] would create inflation.)

Economists also did not foresee the magnitude of the 1970s inflation, and still haven't given a satisfactory explanation fifty years later. The orthodox consensus is that it was caused by the 1974 and 1979 oil shocks, perhaps abetted by food supply shocks. Yet oil was barely 4% of the economy in the 1970s, so even the 500% increase in oil prices from $10 a barrel to $60 can explain at most a 20% increase, not the 100% increase we had. Economists can't really think this inflation was caused by the deaths of anchovies in the late 1970s, as some economists suggested at the time

Credit cards are by now a well-established kind of money surrogate. But new monies are on the way. There are money market cards, which are like debit cards except that it is possible to charge on bonds held in a money market instead of money held in a bank account. Currently, this ends up in a bond liquidation and transfer of money to the recipient's bank account like a debit card. But we could imagine in the future that the correct amount of bonds are transferred directly from the buyer's money market fund to the seller's fund, without any money used at all. Another step is that the buyer might be able to choose the kind of bonds— for example, their maturity bucket—that they charge on. And one could further imagine that there could be stock market charging cards where the buyer could charge on their index stock account without any conventional money payments. Stable coins that are backed by stocks guaranteeing a stable money value are exactly like that. At the moment, there are very few issuers of stable coins. But in the future we could imagine that any holder of a stock index could issue stable coins. Stable coins backed by treasuries or stock indices, or commodities like gold, are another potential future money.

AN AGENT-BASED MODEL OF INFLATION

Prices refer to transactions between two instruments. Yet there are no transactions in most inflation models. In real life, nothing is purchased without putting up either money or a promise in exchange. The promises end in payments, usually of money, except when payments are netted. In the future we can imagine payments in Bitcoin, stable coin, or liquid securities.

The ratio of means of payment to transactions would seem to be related to price levels and interest rates. That is why the Fed injects money into the system every Christmas to maintain a stable price level. If this ratio matters at Christmas, why not every other day?

Not only does the total amount of liquidity matter to prices, but so does how it is divided among agents and what their desired purchases are. A key factor in COVID inflation was the transition of demand from services to the goods sector. If a model does not explicitly include different sectors, it cannot possibly predict the effects of such demand shocks. Supply-chain disruptions are also said to have been a crucial driver of COVID inflation. Needless to say, without an explicit model of supply chains, substitute inputs, and the elasticity of final demands, it is impossible to predict the effects of a supply bottleneck on prices.

A truly satisfactory model of price levels must keep track of who is buying what and with what (money or credit or new money); that is, it must keep track of transactions at a finely disaggregated level. Such a model would not only follow the value and quantity of purchases, but also the means of payment, including the type of credit and the eventual delivery. One example of such a model is Anderson *et al.*'s study (2023) of purchases in Denmark in 2023.[11] Such agent-based models are becoming entirely feasible. Even little hedge funds can follow mortgage decisions of almost every household. Why can't the Fed follow who has money and how many transactions various agents make?

The Anderson *et al.* model shows it is possible to monitor the money transactions of all small entities in a whole economy.[12] One trouble with the Anderson model is that it keeps track of goods and the flow of value but does not really explain where the money goes. For example, if somebody finds a krone, it ends up getting absorbed by the rest of the world, in exchange for foreign goods. Why do the foreigners keep the money at the end? What would happen to the krone in a closed economy? Who would hold it in the end?

This raises an old question, often called the Hahn problem (Hahn 1965). How can money in positive supply have value in a finite-horizon equilibrium model? In the last period, nobody will want to hold it, so its value would apparently be zero. But then nobody would want to end up with it at the end of the second-to-last period. Working backwards, it would seem that money could never have value even if cash was necessary for trade. The economy would break down to autarky.

For this reason economists have embraced infinite-horizon economies, so there is no last period. Unfortunately, then detailed records of agent-based transactions then become unmanageable. I solved the problem of how to model money in a finite-horizon closed economy with Pradeep Dubey (1992; 2003). We explained the value of money by introducing a central bank that can make loans in total of size M, and enforce their collection. Far from propping up the value of money by giving goods for money,

---

[11] The model is a little short on the means of payment data, apparently presuming all purchases are made with cash.

[12] The data do not go down to the actual individuals but rather to neighborhoods and so on.

as the foreigners do in Anderson *et al.*, the central bank injects more money M into the economy, in addition to the money m held free and clear by private agents.[13] The bank does not have anything to offer in exchange for money. It does, however, have the power to lend money to voluntary borrowers, and to enforce the collection of the ensuing debts. That collection power, together with the necessity of trading via money, guarantees money has value, provided the potential gains to trade from autarky are big enough. The crucial idea is that agents, who do not initially owe the bank anything, desire to spend more money than m and so borrow at a positive interest rate r and thereby incur debts $(1 + r)M$ to the bank, that exceed what they borrowed M. They can only repay this by sending the bank their own endowments m of money, resolving the puzzle of who holds the money at the end.

The model allows for an arbitrary number of heterogeneous agents and commodities with arbitrary sectors and supply chains while still maintaining equilibrium in which money has value in a closed economy with a finite horizon. Every transaction is accounted for, including the amount of cash spent on each purchase and precisely how the agent got the cash to spend (out of endowment or by borrowing). Because the time horizon can be taken as short, it is computationally feasible to find equilibrium.

Anderson *et al.* show that such a model could be matched to realistic agents. I propose that it be done with the US economy.

In the model, the qualitative effects of monetary and fiscal policy are described by theorems, with exactly the same generality as classical Walrasian general equilibrium. A transfer of $1 to households of printed money that increases m, like the Trump–Biden COVID checks, will raise interest rates and prices. In fact, as the transfers grow in size, there will always come a finite point

---

[13] Gurley and Shaw's (1960) inside money and outside money are related to our M and m, respectively.

where the prices go to infinity in what we called a hyperinflation. Printing money and giving it to people, what Milton Friedman (1969) called "helicopter money," is thus very inflationary.

Lending $1 (increasing M by $1) whose repayment comes due at the end of the horizon will lower interest rates and raise prices, but the price effect is less than with helicopter money. Furthermore, if the loan comes due before the horizon (so that the money that entered the system will be repaid and leave the system before the horizon), then there will be almost no inflationary effect. This contrast in the inflationary effect of helicopter money versus temporary injections aligns with the inflation after the 2020–2021 COVID transfers and the noninflation after the quantitative easing that began in 2010.[14]

Equilibria in this model can be computed with fully rational agents if there aren't too many, as in classical general equilibrium. If the level of individual detail gets too fine, one has to switch to an agent-based behavioral model. Presumably that will display the same qualitative features as the equilibrium model, perhaps with some twists.

AN AGENT-BASED MODEL OF CREDIT CARDS

General-purpose credit cards are perhaps the most ubiquitous financial innovation in the last fifty years, yet they have been completely ignored by economists. I am not aware of a single course in an economics department that analyzes credit cards beyond noting that the interest rates they charge are very high.

In Dubey and Geanakoplos (2010) we built a model with credit cards in which agents voluntarily choose which kind of money (cash or credit card) to use for which kind of purchase.

---

[14]Recall that, in the massive quantitative easing program beginning in 2010, the Fed made it clear that once the bonds were paid down it would revert to its former lower level of balance sheet.

In equilibrium, credit card markets and money markets all clear. We proved that money and credit cards can coexist in equilibrium, even if every buyer finds credit cards more convenient than cash. We showed that in equilibrium the credit card price of each good would be higher than the cash price, as it is in practice when no law mandates that the prices be the same. Gas stations, for example, give a discount for cash purchases. Department stores are forced to quote the same prices, but credit card purchasers who pay late are charged an exorbitant interest, partly to defray the default costs but also to raise the effective credit card price. Despite the difference in price, we showed that the invention of credit cards improves trading efficiency.

Our main conclusion is that the introduction of credit cards causes a massive increase in the price level, of around 100%, on the order of the 1970s inflation when credit cards emerged in full use, without affecting interest rates. This flies completely counter to the modern macro orthodoxy, which assumes that interest rates determine price levels and that the quantities of money can be ignored. No model of inflation that I am aware that is in use by the Federal Reserve System takes any note of credit cards. The model also indicates that monetary policy cannot thwart the effect on price levels of credit cards without destroying their efficiency benefits.

NEW MONIES AND INFLATION

In my 2024 Klein Lecture at the University of Pennsylvania, I described how Pradeep Dubey and I extended our our analysis of monies and inflation to the new monies I alluded to above. We built a model that includes all these innovations and we asked a fundamental question: Does the march to more monies create inflation? Does it foreshadow the inevitable nonviability of money? If credit becomes so easily available, how could money play any role?

Our main contribution was to provide a framework and parsimonious notation for a model with many monies, like fiat money, credit cards, debit cards, money market cards, stable coins, and a host of potential future monies. Surprisingly, we proved that the viability of old-fashioned cash and bank money is not destroyed by any of these new monies except stable coins backed by commodities, but that inflation will result without Fed action.

Standard macroeconomic models nowadays ignore conventional money, and definitely new monies. The new macro models presume that everything is determined by interest rates. One of our goals was to show that this is wrong. Changes in the stocks of money and access to credit will have real effects, and price-level effects, even if the Fed holds the interest rates constant. If the Fed remains passive, inflation is inevitable.

Once credit cards and other monies are added, the model keeps track of which purchases are funded by which monies. Agents voluntarily choose which kind of money to use for which kind of purchase. In equilibrium, markets all clear. Again, we worked with a very stylized equilibrium model. It goes from a cash economy to new monies, without analyzing the dynamic transition. It is important to see what new features arise in an agent-based model where agents don't have perfect foresight about future monetary developments and to simulate how an inflation might get started and pick up steam.

Booms and Busts via the Leverage Cycle

I have always been curious about what causes booms and busts in economic markets. After living through several of them as a

Wall Street principal, I found what I think is the answer in the years 1997–2003. I called it the leverage cycle.[15]

Phrases like the *business cycle* or the *credit cycle* were commonplace before then. These marked the ebb and flow of activity (like output or employment) or of borrowing, and usually focused attention on the central role of interest rates and investor confidence. I wanted to emphasize a new phenomenon: the credit terms apart from interest that drive borrowing, like collateral or credit score.

Traditional macroeconomists since John Maynard Keynes have attributed economic booms and busts to excessive or insufficient demand. The very language economists and journalists use to describe these episodes invariably refers to investors' animal spirits, irrational exuberance, risk appetite, or precautionary savings. According to the traditional theory, these ups and downs can be smoothed over by raising the interest rate when demand is too high, and lowering the interest rate when demand is too low.

The trouble with this demand-centric and riskless interest-rate-centric view of macroeconomics is that it leaves unanswered what we mean by tight credit, if not just a high interest rate. When businesspeople talk about tight credit, they don't mean that the riskless interest rate set by the Fed is too high. Default, and especially lenders' fear of default, is what is missing in the traditional macroeconomics theory. During a crisis, investors do not stop demanding loans at the riskless interest rate. If anything, they want more. They cannot get a loan of the size they want at the riskless rate, or anything close to it, because lenders are afraid they might default.

---

[15] See Geanakoplos (1997, 2003, 2010b, 2010a, 2014, 2016, 2019, 2024), from which much of this section is taken. See also Fostel and Geanakoplos (2008, 2012a, 2012b, 2014, 2015, 2016).

The leverage cycle places the lenders on center stage, equal at least to the investors in driving booms and busts. Busts, and especially crashes, are typically accompanied by increased uncertainty. When uncertainty rises, investors see more downside risk, but also more upside potential. Lenders, on the other hand, don't share in the upside; they see bigger downside losses. Lenders therefore have a bigger incentive than investors to change behavior. They ask for more collateral, or equivalently, they withdraw funds. In the leverage cycle, the bust comes not from the panic of investors, but the margin calls of lenders at the moment debts come due and might be rolled over.

There are two inspirational precursors of the leverage cycle, William Shakespeare and Hyman Minsky (1977). The title of William Shakespeare's *The Merchant of Venice* trumpets the investor Antonio (or possibly Bassanio). The real leading character, however, is Shylock, the moneylender. Like many high schoolers of the time, I was forced to read over a dozen Shakespeare plays; the one I felt I understood differently from the critics was this one. It is, to be sure, a comedy about weddings, and it does explore antisemitism, as the critics said. It is also a commentary about legal contracts. The plot, however, is driven by lending terms and collateral, the 3000-ducat loan Bassanio takes from Shylock that Antonio guarantees. The characters debate the interest rate Shylock is going to charge for five pages. Bassanio and Antonio point out that, as Christians, they would not charge interest. Yet when I walked out of the Broadway performance of the play, featuring Al Pacino as Shylock, I asked everybody I saw if they remembered the interest rate Shylock and Antonio and Bassanio eventually agreed on just an hour earlier. Not one remembered. Yet they all remembered the pound of flesh collateral. Shakespeare made the collateral more memorable than the interest because he thought it was more important, just as he made Shylock more memorable than the titular investor because he thought the lender was more important. The high drama of the play develops with a rumor that Antonio's boats have sunk, which turns out to be false, followed by Shylock calling in his loan without rolling it over, like a margin call, and claiming the collateral when the payment can't be met. The play ends with the question of whether it is right to forgive an unpaid promise.

Hyman Minsky is known for saying, as paraphrased by James Tobin, that leverage is the Achilles' heel of capitalism. He warned that borrowing goes through three stages he called hedge, speculative, and Ponzi. Corporations that owed more in interest payments than they earned in current profits were engaging in the last Ponzi stage of borrowing that would eventually lead to a moment where it became recognized that they couldn't repay their debts; unable to get more loans they would go bankrupt, possibly crashing the whole system if there were too many of them. He did not mention collateral or loan-to-value ratios for assets such as housing. Like Shakespeare, he did not have a mathematical model.

The first idea of the leverage cycle is that the purchases of a great many assets come partly, and many times mostly, with borrowed money, using the asset itself as collateral. Nobody paid $1 million in cash in today's dollars for a tulip in the 1600s, or $1.3 million in cash for a taxi medallion in 2015.

Second, how much is borrowed is mediated by lending terms like collateral and credit score (FICO for individual borrowers or credit grade for corporations). The relationship between the interest rate and the credit terms is called the *credit surface*. The credit surface emerges in equilibrium, which can be described in a mathematical general-equilibrium model. Figure 2 is a hypothetical picture of the interest rate on loans as a function of the loan-to-value (LTV, the ratio of the loan

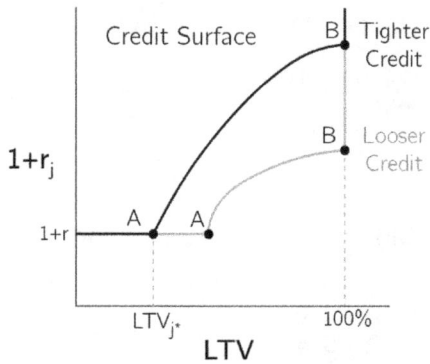

Figure 2. Theoretical credit surface when collateral has two future values. *(Source: Fostel and Geanakoplos 2015)*

amount to the value of the collateral). Figure 3 is the actual picture of interest depending on LTV and FICO credit score for FNMA and Freddie Mac loans in the second quarter of 2006.

Third, an important shaper of the credit surface is perceived uncertainty. When uncertainty increases, the credit surface rises and steepens. If lenders perceive more uncertainty or volatility to future collateral values, they will feel their

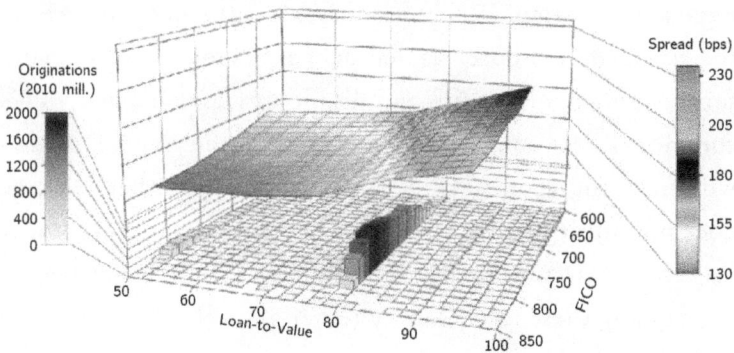

Figure 3. 2006Q2, 30-year conventional purchase mortgages. *(Source: Geanakoplos and Rappoport 2019 using Black Knight Financial Services and BLS)*

Good News, Calm Economy

Low Volatility ➡ Flat Credit Surface ➡ High Leverage ➡ High Price

Bad News, Uncertain Economy

High Volatility ➡ Steep Credit Surface ➡ Low Leverage ➡ Low Price

Figure 4. Logic of the leverage cycle from Geanakoplos (2003, 2010a). *(Source: Geanakoplos 2024)*

loans are more at risk and ask for higher interest rates for the same credit terms, so the surface will shift up. Furthermore, the increased uncertainty matters more for riskier loans, so the credit surface also steepens. *Tighter credit* means a higher and steeper credit surface. As a result of the tighter credit, borrowers will end up choosing to borrow less and at lower LTV, that is, using less leverage.

Fourth, and most importantly, more leverage leads to higher asset price. When homeowners can borrow 97% of the value of a house, as happened in 2005, the demand for housing will go up and house prices will soar. This common-sense conclusion contradicts the Nobel Prize-winning Modigliani–Miller theorem asserting that leverage does not affect prices, which it turns out only applies under very special circumstances that don't include home mortgages.

Putting these four ideas together gives the static leverage cycle described by the two rows in figure 4.

Fifth, crashes occur when scary bad news arrives at the top of the leverage cycle with high leverage and high prices, provided that a significant number of loans are coming due.

The (little bit of) bad news lowers everybody's valuations of the collateral and the scary news substantially increases uncertainty, lurching the economy from the top right to the bottom left of figure 4 above. At that moment asset prices go down because of the bad news, because the highest-valuation agents take steep losses when they are unable to meet their margin calls and are forced to sell, and because the new buyers cannot borrow much money and have to pay with their own cash.[16]

The price can fall much farther than the bad news reduces anybody's valuation. If there is heterogeneity in the buyers' valuations, the asset will start in the hands of those who value it most. The marginal buyer, who is on the cusp of buying or not buying and whose valuation therefore sets the price, will be a high-valuation buyer, and the price will start high. Many of those high-valuation buyers will take steep losses when they are forced to sell after the news and won't be able to buy again. Each new buyer will not be able to buy as much as each old buyer did because they won't be able to borrow as much. Hence, there will be a long interval of new buyers, starting at the top just below the old marginal buyer and stretching a long way down to the new marginal buyer, who will have a much lower valuation than the original marginal buyer. The price falls much more because of the change in marginal buyer than because of the bad news affecting every valuation. See figure 5 below.

---

[16]More precisely, the credit surface steepens because of the increased uncertainty. Leveraged owners whose loans are coming due receive margin calls because (1) the price is falling (from the bad news) *and* (2) because they are limited to borrowing a smaller percentage of the (lower) price (on account of the steeper credit surface), if they want to reborrow at the same interest. In sum, they must pay down part of their old loan. Those who lack the cash are forced to sell. The new buyers cannot borrow nearly as much as the original buyers because the credit surface is tighter.

Marginal Buyer Theory of Price

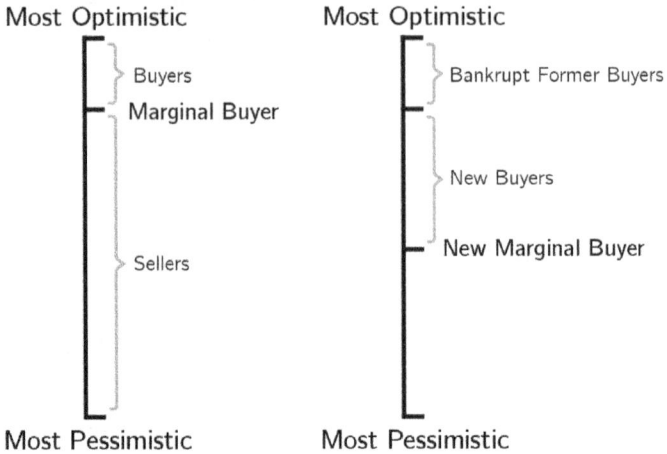

Figure 5. Leverage cycle crash of the marginal buyer. *(Source: Geanakoplos 2010a)*

Sixth, once the prices are much lower, many owners whose loans have not come due are underwater. They will not make improvements to their houses or buildings. They may even stop paying their mortgages or taxes, eating away at the value of the collateral. If the underwater owners cannot be forced out quickly, it may be better to partially forgive some of their debt, because then forgiveness helps the borrowers and the lenders, as Portia said in *The Merchant of Venice*.

SOME RECENT EXAMPLES OF THE LEVERAGE CYCLE IN ACTION

The newspapers are full of stories illustrating the leverage cycle, once the reader knows how to interpret them. Let me begin by recounting my conversation with an elderly New York cabdriver I hailed near my Harlem home just a few months ago. I asked him if he owned his own taxi medallion (the permit that allows taxi drivers to legally operate in New York City) or whether he was employed by the medallion owner. He replied with some pride that he owned his.

"If you don't mind my asking," I said, "how much did you pay for your medallion?"

He said just over $300,000.

"If you don't mind my asking," I continued, "how could you have afforded that?"

He said that a bank got his name as a driver of cabs and called him offering to lend 95% of the value of the medallion if he let them hold it as collateral. He figured that for just $15,000 (5% of $300,000), which he had, plus the borrowed money, he could own his medallion and be his own man. He took the chance.

"What happened to the price of the medallion?" I asked.

"It went up to $1.3 million in 2014 or '15," he replied.

"You could have told your grandkids that you were a millionaire."

"I did tell them."

I replied, "I suppose you sold the medallion." He said he hadn't.

"Why not?" I asked.

"I thought the price would keep going higher."

"What is it worth now?" I asked.

"$140,000," he said.

I apologized as I started to laugh.

"No need to apologize," he said, adding that he laughs himself, when he isn't crying.

"How much could you borrow now on the $140,000 medallion?"

"Not a penny," he replied.

I then asked how he managed to repay the $285,000 loan from the bank. He said he couldn't—but the city bailed him out.

This story fits the leverage cycle almost perfectly. The banks initially figured that the price of medallions was rock-solid. The city rarely increased their number, freezing supply, and more and more New Yorkers and tourists were taking more rides. Then Uber was invented. Scary bad news for the cab business. For New York cab drivers the news wasn't very bad, because the number of cabs was already so limited. The investor–cab drivers like my old friend kept driving. Their incomes barely went down, at most by 10%. The livery cab business was destroyed, but that is another story. The real significance of Uber is that it created much more uncertainty for the lenders. The downside potential for the taxicab business was now much greater. Though the investor–drivers hung in, the lenders withdrew their lending. Now nobody can borrow using medallions as collateral. The price collapsed, and unless lending returns, it is unlikely to rise much. In the end, the loans had to be forgiven.

In November 2022, the legendary cryptocurrency trader Sam Bankman-Fried's US$32 billion Alameda–FTX crypto empire collapsed, apparently out of the blue, as crypto currencies declined.[17] Bankman-Fried was convicted of stealing money from the depositors at his cryptocurrency bank FTX in order to finance investments at his crypto-trading firm Alameda. Why would somebody with that much money need to steal more? Because his crypto investments in Alameda were having a rocky time as crypto values fell. His lenders asked for their money back. Unable to find the funds quickly, he "borrowed" the money from the depositor funds in his

---

[17] As a kid, Bankman-Fried had once vacationed in France with his academic parents and my family.

FTX crypto bank, apparently figuring on repaying them once the markets settled. Eventually crypto did come back and the depositors were repaid in full, but not before Bankman-Fried's actions were exposed. We see again the familiar leverage cycle story where the bad news scares the lenders more than the investors, and causes the lenders to withdraw their funds. Only then do the investors react.

In September 2022, just after Prime Minister Liz Truss announced a tax cut, the British pension system nearly imploded, leading the British central bank to intervene and the newly elected Truss government to fall just six weeks after taking office. The pension liabilities owed by firms had been hedged by holding very long maturity government bonds called gilts, which had been purchased on margin. Truss's determination to cut taxes and possibly increase government debt caused gilt prices to fall and interest rates to rise. This scared the lenders about the future value of gilts. They made margin calls on the gilt holders, who were forced to sell. Unfortunately, their long-dated pension liabilities made the pension system almost the unique natural buyers for very long-dated gilts. As the gilts fell into the hands of those who valued them less, the drop in the valuation of the marginal buyer, and thus the price, was steep.

In early 2020, the Dow Jones plunged 18% and the mortgage markets and especially real estate investment trusts (REITs) completely froze, necessitating the biggest Federal Reserve intervention in history. Of course COVID-19 darkened investors' moods, but no dramatic new information caused investors to suddenly panic in March 2020: The long-term prognosis for mortgage security cash flows did not fall much. What happened was that the lenders saw more potential (not yet actual) downside and dramatically altered the LTV on loans to REITs, in many cases refusing to lend at all. REIT stock market values collapsed by about 80%. Then, a few months later, they came roaring back when LTVs were restored by the Fed intervention starting March 23, 2020.

In 2007, the subprime mortgage market collapsed. Over the next two years the Dow Jones plunged 45% and the Great Recession took hold. In retrospect, the only noteworthy preceding news seemed to be the subprime delinquency reports, which showed delinquencies in the range of 4–5% instead of 2%. The delinquency increase by itself was not very big. But it scared lenders into thinking there might possibly be much worse to come. They drastically curtailed mortgage leverage and repo leverage. In the 2000s house prices and mortgage security prices soared as leverage rose for housing (correctly measured to take into account the new kinds of mortgages) and for repo lending on mortgage securities, then plunged in 2007 when leverage collapsed, then rose again post-2010 as leverage returned. [18]

Two very recent examples are the boom and decline of the office space real-estate market in many US cities, which as of May 2025 has not turned into a crash because many of the long-term loans have not come due. The Chinese real-estate collapse following massive leverage-based building has been as bad as the 2007–2010 American crisis, and it could have been much worse if the government had not absorbed so much of the losses.

Other examples are the sudden crash of the famous American hedge fund Long Term Capital in 1998 on the heels of the late 1997 Asian market bond crisis, and the derivatives crisis of 1994 that led to the closing of Kidder Peabody. More famous examples include the 1980s Japan boom and crash, the 1920s Florida land boom and crash, and the 1637 tulip bulb

---

[18] See Geanakoplos (2003, 2010b, 2010a, 2016, 2019).

craze, in which tulips reached a price of nearly \$1 million in today's dollars before suddenly crashing.[19]

I end with the very recent 2023 collapse of Silicon Valley Bank (SVB), to contrast the leverage-cycle theory of booms and busts with the panic theory of crashes. As the Fed raised interest rates to fight the COVID inflation starting in March 2022, the treasuries held by SVB dropped in value. These securities were held in a special hold-to-maturity bucket that did not have to be marked to market. So only a very attentive depositor would have been immediately aware that the value of SVB assets was gradually sinking below the value of its deposit liabilities. As the tech depositors in SVB bank eventually became aware of this, they pulled their money out and the bank collapsed, requiring government intervention.

According to the panic theory, crashes are the result of panics like bank runs, in which depositors run because they are worried that if they don't get their money out first, others depositors will take it themselves. According to this theory, the main goal of government should be to restore confidence, thereby moving the economy from a bad equilibrium to a good one. The theory has one central similarity with the leverage cycle in that it focuses attention on the lenders, or at least the depositors in banks and money market funds. In both theories the lenders run away with, or substantially reduce, their loans in the crisis.

There are, however, important differences between the leverage cycle theory and the panic theory of crashes. In the leverage cycle, lenders increase margins (i.e., reduce their loans against the same collateral) because the risk that the *collateral* might not cover the old loan amount has increased. As it pertains to banks, the leverage cycle could be called a theory of *insolvency*

---

[19] Of course, some crashes occur because the world changes, and have nothing to do with lending, like the tech bust in 2000.

runs. Depositors regard the assets of the bank as collateral for their loans, and when they become aware that even with an orderly liquidation there is a risk that those assets may become insufficient to pay off their deposits, they will withdraw their money. By contrast, a panic bank run can occur even if the bank assets are certainly worth more than the deposits in an orderly liquidation. The panic run is triggered once depositors worry that *other depositors* will take their money out of the bank, forcing a disorderly liquidation of the assets. In my opinion, the SVB bank crash was the result of a leverage cycle potential insolvency, not a panic run.

The policy implications of the two theories regarding banks are very different. The leverage cycle theory of insolvency runs embraces the positive role of transparency. The idea is that, if the value of bank assets versus liabilities is public, pressure from depositors or regulators will force the bank to raise capital to restore safer LTV levels. Insolvency rarely comes all at once; it is generally a gradual process during which assets lose value a little bit at a time, say, because the Fed is raising interest rates or unemployment is rising. There is plenty of time to act before actual insolvency sets in, provided that people are aware of the growing problem. Bank-run theorists argue the opposite, suggesting that opaqueness is a virtue because too much information might make depositors nervous and lead them to panic. Thus, banks are not required to mark many of their assets to market, which sometimes hides dwindling asset values. According to the leverage theory, efforts to forestall a panic run may make it more likely that there will be an insolvency run.

The main difference is that the leverage cycle applies to any situation involving collateral, whether or not it is connected to banks. The run-to-get-out-first story really only applies to banks. Collateral like houses, unlike bank assets, is reserved for a single lender. No other lender can take my collateral by running. A margin call is the rational reaction to increased down risk, not an arbitrary panic that can be quieted by reassuring words that other depositors will not withdraw their loans.

THE LEVERAGE CYCLE AND AGENT BASED MODELS

Agent-Based Data

The leverage cycle is tailor-made for agent-based modeling. It ~ 125 ~ requires heterogeneity among the agents, so some become lenders and others become borrowers, and so that there is a wide dispersion in valuations of the collateral among potential marginal buyers. It requires information about each agent's debts and assets, including the network of who owes what to whom. The reason is that there can be chain reactions in defaults. If A defaults against B, then B may not have the funds to pay C. Even if all the loans are collateralized, C may be forced to sell their collateral to pay their loan when it comes due, which might lower the value of D's collateral and cause D to default against E.

After the 2008 financial crisis, Congress created the Office of Financial Research (OFR) as a third leg to balance and check the Federal Reserve and the Office of the Treasury, which Congress felt had failed to properly regulate the American economy leading up to the 2007–2010 financial crisis. One of the OFR's important missions was to monitor precisely the sorts of variables that underlie the leverage cycle and were discussed at the end of the last section. To that end it was given unlimited subpeona power to get whatever information it believed was vital.

Doyne Farmer and I realized that the OFR offered a once-in-a-lifetime opportunity to marshal the data and the manpower to create the agent-based model of the US financial system that we had been envisaging. Doyne paid a small fortune to take up an Obama fundraising initiative that gave him a twenty-minute face-to-face meeting with Obama, which he used to suggest me as head of the OFR. Whether through Doyne's efforts or otherwise, I was indeed interviewed for the job by Jeffrey Goldstein, the undersecretary of the Treasury. The interview started off badly and then got worse. Goldstein began by telling me that since Robert Engel and Robert Shiller (both Nobel Prize winners) had turned down the job, there was really nobody worthy left to hire. I had figured that since Goldstein was a Yale economics PhD and a friend of my classmate Larry Summers, who had coached me about the interview, that he would begin on a friendlier note. I asked him if the OFR was really going to have independence from the Fed and the Treasury, especially in view of the fact that it was apparently going to be housed in the basement of the Treasury building. Goldstein asked me if I was calling him a liar. I never figured out whether or not he had a gruff but good sense of humor. I didn't get the job.

The OFR has not in fact grown into the important third leg that its backers in Congress envisaged. The dream that the OFR would create a realistic model of the agents in the American (and world) financial system remains unfulfilled. At the same time, enormous progress in computation and artificial intelligence makes the goal more realistically achievable outside the OFR. Perhaps 2011 was too early, and the time is right now.

Rational Expectations?

The leverage cycle was conceived of in a general-equilibrium setting with rational expectations. One of the points of the theory was to show that, despite the rationality of the agents, they could repeatedly get themselves into leverage-cycle crashes and booms.[20] The crucial optimizing behavior is that lenders should steepen

---

[20] In stark contrast to the Pareto optimality of Arrow and Debreu's complete markets general equilibrium.

the credit surface if they anticipate higher volatility in future asset prices (or, more precisely, worse downside risks), and that borrowers should then use less leverage. That behavior can easily be mimicked by a behavioral rule that makes leverage fall when trailing volatility climbs. In practice, volatility is clustered, so past volatility is generally a very good predictor of forward volatility. The heart of the leverage cycle can be captured by an agent-based model.

A defining characteristic of the leverage cycle is that it depends on anticipations of second moments, not just first moments. Changes in volatility, or in anticipated volatility, change leverage and therefore collateral prices. The leverage cycle is thus driven by expectations, but it does not rely on irrational expectations. Nor does it rule them out. It might well be that lenders might become irrationally exuberant or pessimistic about the direction collateral prices will move. That would increase the amplitude of the leverage cycle.[21]

Dynamic Leverage Cycle

The leverage cycle creates fascinating dynamics that can be explored via simulation. Indebted agents who get margin calls are forced to sell precisely when they would most like to buy. In a standard Walrasian model with equal endowments, one would expect a monotonic relation between valuations and purchases. The standard monotonicity gets reversed with a margin call, when the highest valuation agents become forced sellers.

Furthermore, volatility or uncertainty is a key driver of the leverage cycle. When there is scary bad news, that is, an increase in uncertainty at the same time as bad news, then leverage drops (margin requirements rise) exactly at the moment prices are falling

---

[21]See Bordalo, Gennaioli, and Shleifer (2018) for a sophisticated treatment of expectations.

anyway. This magnifies the drop in prices, as I explained earlier. In Stefan Thurner, Farmer, and Geanakoplos (2012) and Sebastian Poledna *et al.* (2014) we showed that in agent-based simulations in which banks automatically increase margin requirements when trailing volatility increases, independent Gaussian shocks to valuations lead to price changes that have fat tails and display clustered volatility, and that "prudently" responding to increases in volatility with sharper restrictions on leverage could paradoxically cause the system to be more unstable. Christoph Aymanns and Farmer (2015) showed that large asymmetric cycles could be self-generated when investors follow a value-at-risk rule that induces less leverage when volatility goes up. For a two-agent economy, they develop theorems characterizing the dynamics as well as simulations.

Housing

Housing is one of the archetypal assets to which the leverage cycle applies. The leverage cycle as described above has been shown to be an informative way of looking at housing prices. But there is so much more to housing that only an agent-based model can capture. Housing may be the most fruitful place of all for agent-based modeling. Much progress has been made, but we are still just scratching the surface. Understanding what drives housing and homeowner behavior has at least three vital benefits.

First, it is now universally recognized that real estate and housing bubbles played an important role in the 2007–2010 financial crisis and in many others besides. No monitoring of systemic risk can pretend to be satisfactory if it does not include a model of the housing market. Are housing bubbles caused by low interest rates, irrational exuberance, low debt-to-income or FICO standards, too much cash-out refinancing, or too much leverage?

Second, prudential policy in real estate also requires an understanding of the role of forgiveness. I argued that the best way out of the 2008 crisis was to write down principal on subprime housing loans that were underwater (see Geanakoplos and Koniak 2008, 2009; Fostel and Geanakoplos 2012b), on the grounds that the loans would not be repaid anyway, and that, taking into account foreclosure costs, lenders could get as much or almost as much money back by forgiving part of the loans, especially if stopping foreclosures were to lead to a rebound in housing prices. Unfortunately, there was very little forgiveness. Afterward, policy dramatically shifted. Nowadays, homeowners who miss payments on loans routinely see their coupons reduced or delayed. It is not only important to see whether this leads to better recoveries for the lender, but also to know what the implications of avoiding foreclosures are for house prices in general.

Third, housing is also one of the most important and lucrative industries in the world. Brokers and traders would pay through the nose to learn which neighborhoods were going to appreciate and which decline.

Conventional economic analysis attempts to answer these kinds of questions by building equilibrium models with a representative agent, or a very small number of representative agents. In my view, answers can only be given by an agent-based model, that is, a model in which we try to simulate the behavior of literally every household in each city, taking into account the characteristics of all the existing buildings one by one. Just how many more people would want to buy a house of a particular kind (e.g., three bedrooms in a neighborhood with good schools) because they could leverage more easily, because interest rates were lower, or because they were forced to move when their neighborhoods burned down?

Conventional thinking suggests that an agent-based model of the housing market is an impossibly ambitious task. We would need too much data; it all depends on arbitrary behavioral rules, each of which depends on too many parameters to estimate reliably. Without the discipline of equilibrium, expectations cannot be pinned down. As the world changes, what seemed like appropriate behavioral rules will be revealed to be crazy, and so on.

Against this we have the argument that the agent-based approach brings a new kind of discipline because it uses so much more data. For example, records of all bids and offers through brokers over ten years in a city might enable us to say with high confidence that a when a homeowner decides to put their house on the market, they list an offering price that is somewhat above (on average a bit less than 10%) the average of other comparable sales in the past six months. Every few months that a house does not sell, the homeowner reduces the asking price by some percent (e.g., 3–5%). Each behavioral rule must pass a basic plausibility test on its own (which is crucial in any model). When the whole model is run with its multiplicity of behavioral rules, the agent-based approach allows for many more checks, including vacancy rates; time on market; number of renters versus owners; ownership rates by family status, ethnicity, age, sex, wealth, and income; as well as the housing prices used in standard models. And one can check if the same or similar behavior works across dozens of different cities.

A large team of us worked on a housing model for the Washington, DC, area (see Geanakoplos *et al*. 2012; Axtell *et al*. 2014). Since then, part of the team and others led by Doyne Farmer have applied a similar model to London and other cities. These models reproduced some of the dynamics of historical housing prices and gave answers to questions like: How far will prices ripple through the city if a bunch of Russian oligarchs start buying high-end real estate? The models are still too stylized to make neighborhood predictions, for example. And the effects of foreclosures versus modifications on housing prices are still not completely resolved.

Conclusion

I began by describing an aggregate prepayment model that I first formulated by adding over many types of heterogeneous, rational agents, even though I had absolutely no information about agents at the individual level. This was, to the best of my knowledge, the first (optimizing) agent-based prepayment model. The optimizing behavior of each type included a few surprises, like slowing when volatility increased. Nonetheless, each rational individual behavior could be mimicked by a behavioral rule in an agent-based model that proved very reliable for a number of years.

Next I described a general-equilibrium model of monies and inflation. This model differs in many ways from the standard neo-Keynesian and Federal Reserve models of inflation in being fully in the general-equilibrium tradition yet in a finite horizon, with no artificial utility for money. Its main virtue is that it naturally includes the different monies, like cash and credit cards, that have already emerged, as well as others that are soon to arrive, but are completely absent in standard models. Because it is finite horizon, it is in principle computationally compatible with many of the details of the actual flow of funds in the economy between different sectors and agents.

To be truly realistic, I argued, there is no substitute for an agent-based model. Agent-based modeling of inflation is just beginning. The usual question is whether there are plausible behavioral rules that mimic the rational agents of the general-equilibrium model. As I mentioned earlier, the liquidity trap is one feature of the equilibrium model that seems to mirror actual economies (like Japan for decades around 2000 and the United States after the quantitative easing in 2010). When the central bank injects money that everyone expects will be pulled out later, even ten years later, the equilibrium model predicts there should be no inflationary effect, whereas if it is thought to be permanently injected, it will indeed produce a temporary inflation. What behavioral rules will lead to that? We shall see!

Next I described my leverage-cycle theory of booms and busts. Again, it was conceived in a general-equilibrium setting with rational expectations. Once again the behavior of the agents can be mimicked by behavioral rules. The optimizer sets leverage according to anticipated volatility, and anticipated volatility is highly correlated with current volatility, so there is an obvious behavioral rule that is pretty close to optimal.

Finally, housing is perhaps the leading example of the leverage cycle. Yet a fully realistic and more interesting model of housing would require the details of an agent-based model. I started to build one with many coauthors, and now many others are moving much farther ahead. There is still much to be done.

All four examples illustrate the symbiosis between general-equilibrium modeling and agent-based modeling that I described in the introduction. ❧

REFERENCES

Andersen, A., K. Huber, N. Johannesen, L. Straub, and E. T. Vestergaard. 2023. *Disaggregated Economic Accounts.* NBER Working Paper no. 30630. https://doi.org/10.3386/w30630.

Axtell, R., J. D. Farmer, J. Geanakoplos, P. Howitt, E. Carrella, B. Conlee, J. Goldstein, *et al.* 2014. *An Agent-Based Model of the Housing Market Bubble in Metropolitan Washington, D.C.* https://doi.org/10.2139/ssrn.4710928.

Aymanns, C., and J. D. Farmer. 2015. "Dynamics of the Leverage Cycle." *Journal of Economic Dynamics and Control* 50 (C): 155–179. https://doi.org/10.1016/j.jedc.2014.09.015.

Benmelech, E., N. Kumar, and R. Rajhan. 2022. *The Secured Credit Premium and the Issuance of Secured Debt.* NBER Working Paper no. 26799. https://doi.org/10.3386/w26799.

Bordalo, P., N. Gennaioli, and A. Shleifer. 2018. "Diagnostic Expectations and Credit Cycles." *The Journal of Finance* 73 (1): 199–227. https://doi.org/10.1111/jofi.12586.

Brunnermeier, M. K., and L. Pedersen. 2009. "Market Liquidity and Funding Liquidity." *Review of Financial Studies* 22 (6): 2201–2238. https://doi.org/10.1093/rfs/hhn098.

Carroll, C. 1997. "Buffer-Stock Savings and the Life-Cycle/Permanent Income Hypothesis." *Quarterly Journal of Economics* 112 (1): 1–55. https://doi.org/10.1162/003355397555109.

Dubey, P., and J. Geanakoplos. 1992. "The Value of Money in a Finite Horizon Economy: A Role for Banks." In *Economic Analysis of Markets: Essays in Honor of Frank Hahn,* edited by P. Dasgupta, D. Gale, O. Hart, and E. Maskin, 407–444. Cambridge, MA: MIT Press. https://doi.org/10.7551/mitpress/2581.003.0022.

———. 2003. "Inside and Outside Fiat Money, Gains to Trade, and IS-LM." *Economic Theory* 21 (2–3): 347–397. https://doi.org/10.1007/s00199-002-0296-5.

Farmer, J. D. 2024. *Making Sense of Chaos: A Better Economics for a Better World.* New Haven, CT: Yale University Press.

Fostel, A., and J. Geanakoplos. 2008. "Leverage Cycles and the Anxious Economy." *American Economic Review* 98 (4): 1211–1244. https://doi.org/10.1257/aer.98.4.1211.

Fostel, A., and J. Geanakoplos. 2012a. "Tranching, CDS, and Asset Prices: How Financial Innovation Can Cause Bubbles and Crashes." *American Economic Journal: Macroeconomics* 4 (1): 190–225. https://doi.org/10.1257/mac.4.1. 190.

———. 2012b. "Why Does Bad News Increase Volatility and Reduce Leverage?" *Journal of Economic Theory* 147 (2): 501–525. https://doi.org/10.1016/j. jet.2011.07.001.

———. 2014. "Endogenous Collateral Constraints and the Leverage Cycle." *Annual Review of Economics* 6 (1): 771–799. https://doi.org/10.1146/annurev-economics-080213-041426.

———. 2015. "Leverage and Default in Binomial Economies: A Complete Characterization." *Econometrica* 83 (6): 2191–2229. https://doi.org/10.3982/ ECTA11618.

———. 2016. "Financial Innovation, Collateral, and Investment." *American Economic Journal: Macroeconomics* 8 (1): 242–284. https://doi.org/10.1257/ mac.20130183.

Friedman, M. 1969. *The Optimum Quantity of Money and Other Essays.* Chicago, IL: Aldine Publishing Co.

Geanakoplos, J. 1997. "Promises, Promises." In *The Economy as an Evolving Complex System II,* edited by W. B. Arthur, S. Durlauf, and D. Lane, 285–320. Reading, MA: Addison-Wesley.

———. 2003. "Liquidity, Default, and Crashes: Endogenous Contracts in General Equilibrium." In *Advances in Economics and Econometrics: Theory and Applications, Eighth World Conference, August 2000,* edited by M. Dewatripont, L. P. Hansen, and S. J. Turnovsky, II:170–205. Cambridge, UK: Cambridge University Press. https://doi.org/10.1017/CBO9780511610257.007.

———. 2010a. "Solving the Present Crisis and Managing the Leverage Cycle." *Federal Reserve Bank of New York Economic Policy Review,* 101–131.

———. 2010b. "The Leverage Cycle." In *NBER Macroeconomics Annual 2009,* edited by D. Acemoglu, K. Rogoff, and M. Woodford, 1–65. Chicago, IL: University of Chicago Press.

———. 2014. "Leverage, Default, and Forgiveness: Lessons of the American and European Crises." *Journal of Macroeconomics* 39 (Part B): 313–333. https:// doi.org/10.1016/j.jmacro.2014.01.001.

———. 2016. "The Credit Surface and Monetary Policy." In *Progression and Confusion: The State of Macroeconomic Policy,* edited by O. J. Blanchard, R. G. Rajan, K. S. Rogoff, and L. H. Summers, 143–153. Cambridge, MA: MIT Press.

———. 2019. "Leverage Caused the 2007–2009 Crisis." In *Systemic Risk in the Financial Sector: Ten Years After the Great Crash,* edited by D. W. Arner, E. Avgouleas, D. Busch, and S. L. Schwarcz, 235–262. Montréal, Canada: McGill–Queen's University Press.

———. 2024. "Leverage Cycle Theory of Economic Crises and Booms." In *Oxford Research Encyclopedia of Economics and Finance.* Oxford, UK: Oxford University Press. https://doi.org/10.1093/acrefore/9780190625979.013.484.

Geanakoplos, J., R. Axtell, J. D. Farmer, P. Howitt, B. Conlee, J. Goldstein, M. Hendrey, N. M. Palmer, and C.-Y. Yang. 2012. "Getting Systemic Risk via an Agent-Based Model of the Housing Market." *American Economic Review* 102 (3): 53–58. https://doi.org/10.1257/aer.102.3.53.

Geanakoplos, J., and P. Dubey. 2010. "Credit Cards and Inflation." *Games and Economic Behavior* 70 (2): 325–353. https://doi.org/10.1016/j.geb.2010.02.004.

Geanakoplos, J., and S. Koniak. 2008. "Mortgage Justice is Blind." October 20, 2008, *The New York Times,* https://www.nytimes.com/2008/10/30/opinion/30geanakoplos.html.

———. 2009. "Matters of Principal." March 4, 2009, *The New York Times,* https://www.nytimes.com/2009/03/05/opinion/05geanokoplos.html.

Geanakoplos, J., and D. Rappoport. 2019. *Credit Surfaces, Economic Activity, and Monetary Policy.* SSRN no. 3428729.

Gilchrist, S., and E. Zakrajšek. 2012. "Credit Spreads and Business Cycle Fluctuations." *American Economic Review* 102 (4): 1692–1720. https://doi.org/10.1257/aer.102.4.1692.

Glaeser, E. L., J. Gottlieb, and J. Gyourko. 2010. *Can Cheap Credit Explain the Housing Boom?* Harvard Working Paper.

Goodman, L. *Housing Credit Availability Index.* Urban Institute. https://www.urban.org/policy-centers/housing-finance-policy-center/projects/housing-credit-availability-index.

Greenwald, D. 2016. *The Mortgage Credit Channel of Macroeconomic Transmission.* MIT Sloan Research Paper no. 5184-16.

Gurley, J. G., and E. S. Shaw. 1960. *Money in a Theory of Finance.* Washington, DC: The Brookings Institution.

Hahn, F. H. 1965. "On Some Problems of Proving the Existence of an Equilibrium in a Monetary Economy." In *The Theory of Interest Rates,* edited by F. H. Hahn and F. P. R. Brechling, 126–135. London, UK: Macmillan.

Haughwout, A., D. Lee, J. Tracy, and W. van der Klaauw. 2011. *Real Estate Investors, the Leverage Cycle, and the Housing Market Crisis.* Federal Reserve Bank of New York Staff Report no. 514. https://www.newyorkfed.org/research/staff_reports/sr514.html.

Kindleberger, C. 1978. *Manias, Panics, and Crashes: A History of Financial Crises.* New York, NY: Basic Books.

Kiyotaki, N., and J. Moore. 1997. "Credit Cycles." *Journal of Political Economy* 105 (2): 211–248. https://doi.org/10.1086/262072.

Lerner, A. 1947. "Money as a Creature of the State." *Papers and Proceedings of the Fifty-Ninth Annual Meeting of the American Economic Association* 37 (2): 312–317. https://www.jstor.org/stable/1821139.

Megiddo, N., and C. Papadimitriou. 1989. "On Total Functions, Existence Theorems and Computational Complexity." *Theoretical Computer Science* 81 (2): 317–324. https://doi.org/10.1016/0304-3975(91)90200-L.

Minsky, H. 1977. "A Theory of Systemic Fragility." In *Financial Crises: Institutions and Markets in a Fragile Environment,* edited by E. D. Altman and A. W. Sametz, 138–152. New York, NY: John Wiley and Sons.

Poledna, S., S. Thurner, J. D. Farmer, and J. Geanakoplos. 2014. "Leverage-Induced Systemic Risk under Basle II and Other Credit Risk Policies." *Journal of Banking & Finance* 42:199–212. https://doi.org/10.1016/j.jbankfin.2014.01.038.

Shleifer, A., and R. Vishny. 2011. "Fire Sales in Finance and Macroeconomics." *Journal of Economic Perspectives* 25 (1): 29–48. https://doi.org/10.1257/jep.25.1.29.

Soros, G. 2009. *The Crash of 2008 and What it Means: The New Paradigm for Financial Markets.* New York, NY: Public Affairs.

Stiglitz, J. E. 1969. "A Re-Examination of the Modigliani–Miller Theorem." *American Economic Review* 59 (5): 784–793. https://www.jstor.org/stable/1810676.

Thurner, S., J. D. Farmer, and J. Geanakoplos. 2012. "Leverage Causes Fat Tails and Clustered Volatility." *Quantitative Finance* 12 (5): 695–707. https://doi.org/10.1080/14697688.2012.674301.

EQUATIONS VS. MAPS: COMPLEXITY, EQUILIBRIUM, DISEQUILIBRIUM

Marco Pangallo, CENTAI Institute

Abstract

A key theme of complexity economics is nonequilibrium/ disequilibrium dynamics. However, it is often not clear what this precisely means, leading to confusion about the differences and relative advantages of, for instance, general equilibrium and nonequilibrium agent-based models. In this chapter, I first review and survey economists on what equilibrium means. Answers range from equilibrium being a problem-specific concept with no general meaning to general definitions of equilibrium that may not be consistent with one another. Given this lack of consensus, I provide a novel definition. This definition is a technical, practical one: Equilibrium is when model variables are obtained by solving equations; nonequilibrium is when variables are just obtained as a map of other variables. Equilibrium and nonequilibrium can coexist in the same model, and nonequilibrium dynamics may converge to equilibrium. The main value added of our definition is that it clarifies the relative advantages of both equilibrium and nonequilibrium models. On the one hand, solving equations allows economists to determine variables without specifying an explicit process, making equilibrium models more parsimonious. On the other hand, nonequilibrium models such as agent-based models can easily accommodate many complexity-related ideas that are more difficult to include in equilibrium models, such as heterogeneity, nonlinear dynamics, and complex network structure.

Introduction

> *As the model has developed so far, all the actions of*
> *decision-makers are based on previously established*
> *values of the variables influencing the decision: there*
> *is no simultaneity whatever. Thus the model is*
> *never "solved."... This means that decision-makers'*
> *initial plans may be frustrated, and they may have*
> *to fall back to other plans. In this sense, the model*
> *depicts disequilibrium situations.*

— Barbara Bergmann (1974), pioneer of microsimulation

Complexity economics is the study of the economy as a complex
system. Its scope is much more general than the concepts
of equilibrium or nonequilibrium. Indeed, complexity-related
ideas such as networks, phase transitions, criticality, scaling,
and so on are conceptually independent of equilibrium.
Moreover, as far as nonlinear dynamics and chaos are typically
associated with complexity, they also occur in equilibrium
models (Boldrin and Woodford 1990). In addition, many
scholars routinely working with equilibrium models may
nonetheless be attracted by ideas from complexity science such
as self-organized criticality (Krugman 1996).

However, several scholars view nonequilibrium as a key
component of complexity economics. For instance, Brian Arthur
(2014) writes: "Complexity economics builds from the proposition that the economy is not necessarily in equilibrium. . . .
Where equilibrium economics emphasizes order, determinacy,
deduction, and stasis, complexity economics emphasizes contin-
gency, indeterminacy, sense-making, and openness to change. . .
. Equilibrium economics is a special case of nonequilibrium and
hence complexity economics, therefore complexity economics is
economics done in a more general way." Similarly, Alan Kirman
(2010) argues: "Out-of-equilibrium dynamics are not a central is-

sue in economics and Gerard Debreu said explicitly that their analysis was too difficult and that was why he had never ventured in
that direction. So, the most interesting aspects of economics if the
economy is viewed as a complex interactive and adaptive system,
are absent in macroeconomic models based on the General Equilibrium view." Similar statements can be found in several other
references (Simon 1996; Epstein and Axtell 1996; Axelrod 2006).

These definitions consider equilibrium as a situation of stasis.
But stasis is incompatible with models that display endogenous
dynamics, even chaos, in an equilibrium framework (Boldrin and
Woodford 1990): It is hard to imagine that an economy displaying
substantial oscillations is "static." Yet, this tension is just due to
a different use of the word "equilibrium." This is by no chance:
Equilibrium is probably the word that has the most different
meanings in economics. In fact, equilibrium is often used in a
nontechnical way, to mean that one is thinking about the various
feedbacks that affect an outcome of interest, without reference to
any particular model.

Thus, the first goal of this chapter is to put some order into
the various definitions of equilibrium. We first review a number of
equilibrium concepts and general definitions from the literature,
and then ask economists if they can provide their own general
definition of equilibrium. While some think that none exists,
others disagree, and there is no consensus on what the general
definition may be.

Our next goal is to provide a technical, practical definition
of equilibrium that we think is particularly useful to clarify
the main mathematical difference between equilibrium and
nonequilibrium models, such as general equilibrium and agent-
based models. This definition clarifies the advantages and
disadvantages of both types of models. We hope that this
contribution will help economists working with equilibrium

models better understand what makes nonequilibrium, agent-based models different. At the same time, we argue that the community working with nonequilibrium models should be more careful in the use of this term, highlighting in which precise technical sense their models are nonequilibrium.

Some Definitions of Equilibrium

In economics there are hundreds of definitions of equilibrium. In the following, we review a few important ones, highlighting that they are not all static and setting the stage for our own definition.

~ 141 ~

Static equilibrium concepts. A standard notion is Walrasian, or competitive, equilibrium where demand equals supply. At the price where demand and supply curves intersect, buyers and sellers trade a quantity that maximizes their utility. The main limitation of Walrasian equilibrium is that many important markets, such as the labor market, do not clear, as evidenced by the existence of unemployment. For this reason, equilibrium unemployment theories (Pissarides 2000) have been proposed that take into account the cost to both firms and workers to find a good matching on the job market. Here, equilibrium means multiple things, including that the profit to the firm from a new worker equals the hiring cost, that the wage and job creation curves intersect in the wage-tightness plane, that the job creation and Beveridge curves intersect in the unemployment–vacancy plane, and more.[1] But, in general, "the aggregate equilibrium state is one where firms and workers maximize their respective objective functions, subject to the matching and separation technologies, and where the flow of workers into unemployment is equal to the flow of workers out of unemployment" (Pissarides 2000).

---

[1] The word *equilibrium* is mentioned about eighty times in the first chapter of Pissarides (2000).

Another well-known equilibrium concept is *Nash*. In a Nash equilibrium, no player has an incentive to deviate if they assume that other players do not deviate. It is not enough for players to be rational to choose the Nash equilibrium; they also need to assume that the other players play their equilibrium strategies, so their beliefs must be consistent with equilibrium play. In fact, there is no need for players to be rational, as the concept of *quantal response equilibrium* (QRE) (McKelvey and Palfrey 1995) extends Nash equilibrium to boundedly rational players that use a logit function to choose their strategies depending on their payoffs. Still, at the QRE point, all players' beliefs are mutually consistent.[2]

Dynamic equilibrium concepts. Walrasian equilibrium, as presented so far, is a static concept. However, Ken Arrow (1953) and Gérard Debreu (1959) extended general equilibrium theory to deal with time. In their setup, the future is modeled as a sequence of states of nature that unfold as the nodes in a tree. If the agents know the probabilities to reach any node, it is possible to apply the usual Walrasian concepts to derive equilibrium allocations at all points in time. Another well-known dynamic equilibrium concept in economics is *rational expectations*. In this setting, more than rational, expectations are model-consistent, in the sense that the law of motion implied by the choices made under rational expectations must match the expectations themselves. In practice, by assuming that agents with rational expectations optimize intertemporally, finding rational-expectations equilibria boils down to solving dynamic programming problems, for which there are plenty of analytical and computational techniques available. There are

---

[2]Foster and Young (2001) even argue that there is a tension between rationality and accurate prediction of beliefs, and so equilibrium.

plenty of dynamic equilibrium concepts in game theory as well, including subgame perfect equilibria in extensive-form games and Nash equilibria of history-dependent strategies such as tit-for-tat in iterated games.

Steady state. Given a law of motion, the steady state is the value of the variables at which there is no motion. Economists tend to distinguish between "steady" or "stationary" state and equilibrium, but it is not uncommon to read "steady-state equilibrium." This is usually not thought of as a natural condition of the model but rather as a special case.

It is important to note that most dynamic equilibrium concepts do not constrain the model to be in the steady state. The most standard dynamics in rational expectations equilibria are around the *saddle-point stable* steady state,[3] but there exist plenty of rational expectations models in which steady states are unstable and dynamics follow limit cycles or chaos. In these models, agents expect a chaotic law of motion, and that law of motion comes true. Rational expectations equilibria are not necessarily steady states (Boldrin and Woodford 1990).

Survey

We consider four definitions from the literature and four definitions obtained for this chapter.

> *We may define equilibrium, in economic analysis, as a constellation of selected interrelated variables so adjusted to one another that no inherent tendency to change prevails in the model which they constitute. ... As an alternative definition of equilibrium we*

---

[3] Or around the balanced growth path, which is a steady state up to a constant trend.

may propose mutual compatibility of a selected set of interrelated variables of particular magnitudes. (Machlup 1958)

This takes us into what has become the ruling conception of equilibrium in economics. This is the notion that equilibrium is where (a) agents' decisions are compatible with each other and (b) no agent has any reason to change his or her behavior. The second part of this condition (which might be thought to encompass the first) is often represented by saying that agents are maximizing utility subject to the constraints they face. This, however, introduces the assumption of optimizing behavior, which is not essential to the notion. (Backhouse 2004)

Economic equilibrium, at least as the term has traditionally been used, has always implied an outcome, typically from the application of some inputs, that conforms to the expectations of the participants in the economy. Many theorists, especially those employing the "economic man" postulate, have also required the further condition for equilibrium that every participant be optimizing in relation to those correct expectations. However it is the former condition, correct expectations, that appears to be the essential property of equilibrium at least in the orthodox use of the term. Economic equilibrium is therefore not defined in the same terms as physical equilibrium. The rest positions or damped oscillations of pendulums cannot be economic equilibria nor disequilibria since pendulums have no expectations. (Phelps 2018)

At the most elementary level, "equilibrium" is spoken about in a number of ways. It may be regarded as a "balance of forces," as when, for example, it is used to describe the familiar idea of a balance between the forces of demand and supply. Or it can be taken to signify a point from which there is no endogenous "tendency to change": stationary or steady states exhibit this kind of property. However, it may also be thought of as that outcome which any given economic process might be said to be "tending towards," as in the idea that competitive processes tend to produce determinate outcomes. (Milgate 2018)

I do not think there is any settled all-purpose definition of equilibrium as the term is understood by economists. It is context dependent. (Peyton Young)

Equilibrium just means no [group of] agent(s) have an incentive to deviate from their action given the state of the system. (Alex Teytelboym)

It is obviously a common accusation leveled against economics that the idea of an economy in equilibrium is unrealistic, so any analysis using the concept is inherently flawed. This misunderstands the way I for one—and I think many other economists—use the idea of equilibrium. First, "general equilibrium." The formal Arrow–Debreu or similar construct is largely confined to the boot camp of graduate advance micro and some economic theorists. Most of us use the idea of general equilibrium in a heuristic way in opposition to partial equilibrium, to indicate that a change or intervention in the economy will have consequential effects outside the immediate context of analysis. The specific concept of a Nash equilibrium I have found practically useful in policy contexts: It is a way of considering strategically and within a framework that enforces logically the likely decisions of other parties. For example, identifying the Nash equilibrium proved an accurate prediction of the decisions of other governments on a policy issue I was advising on. More broadly, "equilibrium" is a useful counterfactual thought experiment about likely dynamics. Is the economic system in question tending to converge or diverge? Which way would variables need to move if it were a stable system? Can we compare the equilibrium state of our model to the actuality and deduce likely trends? In short, the concept of equilibrium is a useful tool for analysis, not a descriptor of the economy. (Diane Coyle)

An equilibrium state is one in which there is no inherent tendency to change. A disequilibrium is any situation that is not an equilibrium and hence characterizes a state that contains the seeds of its own destruction in that it will change. This concept goes beyond defining equilibrium as "market-clearing" (supply = demand), but a temporary equilibrium is possible before new forces change it. Stock markets are the classic case of clearing by price adjustments that change the state; the climate system is always in disequilibrium as ocean heat chases air temperature. Even if external shocks are constantly occurring, so that a system is never in equilibrium, the concept of long-run equilibrium path *may be useful. The present is the long-run outcome of the distant past, and long-run relationships often hold "on average" over time (e.g., reversion to the mean). For example, the UK unemployment rate has averaged 5% since 1860 despite large and persistent deviations in some periods. However, equilibria can be unstable: a penny on its edge versus lying flat.* (David Hendry)

A New Definition of Equilibrium

In mathematics, a function or *map* f from a set X to a set Y is an assignment of one element of Y to each element of X. For instance, in discrete-time dynamical systems (including agent-based models) a map takes past values of variables and maps them to current values. Although typically one indicates a map as $y = f(x)$, with $x \in X$ and $y \in Y$, in this chapter we use the notation $y \leftarrow f(x)$. This choice confines the use of the equal symbol to *equations*, which impose that an expression on the left hand side is equal to an expression on the right hand side. For instance, given two functions $f(x)$ and $g(x)$, setting $f(x) = g(x)$ means that $f(x)$ and $g(x)$ must be equal, making it possible to find the value(s) of x for which this is true.

Maps are used every time the modeler assumes causal relations between variables. Equations are used when the modeler wants to impose, for whatever reason, that two quantities are equal. In this chapter, equilibrium corresponds to equations, and I say that a variable x takes its equilibrium value when it is obtained by solving an equation. Conversely, I use the term *nonequilibrium* when variables are obtained just from maps. Of course, it is possible that some variables are determined from maps and other variables are determined by solving equations, implying that some model variables may always take their equilibrium values and others may not. Usually, models specified analytically use a combination of maps and equations, while models specified as computer programs (such as most agent-based models) mostly use maps, as they would need to resort to root-finding algorithms to solve equations. One important case in which one solves equations even if the model specification does not require to do so is to find the steady state. We consider this a special case, corresponding to the steady-state equilibrium of a model that may be nonequilibrium.

   In the following, we present a number of examples, and then discuss a few general issues with our definition.

EXAMPLES

The consumer's problem. Consider first the textbook model derivation. A consumer maximizes a utility function $u(x_1, x_2)$ under the constraint that $y \geq p_1 x_1 + p_2 x_2$, where p_1 and x_1 are price and quantity of good 1, p_2 and x_2 are price and quantity of good 2, and y is the budget. Solving the optimization problem via a Lagrangian leads to a system of two equations $\partial u / \partial x_1 - \lambda p_1 = 0, \partial u / \partial x_2 - \lambda p_2 = 0$, where λ is the Lagrange multiplier. Solving these two equations leads to the Marshallian demand for both goods, $x_1^\star \leftarrow x_1(p_1, p_2), x_2^\star \leftarrow x_2(p_1, p_2)$. As an alternative way to determine consumer demand, suppose that consumers follow the multi-attribute aspiration level heuristic of Richard Selten 1998. This is a version of Herb Simon 1955's satisficing model with multiple incommensurable attributes. In a simplified version of this setting, a consumer may start from aspiration levels A_1, A_2, which means that the consumer accepts to demand any (x_1, x_2) pair such that both $x_1 > A_1$ and $x_2 > A_2$. The aspiration level is feasible if $p_1 A_1 + p_2 A_2 \leq y$. The aspiration levels may be adapted as the consumer is presented with multiple (x_1, x_2) pairs, and at the end the consumer settles on a quantity demanded $(\hat{x}_1, \hat{x}_2)$ that is feasible. Under some assumptions, this may converge to Marshallian demand, but $\hat{x}_1$ and $\hat{x}_2$ are the output of an algorithm, not the solutions of an equation. In conclusion, textbook Marshallian demand is an equilibrium quantity, whereas the heuristic process of Selten 1998 is nonequilibrium.

Walrasian equilibrium, tâtonnement, and the cobweb model. Consider a demand curve $q^D(p_t)$, specifying the quantity that consumers are willing to demand at price p_t, ~149~ and the supply curve $q^S(p_t)$. Under Walrasian equilibrium, p_t is obtained by setting $q^D(p_t) = q^S(p_t)$. Under tâtonnement, one assumes instead that prices increase when demand is larger than supply, and that prices decrease when demand is smaller than supply, leading for instance to a specification $p_{t+1} \leftarrow p_t + \gamma \left(q^D(p_t) - q^S(p_t) \right)$, where γ is a parameter. Although, under most reasonable assumptions, tâtonnement converges to the Walrasian equilibrium over time, it is a nonequilibrium model as it does not assume that any variable is in equilibrium. The case of the cobweb model is the most interesting (Brock and Hommes 1997). Under the cobweb model, consumers have an instantaneous demand curve $q^D(p_t)$, but producers face a lag and so must produce based on their expectation for p_t, which we denote p_t^e. All cobweb models assume Walrasian equilibrium: $q^D(p_t) = q^S(p_t^e)$. Next, expectations, when backward-looking, are obtained as a map from previously computed prices: $p_t^e \leftarrow f(p_{t-1}, p_{t-2}, \ldots)$. If instead expectations are rational, they must be equal to the correct price, that is, $p_t^e = p_t$, corresponding to the Walrasian equilibrium at all times. Therefore, one could say that the cobweb model with backward-looking expectations is half equilibrium and half nonequilibrium, as markets clear but expectations are not necessarily equal to realizations, while in case of rational expectations the model is fully in equilibrium.

Learning in games. Consider a two-action, two-player game, and denote by x the probability that player Row plays action 1 and by y the probability that player Column plays action 1. Due to normalization, x and y fully specify a mixed-strategy profile for both players. Denoting by $\Pi(1, y)$ the expected payoff for Row playing action 1 if Column plays $(y, 1 - y)$ and similarly by $\Pi(2, y)$ the same payoff for action 2, the quantal response equilibrium (McKelvey and Palfrey 1995) for player Row is

$$x = \frac{\exp\left(\beta\Pi(1, y)\right)}{\exp\left(\beta\Pi(1, y)\right) + \exp\left(\beta\Pi(2, y)\right)}, \tag{1}$$

with β denoting a parameter often named intensity of choice. The expression for player Column is similar, and the solution of the two-equations system gives the one or more equilibria. When β is large, the quantal response equilibrium approximates the Nash equilibrium. Suppose now that players repeatedly play the game and use a learning rule named experience-weighted attraction (Camerer and Ho 1999). Under a simplified version of this rule (Pangallo *et al.* 2022), player Row updates the probability x as

$$x_t \leftarrow \frac{x_{t-1}^{1-\alpha} \exp\left(\beta\alpha\Pi(1, y_{t-1})\right)}{x_{t-1}^{1-\alpha} \exp\left(\beta\alpha\Pi(1, y_{t-1})\right) + x_{t-1}^{1-\alpha} \exp\left(\beta\alpha\Pi(2, y_{t-1})\right)}, \tag{2}$$

where α is a memory loss parameter. By our definition, players following this rule are "out of equilibrium." For general α and β, the steady state they may converge to is not a known game-theoretic equilibrium (Pangallo *et al.* 2022). For $\alpha \to 0$, the learning rule reduces to stochastic fictitious play (Fudenberg and Levine 1998), and for some games and learning parameters may converge to quantal response equilibria. For $\beta \to \infty$, it would converge to Nash equilibria. But in many cases, especially in large competitive games, EWA dynamics would not converge to any equilibrium (Pangallo, Heinrich, and Farmer 2019).

The Schelling model. Perhaps the most famous agent-based model in the social sciences is the racial segregation model by Thomas Schelling 1971. This model cannot be specified analytically, and was in fact originally simulated by hand on a checkerboard before it was implemented as a computer program. In the Schelling model, agents located on a squared grid at positions x, y can be of two colors c, Black or White. Every agent i first computes the share of neighbors like him, $s_{i,t} \leftarrow 1/|\mathcal{N}(i)| \sum_{j \in \mathcal{N}(i)} I(c_j, c_i)$, where $\mathcal{N}(i)$ denotes the set of neighbors of i and $I(c_j, c_i)$ takes value one if $c_j = c_i$ and zero otherwise. Then, the agent computes happiness $h_{i,t}$, taking value 1 if the share of similar neighbors is above a threshold ξ, that is $h_{i,t} \leftarrow \Theta(s_{i,t} - \xi)$, where $\Theta(\cdot)$ is the Heaviside function. Finally, if the agent is happy, they do not move; if they are unhappy, they move to a new empty location chosen uniformly at random among empty cells, $x_t, y_t \leftarrow U(\text{empty cells})$, where U denotes the uniform distribution. This procedure is repeated for all agents and for a certain number of time steps. No equation is solved at any step of the agent-based model, but variables are always computed from previously computed variables. Therefore, by our definition the model follows out-of-equilibrium dynamics at all times. For sufficiently small values of ξ, the system reaches a stable configuration where all agents are happy, and this can be considered a steady state.

ISSUES WITH THE PROPOSED DEFINITION

As with every definition, the one proposed in this chapter has caveats and limitations.

Differential equations. If differential equations were considered equilibrium, all continuous-time models would be equilibrium models. It would not make sense to consider the discrete-time version of models as nonequilibrium and the continuous-time version as equilibrium. However, setting the rate of change of a variable equal to an expression is like finding steady states: It is not about market features, beliefs or other economically relevant variables, it is just a condition for analyzing a model. Thus, in the same way that we view steady states as a special sort of equilibrium, we also do not consider the very specification of a differential equation as an equilibrium assumption.

Where does modeling start? Going back to the example on the consumer's problem, if we knew the solution we could have postulated that demand corresponds to Marshallian demand, that is, $x_1 \leftarrow x_1^\star$, instead of solving the optimization equations. Similarly, in standard macroeconomic models with rational expectations, we could consider the log-linearized solution of the model, which boils down to a vector autoregression, as "the model." In both cases, both models would be nonequilibrium according to our definition. This raises the question of where modeling starts. Is "the model" the reduced-form expression that is used to do the plots in the paper, or is it the structural micro-founded set of equations that lead to the reduced-form expressions? Our view is that the structural model is the real model, because whenever the modeler wants to add new features to the model, they should first add the new features to the micro-founded model and then work out the changes in the reduced-form model. We consider a specific example in the next section.

Advantages and Disadvantages

Equilibrium is a disciplining device. Although Christopher Sims 1980 is usually cited for inventing the term "wilderness of bounded rationality," he talked instead about the wilderness of "disequilibrium economics." Yet, just as there are multiple ways to be boundedly rational and only one way to be rational,[4] there are multiple ways to specify nonequilibrium processes but only one set of equilibrium values.[5] This leads to two main advantages of equilibrium models. First, comparability. Two models with the same equilibrium definition that only differ by one assumption (say, having a minimum wage or not) make it possible to understand the impact of that assumption on equilibrium outcomes. It is more difficult to understand the impact of different assumptions across two disequilibrium models with different nonequilibrium processes, because it is hard to tell to what extent differences in the results are due to differences in the assumptions of interest or to differences in the nonequilibrium processes. The second advantage of equilibrium models is that they are more parsimonious. For instance, solving for market-clearing prices requires fewer assumptions than specifying an adjustment process for prices. Simple, more parsimonious models are more tractable, can be solved analytically, and are easier to teach in class (if the students are mathematically skilled).

Nonequilibrium enables much more flexibility. Including complexity-related concepts such as heterogeneity, nonlinear dynamics, and complex network structures is more difficult in equilibrium models. It is not impossible, just more difficult,

~153~

---

[4] If we take rationality to mean optimizing given beliefs and information.
[5] There could be multiple ways to find the equilibrium solution, but in principle all solution methods should converge to the same equilibrium, or to one of a set of multiple equilibria.

and it is even more difficult to include more than one of these concepts at the same time. Let us tackle them one by one.

In the last few years, heterogeneity has become one of the hottest topics in macroeconomics. After realizing that the representative agent assumption was incompatible with several important empirical facts about consumption and income dynamics, representative agent macroeconomic models have been enhanced with heterogeneous agents (Kaplan and Violante 2018).[6] As well understood since the inception of this literature (Krusell and Smith, Jr. 1998), making the distribution of agent characteristics' compatible with rational expectations equilibrium is very difficult from a technical point of view. Despite clever attempts to approximate the equilibrium (Auclert *et al.* 2021), the community working with these models is starting to question the rational expectations equilibrium, as evidenced by the statement in Moll 2024 that "we are spending a lot of intellectual and computational horse power solving a nonsensical problem." Ultimately, the problem with heterogeneous-agent new Keynesian (HANK) models is that one would in principle have to solve an infinite number of equations to deal with heterogeneity. By contrast, heterogeneity has always been at the forefront of agent-based macroeconomics (Fagiolo and Roventini 2017; Dawid and Delli Gatti 2018). As they do not need to satisfy equilibrium conditions, agent-based models can incorporate as much multidimensional heterogeneity as needed. In a recent application on the economic effects of the COVID-19 pandemic, Pangallo *et al.* 2023 replicated the COVID-19 empirical fact that rich households reduced their consumption more than

---

[6]This enhancement occurs at the point of specifying the model, not at the solution step, supporting our view that the solution of the model cannot be considered "the model."

poor households, despite experiencing lower unemployment. This finding was only possible thanks to rich modeling of heterogeneity including the joint distributions of consumption patterns, income, industry, occupation, and the possibility to work from home. Equivalent heterogeneous-agent equilibrium models could not match this level of heterogeneity (Kaplan, Moll, and Violante 2020).

Another concept that is easier to include in disequilibrium models is nonlinear dynamics. Although equilibrium models displaying limit cycles or chaos exist, in many of these models nonlinear dynamics only occur under relatively extreme parameter values (Boldrin and Woodford 1990). This is because it must be optimal for agents to face an oscillating economy, which is unlikely under sufficiently high discount factors (that seem to be motivated empirically). These problems are not faced by the nonequilibrium endogenous business cycles literature, from historical papers (Hicks 1950; Goodwin 1967) to more recent ones (Bonart *et al.* 2014; Asano *et al.* 2021; Pangallo 2024). The lack of equilibrium constraints also makes it easy to obtain nonlinear dynamics in system-dynamics models (Sterman 2000).

Finally, although there exist a lot of equilibrium concepts in the modeling of economic networks, in some cases equilibrium restricts networks to rather simple structures. For instance, using the concept of pairwise stability equilibrium (Jackson and Wolinsky 1996) in an oligopoly setting, Sanjeev Goyal and Sumit Joshi (2003) show that under strategic network formation only complete, dominant-group, and star networks are compatible with equilibrium. In contrast, nonequilibrium network-formation processes make it easier to obtain network features that are empirically supported in large-scale real-world networks, such as heavy-tailed degree distributions (Gualdi and Mandel 2016), clustering, small-world, and assortativity (Jackson and Rogers 2007).

Discussion

"Convergence to the nonlinear solution in just a few iterations"; "aggregate consistency conditions"; "terminal condition for backward iteration"; "Newton-Raphson procedure"; "bisection procedure"; "convergence of value functions"; "the grid on which the model is solved"; "certainty equivalence"; "we approximate the true aggregate state"; "initial guess for the value function." If you read these phrases, you are most likely reading an equilibrium paper, and these words describe the usually complicated algorithm to find the equilibrium. It is very unlikely that you read these words in an agent-based paper, unless the authors are trying to find the steady state or to estimate the parameters. This is because agent-based models compute variables as maps of other previously computed variables, rather than by solving equations. Although this deep difference in the practice of modeling is often discussed at seminars and conferences, it is rarely discussed in print, and when so, it is discussed in passing. We think it is time to dedicate an entire chapter to it.

Do we need to use the words *equilibrium* vs. *nonequilibrium* to distinguish between solving equations vs. specifying maps? Is this just adding yet another meaning to these already abused terms? In fact, I do think that *equilibrium* and *nonequilibrium* are the appropriate terms for this concept. First, there are no clear alternatives. We could distinguish between *fixed-point* vs. *recursive* models, but these terms are often used to characterize the solution to the models, not the models themselves. Similarly, if we differentiated between *simultaneous* vs. *sequential* models, we would confuse the role of time, and using *micro-founded* vs. *reduced-form* would be unfair to many behavioral models that do not rely on optimization and on solving equations but are microfounded in psychological evidence (Artinger, Gigerenzer, and Jacobs 2022). The second reason why we name equilibrium the solution of equations and name nonequilibrium the simple specification of maps is that authors already use this terminology, and we think it is better to clarify what they mean than to propose an alternative terminology.

What does all this have to do with complexity? As stated throughout this chapter, complexity is a much deeper and broader concept than the modeling practice of using equations vs. specifying maps. We could say that assuming that equations hold is a top-down constraint on the system, while letting behaviorally motivated nonequilibrium dynamics potentially reach the equations' solution is more in line with a complex-systems view of the economy. But this statement would be controversial. This chapter proposes instead to think in practical terms: without assuming equilibrium, it is much easier to include concepts such as heterogeneity, nonlinear dynamics, and network structure, which have always been considered crucial to model the economy as a complex system. ❦

Acknowledgments

I would like to thank Brian Arthur, Paul Beaudry, Matteo Bizzarri, François Lafond, and Cars Hommes for helpful discussions around this chapter, and Diane Coyle, David Hendry, Alex Teytelboym, and Peyton Young for providing their own takes on what equilibrium means.

REFERENCES

Arrow, K. J. 1953. "Le Rôle des Valeurs Boursières pour la Répartition la Meilleure des Risques." *Économétrie, Colloques Internationaux du Centre National de la Recherche Scientifique* 11:41–47.

Arthur, W. B. 2014. *Complexity Economics: A Different Framework for Economic Thought.* Oxford, UK: Oxford University Press.

Artinger, F. M., G. Gigerenzer, and P. Jacobs. 2022. "Satisficing: Integrating Two Traditions." *Journal of Economic Literature* 60:598–635. https://www.jstor.org/stable/27183063.

Asano, Y. M., J. J. Kolb, J. Heitzig, and J. D. Farmer. 2021. "Emergent Inequality and Business Cycles in a Simple Behavioral Macroeconomic Model." *Proceedings of the National Academy of Sciences* 118:e2025721118. https://doi.org/10.1073/pnas.2025721118.

Auclert, A., B. Bardóczy, M. Rognlie, and L. Straub. 2021. "Using the Sequence-Space Jacobian to Solve and Estimate Heterogeneous-Agent Models." *Econometrica* 89:2375–2408. https://doi.org/10.3982/ECTA17434.

Axelrod, R. 2006. "Agent-Based Modeling as a Bridge between Disciplines." *Handbook of Computational Economics* 2:1565–1584. https://doi.org/10.1016/S1574-0021(05)02033-2.

Backhouse, R. E. 2004. "History and Equilibrium: A Partial Defense of Equilibrium Economics." *Journal of Economic Methodology* 11:291–305. https://doi.org/10.1080/1350178042000252974.

Bergmann, B. R. 1974. "A Microsimulation of the Macroeconomy with Explicitly Represented Money Flows." *Annals of Economic and Social Measurement* 3 (3): 475–489.

Boldrin, M., and M. Woodford. 1990. "Equilibrium Models Displaying Endogenous Fluctuations and Chaos: A Survey." *Journal of Monetary Economics* 25:189–222. https://doi.org/10.1016/0304-3932(90)90013-T.

Bonart, J., J.-P. Bouchaud, A. Landier, and D. Thesmar. 2014. "Instabilities in Large Economies: Aggregate Volatility without Idiosyncratic Shocks." *Journal of Statistical Mechanics: Theory and Experiment* 2014:P10040. https://doi.org/10.1088/1742-5468/2014/10/P10040.

Brock, W. A., and C. H. Hommes. 1997. "A Rational Route to Randomness." *Econometrica,* 1059–1095. https://doi.org/10.2307/2171879.

Camerer, C., and T. Ho. 1999. "Experience-Weighted Attraction Learning in Normal Form Games." *Econometrica* 67:827–874. https://doi.org/10.1111/1468-0262.00054.

Dawid, H., and D. Delli Gatti. 2018. "Agent-Based Macroeconomics." In *Handbook of Computational Economics,* 63–156. Amsterdam, Netherlands: Elsevier. https://doi.org/10.1016/bs.hescom.2018.02.006.

Debreu, G. 1959. *Theory of Value: An Axiomatic Analysis of Economic Equilibrium.* Vol. 17. Cowles Foundation Monographs. New Haven, CT: Yale University Press.

Epstein, J. M., and R. Axtell. 1996. *Growing Artificial Societies: Social Science from the Bottom Up.* Cambridge, MA: Brookings Institution/MIT Press. https://doi.org/10.7551/mitpress/3374.001.0001.

Fagiolo, G., and A. Roventini. 2017. "Macroeconomic Policy in DSGE and Agent-Based Models Redux: New Developments and Challenges Ahead." *Journal of Artificial Societies & Social Simulation* 20 (1): 1. https://doi.org/10.18564/jasss.3280.

Foster, D. P., and H. P. Young. 2001. "On the Impossibility of Predicting the Behavior of Rational Agents." *Proceedings of the National Academy of Sciences* 98:12848–12853. https://doi.org/10.1073/pnas.211534898.

Fudenberg, D., and D. K. Levine. 1998. *The Theory of Learning in Games.* Vol. 2. Cambridge, MA: MIT Press.

Goodwin, R. M. 1967. *A Growth Cycle.* Cambridge, UK: Cambridge University Press.

Goyal, S., and S. Joshi. 2003. "Networks of Collaboration in Oligopoly." *Games and Economic Behavior* 43:57–85. https://doi.org/10.1016/S0899-8256(02)00562-6.

Gualdi, S., and A. Mandel. 2016. "On the Emergence of Scale-Free Production Networks." *Journal of Economic Dynamics and Control* 73:61–77. https://doi.org/10.1016/j.jedc.2016.09.012.

Hicks, J. R. 1950. *A Contribution to the Theory of the Trade Cycle.* Oxford, UK: Oxford University Press.

Jackson, M. O., and B. W. Rogers. 2007. "Meeting Strangers and Friends of Friends: How Random are Social Networks?" *American Economic Review* 97:890–915. https://doi.org/10.1257/aer.97.3.890.

Jackson, M. O., and A. Wolinsky. 1996. "A Strategic Model of Social and Economic Networks." *Journal of Economic Theory* 71:44–74. https://doi.org/10.1006/jeth.1996.0108.

Kaplan, G., B. Moll, and G. L. Violante. 2020. *The Great Lockdown and the Big Stimulus: Tracing the Pandemic Possibility Frontier for the US.* NBER Working Paper 27794. https://doi.org/10.3386/w27794.

Kaplan, G., and G. L. Violante. 2018. "Microeconomic Heterogeneity and Macroeconomic Shocks." *Journal of Economic Perspectives* 32:167–94. https://doi.org/10.1257/jep.32.3.167.

Kirman, A. 2010. *Complex Economics: Individual and Collective Rationality.* London, UK: Routledge.

Krugman, P. 1996. *The Self-Organizing Economy.* Cambridge, MA: Blackwell Publishers.

Krusell, P., and A. A. Smith, Jr. 1998. "Income and Wealth Heterogeneity in the Macroeconomy." *Journal of Political Economy* 106:867–896. https://doi.org/10.1086/250034.

Machlup, F. 1958. "Equilibrium and Disequilibrium: Misplaced Concreteness and Disguised Politics." *The Economic Journal* 68:1–24. https://doi.org/10.2307/2227241.

McKelvey, R. D., and T. R. Palfrey. 1995. "Quantal Response Equilibria for Normal Form Games." *Games and Economic Behavior* 10 (1): 6–38. https://doi.org/10.1006/game.1995.1023.

Milgate, M. 2018. "Equilibrium (Development of the Concept)." In *The New Palgrave Dictionary of Economics,* Third, 1:3851–3857. London, UK: Palgrave Macmillan.

Moll, B. 2024. *Heterogeneous Agent Macroeconomics: Eight Lessons and a Challenge.* Economic Journal Lecture. Royal Economic Society.

Pangallo, M. 2024. "Synchronization of Endogenous Business Cycles." *Journal of Economic Behavior & Organization* 229:106827. https://doi.org/10.1016/j.jebo.2024.106827.

Pangallo, M., A. Aleta, R. M. del Rio-Chanona, A. Pichler, D. Martín-Corral, M. Chinazzi, F. Lafond, *et al.* 2023. "The Unequal Effects of the Health–Economy Trade-Off during the COVID-19 Pandemic." *Nature Human Behaviour,* 1–12. https://doi.org/10.1038/s41562-023-01747-x.

Pangallo, M., T. Heinrich, and J. D. Farmer. 2019. "Best Reply Structure and Equilibrium Convergence in Generic Games." *Science Advances* 5 (2): eaat1328. https://doi.org/10.1126/sciadv.aat1328.

Pangallo, M., J. B. T. Sanders, T. Galla, and J. D. Farmer. 2022. "Towards a Taxonomy of Learning Dynamics in 2×2 Games." *Games and Economic Behavior* 132:1–21. https://doi.org/10.1016/j.geb.2021.11.015.

Phelps, E. S. 2018. "Equilibrium: An Expectational Concept." In *The New Palgrave Dictionary of Economics,* Third, 1:3857–3860. London, UK: Palgrave Macmillan.

Pissarides, C. A. 2000. *Equilibrium Unemployment Theory.* Cambridge, MA: MIT Press.

Schelling, T. C. 1971. "Dynamic Models of Segregation." *Journal of Mathematical Sociology* 1 (2): 143–186. https://doi.org/10.1080/0022250X.1971.9989794.

Selten, R. 1998. "Aspiration Adaptation Theory." *Journal of Mathematical Psychology* 42 (2–3): 191–214. https://doi.org/10.1006/jmps.1997.1205.

Simon, H. A. 1955. "A Behavioral Model of Rational Choice." *Quarterly Journal of Economics* 69 (1): 99–118. https://doi.org/10.2307/1884852.

———. 1996. *The Sciences of the Artificial.* 3rd ed. Cambridge, MA: MIT Press.

Sims, C. A. 1980. "Macroeconomics and Reality." *Econometrica* 48 (1): 1–48. https://doi.org/10.2307/1912017.

Sterman, J. 2000. *Business Dynamics: Systems Thinking and Modeling for a Complex World.* Boston, MA: Irwin/McGraw-Hill.

SOME REFLECTIONS ON COMPLEXITY ECONOMICS: RESEARCH IN THE SFI SPIRIT

William Brock, University of Wisconsin, Madison,
and University of Missouri, Columbia; and
Cars Hommes, Bank of Canada,
University of Amsterdam, and Tinbergen Institute

Abstract

This chapter discusses some of our thoughts and reflections about complexity research in economics at the Santa Fe Institute and beyond in the last few decades. We cover a wide arrange of complexity economics topics: detecting chaos and nonlinearity in economic data, early warning signals, complex dynamics under bounded rationality and heterogeneous expectations, empirical validation of heuristics switching models, social interactions, experimental macroeconomics, managing positive feedback in complex systems, and climate modeling. We discuss some policy implications and end with a short future perspective.

Introduction

This chapter discusses some of our past research in complexity economics with some reflections on how what we call the SFI spirit has pushed the two of us into research adventures that drive us to this day.

One of us (WB) has written chapters for all three previous volumes of *The Economy as an Evolving Complex System* and has been a fan of SFI's complexity efforts in economics since its beginning. This chapter puts forth some thoughts and reflections on the previous volumes and speculates about future research directions in complexity economics.

The other of us (CH) founded a complexity-research institute, the Center for Nonlinear Dynamics in Economics and Finance (CeNDEF) at the University of Amsterdam twenty-five years ago, inspired by SFI's work. Our dean used to call CeNDEF the Santa Fe Institute at the Amstel. This chapter discusses some CeNDEF complexity research over the years.

We touch upon a wide range of complexity-economics topics: detecting chaos and nonlinearity in economic and financial data; early warning signals in ecology; complex dynamics under bounded rationality and heterogeneous expectations; empirical validation of heuristics switching models; behavioral and experimental macroeconomics; how policy can manage positive feedback in complex systems; and climate modeling.

Nonlinearity and Chaos in Economic Data

Interest in nonlinearity and chaos in various time-series datasets including economic and financial data, flourished during the 1980s and 1990s (Dechert 1996) and continues to this day. Part of the first volume of *The Economy as an Evolving Complex System* (Anderson, Arrow, and Pines 1988) is devoted to this topic (see articles by Brock, Boldrin, Farmer and Sidorowich, Packard, and Ruelle).

The possibility of noisy deterministic chaotic explanations of economic and financial time-series data, as well as time-series datasets from other sciences, attracted research not only in economic theory (e.g., Grandmont 1987), but also in empirical research and formulation of various testing methods.

Methods of testing for chaos were developed by physicists and mathematicians (e.g., Grassberger and Procaccia 1983;

Crutchfield *et al.* 1986; Takens 1981; Theiler *et al.* 1992). George Sugihara and Robert May (1990) took a different approach, using nonlinear forecasting as a way of distinguishing chaos from noise.

Initial studies in economics and finance (Barnett and Chen 1988; Brock and Sayers 1988; Scheinkman and LeBaron 1989) raised interesting questions suggesting the possibility of attractors generated by deterministic dynamical systems that "looked random" but could be reconstructed (despite being pummeled by stochastic noise and unmeasured variables) well enough to give an incremental advantage in prediction over received methods of prediction. Grappling with such a challenge spurred many investigations and attracted a lot of publicity, (e.g., Berreby 1993) as well as generating articles in high-profile, more popular science outlets like *Scientific American* (Crutchfield *et al.* 1986).

Econometricians adapted this work into formal and applied statistical methods for testing for hidden structure where size and power against various alternatives could be evaluated by econometricians using, for example, tools from U-statistics theory (Brock *et al.* 1996; Brock, Hsieh, and LeBaron 1991). Brock's work with Dechert, Scheinkman, and LeBaron (1996) and other explorations into nonlinearity benefited much from interactions with Santa Fe Institute scholars.

The evidence in financial and macroeconomic data, especially financial returns data, for the hypothesis of chaos buffeted with dynamic noise is weak. The evidence for nonlinear dependence in the data, however, is strong, and the BDS test (after Brock, Dechert and Scheinkman) has strong power against many nonlinear alternatives. At the same time economic theorists built new classical general equilibrium models with rational expectations that could generate complex dynamics similar to the chaotic quadratic map (e.g., Grandmont 1987; Benhabib and Day 1982; Boldrin and Woodford 1990). The patterns in the time series generated by these theoretical models, however, looked different from economic and financial data.

Empirical difficulties, both in calibrating new classical nonlinear endogenous business-cycle models to economic data and in finding evidence for low-dimensional chaos in economic and financial time series, thus prevented a full embrace and appreciation of complex nonlinear dynamics in economics in the 1980s and early 1990s (Hommes 2013). More recent complexity-economics research relies not strongly on chaos, but rather on nonlinearities and near-unit root processes that may not generate sustained fluctuations on their own, but provide important nonlinear amplification mechanisms, consistent with behavior observed in laboratory experiments with human subjects and in empirical data, that generate near-unit root (temporary) bubble- and crash-like phenomena (see, e.g., the recent survey in Hommes 2021).[1]

Early Warning Signals of Regime Changes

Are regime changes in ecology and other sciences predictable to some extent? One source of potential regime changes in ecosystems is impending bifurcations caused by slow changing forces that impact the dynamics of an ecosystem, which can cause loss of stability or other structural changes. Yuri Kuznetsov (1995) is a basic textbook on bifurcations in

---

[1] Interestingly, empirical estimates of parameters of nonlinear behavioral heterogeneous expectations switching models, such as those discussed in the "Empirical Validation of Two-Type Switching Model" and "Heuristics-Switching Model" sections of this chapter, typically do not lie in the chaotic parameter range, but are rather close to the first bifurcation towards instability and are thus consistent with near-unit root behavior around a benchmark fundamental.

differential and difference equations where there is a parameter that moves slowly relative to the dynamics of the differential and difference equations under study. Some bifurcations can admit early warning signals (EWS) as explained below.

Consider the difference equation system perturbed by an independent and identically distributed stochastic process $\{e_t\}$:

$$x_{t+1} = f(x_t, a, \epsilon e_{t+1}), \qquad\qquad x_t \in R^n, a \in R. \qquad (1)$$

Suppose a moves slowly relative to the timescale of the difference equation and suppose there is a steady state $\bar{x}(a) = f(\bar{x}(a), a)$. A bifurcation point a^* is where the qualitative behavior of (1) changes as a increases past a^*. Consider the case where the steady state $\bar{x}(a)$ loses local asymptotic stability at a^*. Is there some kind of EWS generated by (1) when a approaches a^*? Under appropriate sufficient conditions a local measure of variance increases as a approaches a^* (see Brock and Carpenter 2010 for a general treatment where x is an n-dimensional state vector and Scheffer *et al.* 2009 for a review of work up to 2009). Closely related to an increase in local variance is an increase in local autocorrelation as a approaches a^* (Scheffer *et al.* 2009).

The general method of analysis is to expand (1) in a Taylor series around the steady state $\bar{x}(a)$ and $\epsilon = 0$, collect the terms in $dx_t \equiv x_t - \bar{x}(a)$ and ϵ, compute the local variance covariance matrix of dx_t using the $n \times n$ derivative matrix $\partial f(\bar{x}(a), a, 0)/\partial x$, and study the behavior of the quadratic form built out of the local variance–covariance matrix as a approaches bifurcation point a^* (see Brock and Carpenter 2010 and the appendix to Biggs, Carpenter, and Brock 2009, as cited therein).

We consider the work on EWS part of the general literature on complexity because it involves studying bifurcations and patterns that can surround some classes of impending bifurcations (but not all). A large literature has grown out of the study of EWS and has found many applications. Application to observational data in search of EWS is fraught with problems (e.g., O'Brien *et al.* 2023). Some of the issues raised by Duncan O'Brien *et al.* (2023) are less of a concern in field studies, where there is more control over noise, observed and unobserved confounders, etc., than in studies using observational data (Carpenter *et al.* 2011).

Complex Dynamics under Heterogeneous Expectations

In a world of bounded rationality there is an abundance of nonlinearity and complexity due to learning and interactions of heterogeneous agents. Our joint route in complexity started with our "rational route to randomness" paper (Brock and Hommes 1997) introducing heterogeneous expectations in a cobweb model, where agents switch between different forecasting rules depending on their relative performance. Agents can choose between a freely available simple rule, naive expectations, and a perfectly rational rule at positive information-gathering costs. This model generates complicated, chaotic price fluctuations with irregular switching between a stable phase of close-to-the-steady-state fluctuations, when the cheap naive rule dominates, and an unstable phase with large price fluctuations with agents at some point switching back to rational expectations, pushing prices back close to the steady state. The paper showed that the system exhibits a "rational route to randomness," that is, a bifurcation route to complex dynamics when agents become more "rational" in the sense that they become more sensitive to differences in fitness and switch faster to better, that is, more profitable, strategies.

In a follow-up paper (Brock and Hommes 1998) we applied the same nonlinear switching mechanism to a standard asset-

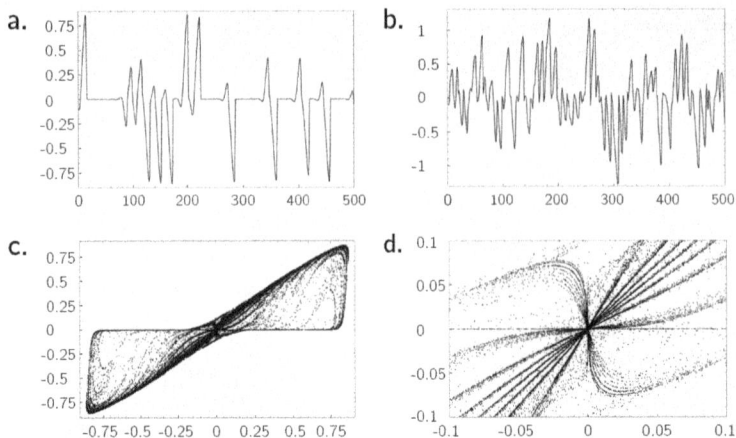

Figure 1. Chaotic (a) and noisy chaotic (b) bubble-and-crash dynamics in deviations from fundamental in behavioral asset pricing model with fundamentalists versus trend-extrapolators. The dynamics are characterized by a strange attractor (c) with fractal structure (see enlargement, d) .

pricing model, with the price derived from myopic mean-variance maximizing agents with heterogeneous expectations. This paper took inspiration from the Santa Fe artificial stock market agent-based simulation model (Arthur *et al.* 1997; LeBaron, Arthur, and Palmer 1999; see also Brock 1997). Our goal was to simplify a rather complex agent-based simulation model and build an analytically tractable framework that could be used for empirical testing. Agents switch between fundamentalists' strategies, assuming that asset prices are fully determined by economic fundamentals, and technical analysts, extrapolating observed patterns in past prices. Switching between strategies is based on the relative strategy performance. This simple behavioral asset pricing model exhibits complicated (chaotic) bubble and crash dynamics of asset prices, as illustrated in figure 1. Along the bubbles, trend-following strategies dominate and amplify prices, with agents switching back to fundamentalists' strategies when prices deviate too far from fundamental value.

Empirical Validation of Two-Type Switching Model

Are these (chaotic) price fluctuation with temporary bubbles and crashes empirically relevant? In H. Peter Boswijk, Hommes, and Sebastiano Manzan (2007), we estimated a simple two-type switching model using yearly price-to-dividend or price-to-earnings S&P500 data. The simple two-type switching model is of the form

$$R^* x_t = n_t \phi_1 x_{t-1} + (1 - n_t)\phi_2 x_{t-1} + \epsilon_t$$
$$R^* = \frac{1+r}{1+g} > 1, \tag{2}$$

where x_t is the price deviation from the fundamental benchmark (discounted sum of expected future dividends or earnings); n_t and $1 - n_t$ are the time-varying fractions of the two types, fundamentalists and trend-followers; $\phi_1 x_{t-1}$ and $\phi_2 x_{t-1}$ are the forecasted deviations x_{t+1} by types 1 and 2; ϵ_t is a noise term; and R^* depends on the interest rate r and the growth rate g of dividends or earnings.

Boswijk, Hommes, and Manzan (2007) find *behavioral heterogeneity* in the data, that is, the estimation supports two type of rules: fundamentalists using a mean-reverting rule with $\hat{\phi}_1 = 0.762$ and trend-followers with a coefficient $\hat{\phi}_2 = 1.135$, believing that the price deviation from fundamental will increase. Notice that the pricing equation (2) of the two-type model is in fact a nonlinear AR(1) model with a time varying AR(1) coefficient, because the fractions n_t and $1 - n_t$ are time-varying and depend on past relative performance and prices. Define the *market sentiment* or the *exuberance index*

$$\phi_t = \frac{n_t \phi_1 + (1 - n_t)\phi_2}{R^*}. \tag{3}$$

When the fundamentalists dominate $\phi_t < 1$ and there is mean reversion in prices. If on the other hand $\phi_t > 1$, the market is explosive and a (temporary) bubble may arise due to coordination on trend-following behavior. In this behavioral asset pricing model, bubbles are triggered by shocks to fundamentals and amplified by the nonlinear switching mechanism to trend-extrapolation.

In Hommes and in't Veld (2017), we estimated a two-type switching model using quarterly data 1950–2016 and obtained similar results, as illustrated in figure 2. The market sentiment fluctuates between 0.95 and 1.01, so price deviations from fundamental follow a near-unit root process. In the late 1990s and early 2000s, the exuberance index exceeds 1 during the dot-com bubble. The behavioral story of the model is that the dot-com bubble was triggered by a positive shock (good news about a new internet technology) and subsequently amplified by trend-following behavior.

EXUBERANCE IN HOUSING MARKETS

Following a similar approach, Wilko Bolt *et al.* (2019) estimated a two-type housing model with fundamentalists versus trend-followers using housing prices from OECD countries. Many of these countries exhibit housing bubbles lasting several years, sometimes more than a decade. Ugochi Emenogu, Hommes, and Mikael Khan (2021) applied the same model to housing prices of Canadian cities. They computed a *heatmap* for housing prices in Canadian cities as shown in figure 3. Medium gray indicates a relatively calm housing market with housing price exuberance index (HPEI) below 0.95; light gray indicates HPEI between 0.95 and 1. Dark gray indicates that HPEI exceeds 1, implying higher risk for a bubble to appear. Over the years

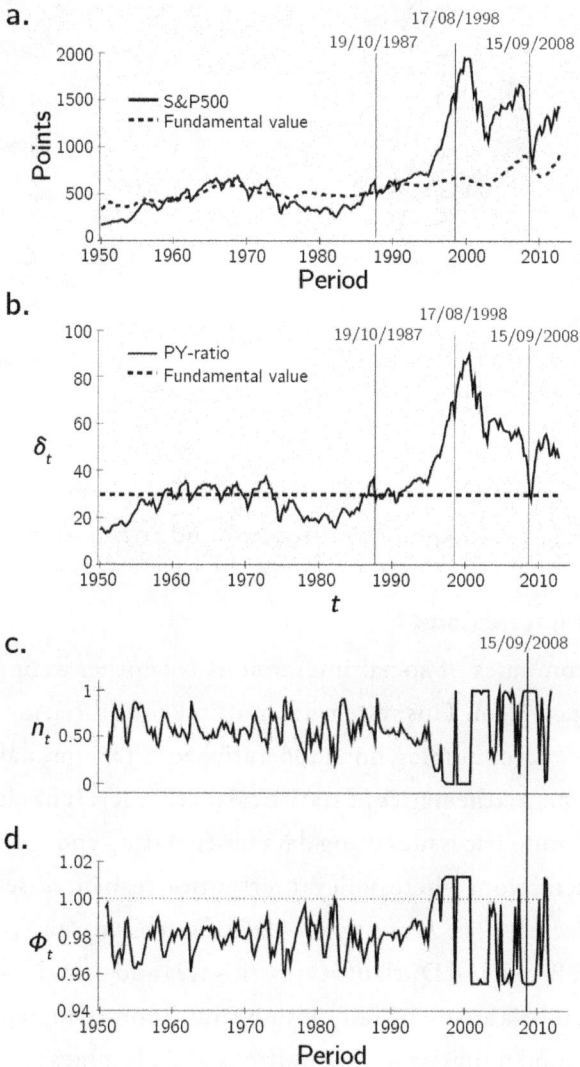

Figure 2. (a) S&P 500 together with benchmark fundamental (discounted sum of expected future dividends, dotted line). (b) Price-to-dividend ratio of S&P 500 and its fundamental value around 30 (dotted line). (c) Estimated fractions n_t of fundamentalists. (d) time-varying market sentiment or exuberance index.

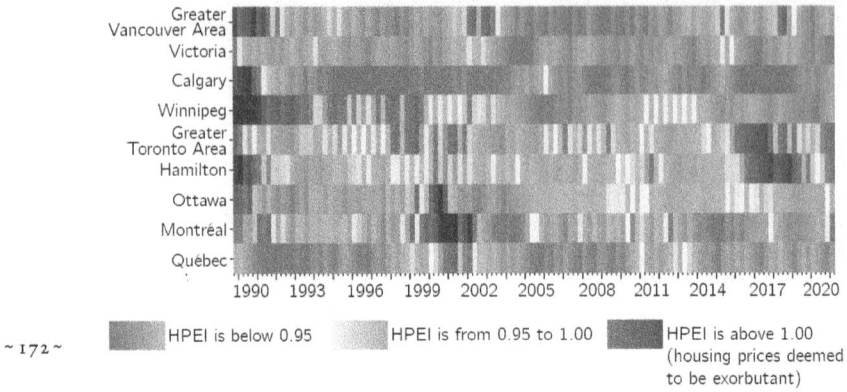

Figure 3. Heatmap of housing price exuberance index (HPEI) measuring the stability or instability of housing prices in nine Canadian cities.

several Canadian cities have been identified as exuberant, for example, Toronto and Hamilton, around 2016–2017.

Social Interactions

The economics of social interactions is another example of a complex system. This system has multiple equilibria, ingredients of network economics, noncooperative equilibrium analysis, and even some mathematics of statistical mechanics. It has interesting econometric issues using data in separating endogenous social interactions where policy interventions can increase welfare of society. In *The Economy as an Evolving Complex System III* (2006), Brock and Durlauf review this area and show how multiple equilibria arise naturally. Using a multinomial logit discrete-choice model similar to that in Brock and Hommes (1997), the probability of choosing choice k out of L choices is given by

$$p(k) = \exp(\beta h_k + \beta J p^e(k)) / \sum_{j=1}^{L} \exp(\beta h_j + \beta J p^e(j)), \tag{4}$$

$$k = 1, 2, \cdots, L.$$

Here h_k, $\beta \geq 0$, $J \geq 0$ and $p^e(k)$ are utility of choice k when $J = 0$, intensity of choice, social interaction parameter, and point expectation held by the representative agent on the probability of choice k. We impose rational point expectations, $p^e(k) = p(k)$, $k = 1, 2, \cdots, L$, apply Brouwer's fixed-point theorem to obtain the equilibrium equations

$$p(k) = \exp(\beta h_k + \beta J p(k)) / \sum_{j=1}^{L} \exp(\beta h_j + \beta J p(j)), \tag{5}$$

$$k = 1, 2, \cdots, L.$$

It is obvious that the solution to (5) is unique when $\beta = 0$, and the solution is unique for all $\beta \geq 0$, when $J = 0$. In Brock and Durlauf (2006) we show that, if the product βJ is large enough, multiple equilibria appear.[2]

In Brock and Durlauf (2006) we show how difficult the econometric identification problem of policy-relevant social interactions is in using observational datasets because of unobserved variables, group selection bias, and other confounders. Brock and Durlauf (2006) and its references give a reader's guide to the voluminous literature in this area as well to the policy relevance of positive social interactions and the existence of social multipliers that can be exploited by policy.

Experimental Macro

Laboratory experiments with human subjects have become a standard tool to test macroeconomic theory in the lab; see for example the extensive surveys by John Duffy (2016), Jasmina Arifovic and Duffy (2018), and Hommes (2021). A macro experiment is a group experiment, where individual and aggregate behavior and their interactions are tested simultaneously in a controlled experimental environment. A

---

[2]Refer to Brock and Durlauf (2006) for the details.

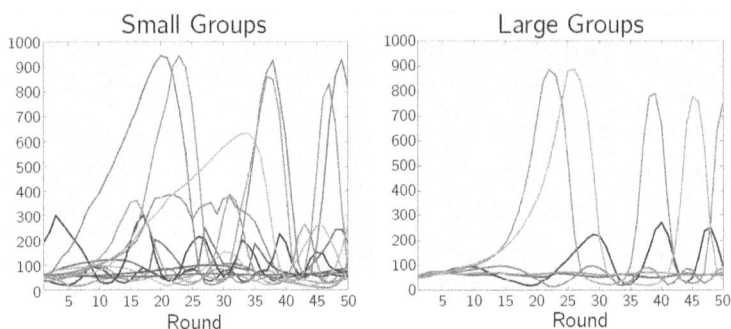

Figure 4. Experimental asset markets in LtFEs (Hommes et al. 2021). Often, experimental markets do not converge to the rational expectations fundamental price (here 65), but rather exhibit bubble and crash dynamics. Coordination of expectations on large bubbles and crashes arises in small groups of six subjects (left plot) as well as in large groups of up to 100 subjects (right plot).

macro experiment thus collects individual as well as aggregate data from the lab. For example, expectation formation has been extensively tested in *learning-to-forecast experiments* (LtFEs), where the subjects' only task is to forecast a price for, say, fifty periods in an environment with expectations feedback, where realized price depends on the average or median forecast of a group of individuals. Key questions in the LtFEs are: (1) Does the price converge to a rational expectations equilibrium? and, (2) if not, which alternative theory of expectations explains individual and aggregate laboratory data? In many of these experiments prices do not converge to rational equilibrium, but rather fluctuate and follow bubble-and-crash patterns (e.g., C. H. Hommes *et al.* 2005). Some examples are shown in figure 4. In Hommes, Anita Kopányi-Peuker, and Joep Sonnemans (2021), we show that the coordination on a bubble-and-crash pattern is robust against group size and occurs not only for small groups of six to ten subjects, but also for large group sizes up to 100.

Positive versus Negative Feedback

The type of expectations feedback in the system is an important driver of its stability or instability. Positive (negative) feedback means that the price increases (decreases) when average forecast increases. Positive feedback corresponds to an environment with strategic complementarity, and negative feedback to an environment with strategic substitutability. Positive feedback is typical in a speculative-asset market, where price increases when investors become more optimistic. Negative feedback is common in supply-driven commodity markets where optimistic producers produce more, leading to lower market prices.

~ 175 ~

Peter Heemeijer *et al.* (2009) ran simple LtFEs with positive versus negative feedback using the linear pricing rules:

negative feedback: $\qquad p_t = 60 - 0.95(\bar{p}_t^e - 60) + \epsilon_t$

positive feedback: $\qquad p_t = 60 + 0.95(\bar{p}_t^e - 60) + \epsilon_t,$

where $\bar{p}_t^e$ is the average forecast of a group of subjects and ϵ_t is a small noise term. The linear pricing rules have the same RE equilibrium 60 and only differ in the sign of their slopes (or eigenvalues) +0.95 vs −0.95.

Figure 5 shows that the behaviors under positive versus negative feedback are strikingly different. Under negative feedback, the market is stable, with prices and expectations converging to the rational outcome 60. Under positive feedback, prices fluctuate and expectations coordinate on trend-following behavior. Which theory of expectations can explain the different behaviors of positive versus negative feedback systems? The fact that the behavior in the lab is different for different treatments suggests that heterogeneity is important for expectations.

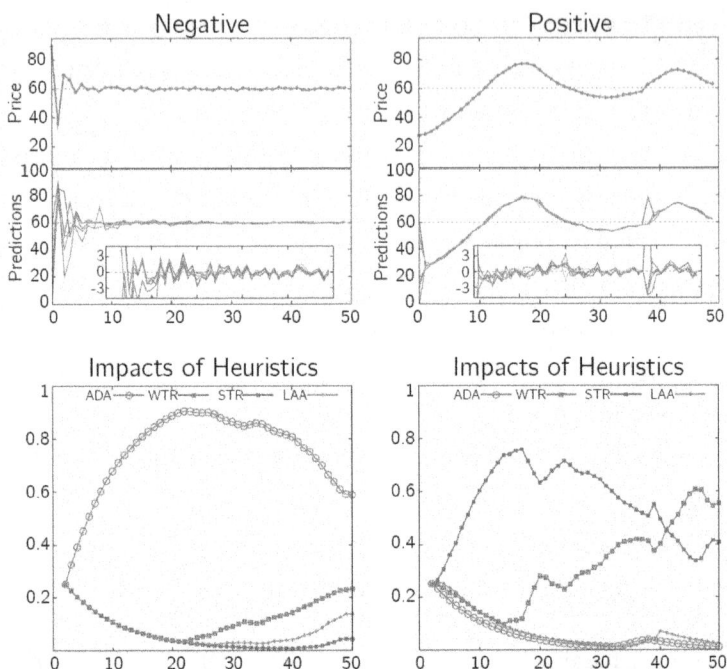

Figure 5. Negative feedback experiments (top left plots) are stable and converge quickly to the rational outcome. Positive feedback experiments (top right plots) are unstable, with drifting expectations and fluctuating prices. The heuristics-switching model shows that subjects coordinate on adaptive expectations under negative feedback (bottom left plot). In contrast, under positive feedback (bottom right plot) coordination on trend-extrapolating behavior amplifies price fluctuations.

Heuristics-Switching Model

In Mikhail Anufriev and Hommes (2012) we proposed a heuristics-switching model (HSM), an extension of the nonlinear switching mechanism of Brock and Hommes (1997), to explain the different types of behavior observed in laboratory experiments: monotonic convergence, oscillatory convergence, explosive behavior and persistent oscillations. The idea of the HSM is that participants choose from four heuristics to make their predictions in each time step. The heuristics are an adaptive heuristic (ADA), a weak trend-following rule (WTR), a strong trend-following rule (STR) and a learning anchor and adjustment rule (LAA):

$$p_{1,t+1}^e = 0.65p_{t-1} + 0.35p_{1,t}^e \qquad \text{(ADA)}$$

$$p_{2,t+1}^e = p_{t-1} + 0.4(p_{t-1} - p_{t-2}) \qquad \text{(WTR)}$$

$$p_{3,t+1}^e = p_{t-1} + 1.3(p_{t-1} - p_{t-2}) \qquad \text{(STR)}$$

$$p_{4,t+1}^e = 0.5(p_{t-1}^{av} + p_{t-1}) + (p_{t-1} - p_{t-2}). \qquad \text{(LAA)}$$

Agents switch between these four heuristics based upon their relative performance. They use the same measure that is used to determine the payoff of a forecast:

$$U_{t-1,i} = \frac{100}{1 + |p_{t-1} - p_{t-1,i}^e|} + \eta U_{t-2,i}, \qquad (6)$$

where $p_{t-1,i}$ is the previous forecast of rule i and η is a memory parameter. The HSM uses a discrete-choice model with asynchronous updating to determine what heuristics the participants will use in the next time step. The intensity of choice parameter β determines how fast participants will switch to the most successful rule, and the parameter δ determines how many participants are switching strategies in an individual time step. The fraction of participants using strategy i to make their next prediction is then given by

$$n_{t,i} = \delta n_{t-1,i} + (1 - \delta)\frac{e^{\beta U_{t-1,i}}}{\sum_j e^{\beta U_{t-1,j}}}. \qquad (7)$$

HSM APPLIED TO POSITIVE VERSUS NEGATIVE FEEDBACK EXPERIMENT

The HSM can now be applied to the positive–negative feedback experiment discussed before. The result is illustrated in figure 5 (bottom plots). Under negative feedback, the HSM shows that subjects coordinate on adaptive expectations (bottom left plot). In contrast, under positive feedback, (bottom right plot), coordination on trend-extrapolating behavior amplifies price fluctuations. The behavioral HSM provides a clear explanation of why negative feedback markets are stable and positive feedback markets are unstable. Under negative feedback, trend-following rules perform poorly and agents coordinate on adaptive expectations leading to stable dynamics. In contrast, under (strong) positive feedback, trend-extrapolating rules perform well and coordination on trend-following behavior amplifies fluctuations.

In Te Bao and Hommes (2019), we ran an experimental housing market with positive feedback due to speculative demand for housing and negative feedback due to supply of housing. We showed that the housing market is stable as long as the overall positive feedback is weak (an eigenvalue of $\lambda = 0.7$). When the positive feedback becomes stronger (eigenvalue $\lambda = 0.85$), the experimental housing market fluctuates and eventually shows coordination on trend-extrapolative behavior and (explosive) housing bubbles under strong positive feedback (eigenvalue $\lambda = 0.95$).

Policy: Managing Positive Feedback

These results suggest an important role for policy in managing the instability of complex economic systems. If negative feedback leads to stable behavior and (strong) positive feedback to unstable behavior, policy should in general add negative feedback to the system to prevent coordination on trend-extrapolating behavior. We illustrate this by discussing an experiment by Tiziana Assenza *et al.* (2021) in the New Keynesian (NK) framework. The NK model is given by the simple three-equations system:

$$y_t = y_{t+1}^e - \varphi(i_t - \pi_{t+1}^e) + \epsilon_t \quad \textbf{output gap}$$

$$\pi_t = \lambda y_t + \rho \pi_{t+1}^e + v_t \qquad\qquad \textbf{inflation}$$

$$i_t = \text{Max}\{\overline{\pi} + \phi_\pi(\pi_t - \overline{\pi}), 0\} \quad \textbf{monetary policy rule}$$

Here y_t is the output gap, π_t inflation, and i_t the interest rate set by the central bank. The monetary policy rule is an inflation-targeting rule with inflation target $\overline{\pi}$ and policy coefficient ϕ_π. An increase of ϕ_π means a more aggressive inflation targeting rule.

~179~

The NK model exhibits positive feedback from expectations of inflation π_{t+1}^e and output gap y_{t+1}^e, but negative feedback from the monetary policy inflation-targeting rule. Assenza *et al.* (2021) tested in the lab whether a more aggressive inflation targeting rule could stabilize expectations, inflation, and output when expectations are formed in the lab. Some of their results are shown in figure 6.

Under weak inflation targeting ($\phi_\pi = 1$), the system is highly unstable and inflation explodes. Under strong inflation targeting ($\phi_\pi = 1.5$), the system is stable and inflation and output gap converge to rational expectations equilibrium. Adding more negative feedback through a more aggressive inflation-targeting rule thus stabilizes inflation (and output gap).

The HSM fits the individual and aggregate experimental data well. What is the behavioral explanation of the HSM? Under weak inflation-targeting agents coordinate on a strong trend-extrapolating rule amplifying fluctuations and inflation (and output gap) explode. Under a stronger inflation-targeting policy, trend-following behavior can not survive competition with other heuristics, agents abandon trend-following rules, and rather coordinate on adaptive expectations so that inflation (and output gap) stabilize.

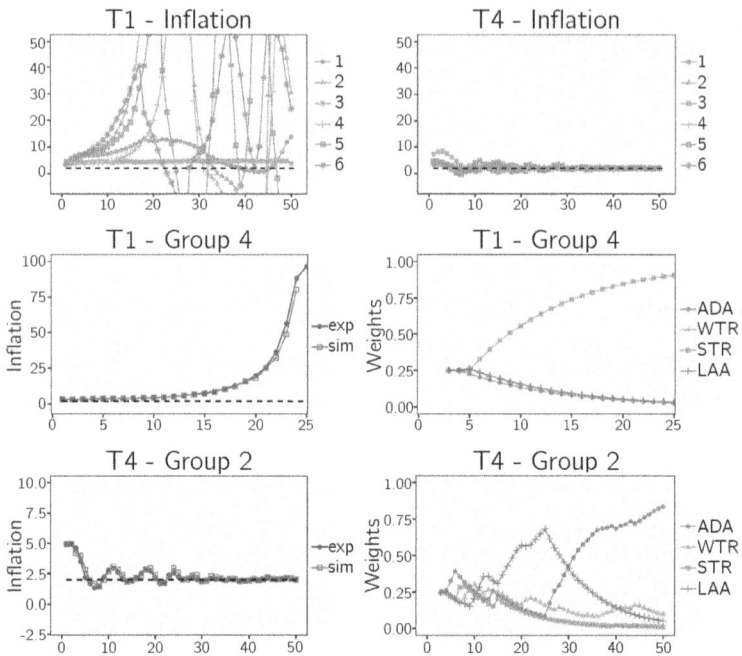

Figure 6. A stronger inflation-targeting rule adds negative feedback to the system, preventing coordination on trend-extrapolating behavior and thus stabilizing the NK system (Assenza *et al.* 2021).

Climate Modeling

An example of a complex system generating headlines is the climate system. The majestic review of Michael Ghil and Valerio Lucarini (2020) opens with the sentences, "The climate is a forced, dissipative, nonlinear, complex, and heterogeneous system that is out of thermodynamic equilibrium. The system exhibits natural variability on many scales of motion, in time as well as space, and it is subject to various external forcings, natural as well as anthropogenic."

Their review covers chaos, critical transitions, bistability, Hopf bifurcations, teleconnections, homoclinic and heteroclinic bifurcations, bifurcation trees, and much more, all germane to the understanding of climate dynamics at a range of spatial and temporal scales.

Like climate scientists, climate economists work with a hierarchy of models ranging from simple energy-balance models to spatial diffusion models illustrating poleward energy transport, but not yet encapsulating the full range of complexity. Climate economists have yet to build economic models that build in the spatial and temporal dynamics respecting the complexity as reviewed by Ghil and Lucarini (2020). We would like to tackle this challenge ourselves. In view of space limitations, we do not review the received work by climate economists here, but mention just a few here (William Nordhaus, Yongyang Cai, Kenneth Judd, Thomas Lontzek, Michael Barnett, Lars Hansen, Anastasios Xepapadeas).

A major concern as the world tries to design policies to cope with climate change are the multiple layers of uncertainty that appear in the coupled economic and climate system. An example of recent work that accounts for multiple layers of uncertainty in addition to risk (where probabilities are known) is Barnett, Brock, and Hansen (2022). Accounting for multiple layers of uncertainty strengthens the case for strong policy action now rather than waiting until we learn more about the unknown unknowns in the coupled climate and economic dynamical system.[3]

Future Perspective

We have discussed a variety of examples of complexity economics research, and there are many more not covered here. SFI has played a key role in generating complexity ideas and providing us with the SFI spirit driving us.

---

[3] For more discussion of climate (change) models, see part IV of this volume, Climate & Sustainability.

Here, we have emphasized rather stylized models of complexity that are (partly) analytically tractable. More complex simulation models, such as agent-based models and climate-change models, are an important complementary tool of complexity to study empirically relevant phenomena. For example, C. H. Hommes *et al.* (2025) and C. Hommes *et al.* (2026, ch. 14 in this volume) use a detailed ABM of the Canadian economy at scale 1:100 for monetary policy analysis. For more on ABMs in macro and central banking, see, for example, Dawid *et al.* (2026, ch. 13 in this volume) and Borsos *et al.* (2026, ch. 18 in this volume). We conclude by saying that both stylized models and detailed ABMs should become part of the complexity toolbox for policy analysis. ❦

Acknowledgments

We would like to thank Doyne Farmer, Marco Pangallo, and participants of the 2023 SFI Workshop, "Complex-System Approaches to Twenty-First-Century Challenges: Inequality, Climate Change, and New Technologies" for stimulating discussions and helpful feedback on an earlier draft of this chapter. Opinions expressed in this chapter are those of the authors and do not necessarily reflect those of the Bank of Canada or its staff. Any remaining errors are ours.

REFERENCES

Anderson, P. W., K. Arrow, and D. Pines, eds. 1988. *The Economy As An Evolving Complex System.* Boston, MA: Addison-Wesley.

Anufriev, M., and C. H. Hommes. 2012. "Evolutionary Selection of Individual Expectations and Aggregate Outcomes." *American Economic Journal-Micro* 4 (4): 35–64. https://doi.org/10.1257/mic.4.4.35.

Arifovic, J., and J. Duffy. 2018. "Heterogeneous Agent Modeling: Experimental Evidence." In *Handbook of Computational Economics: Heterogeneous Agent Modeling,* edited by C. H. Hommes and B. LeBaron, 4:491–540. Amsterdam, Netherlands: North-Holland.

Arthur, W. B., J. H. Holland, B. LeBaron, R. Palmer, and P. Taylor. 1997. "Asset Pricing under Endogenous Expectation in an Artificial Stock Market." In *The Economy as an Evolving Complex System II,* edited by W. B. Arthur, S. N. Durlauf, and D. A. Lane. Reading, MA: Addison-Wesley.

Assenza, T., P. Heemeijer, C. H. Hommes, and D. Massaro. 2021. "Managing Self-Organization of Expectations Through Monetary Policy: A Macro Experiment." *Journal of Monetary Economics* 117:170–186. https://doi.org/10.1016/j.jmoneco.2019.12.005.

Bao, T., and C. H. Hommes. 2019. "When Speculators meet Suppliers: Positive Versus Negative Feedback in Experimental Housing Markets." *Journal of Economic Dynamics and Control* 107:103730. https://doi.org/10.1016/j.jedc.2019.103730.

Barnett, M., W. Brock, and L. P. Hansen. 2022. "Climate Change Uncertainty Spillover in the Macroeconomy." *NBER Macroeconomics Annual* 36 (1): 253–320. https://doi.org/10.3386/w29064.

Barnett, W., and P. Chen. 1988. "Deterministic Chaos and Fractal Attractors as Tools for Nonparametric Dynamical Econometric Inference: With an Application to the Divisia Monetary Aggregates." *Mathematical and Computer Modelling* 10 (4): 275–296. https://doi.org/10.1016/0895-7177(88)90006-4.

Benhabib, J., and R. H. Day. 1982. "A Characterization of Erratic Dynamics in, the Overlapping Generations Model." *Journal of Economic Dynamics and Control* 4:37–55. https://doi.org/10.1016/0165-1889(82)90002-1.

Berreby, D. 1993. "Chaos Hits Wall Street." *Discover Magazine,* https://www.discovermagazine.com/the-sciences/chaos-hits-wall-street.

Biggs, R., S. R. Carpenter, and W. A. Brock. 2009. "Turning Back from the Brink: Detecting an Impending Regime Shift in Time to Avert It." *Proceedings of the National Academy of Sciences* 106 (3): 826–831. https://doi.org/10.1073/pnas.0811729106.

Boldrin, M., and M. Woodford. 1990. "Equilibrium Models Displaying Endogenous Fluctuations and Chaos." *Journal of Monetary Economics* 25:189–222. https://doi.org/10.1016/0304-3932(90)90013-T.

Bolt, W., M. Demertzis, C. Diks, C. H. Hommes, and M. van der Leij. 2019. "Identifying Booms and Busts in House Prices Under Heterogeneous Expectations." *Journal of Economic Dynamics & Control* 103:234–259. https://doi.org/10.1016/j.jedc.2019.04.003.

Borsos, A., A. Carro, A. Glielmo, M. Hinterschweiger, J. Kaszowska-Mojsa, and A. Uluc. 2026. "Agent-Based Modeling at Central Banks: Recent Developments and New Challenges." In *The Economy as an Evolving Complex System IV,* edited by R. M. del Rio-Chanona, M. Pangallo, J. Bednar, E. D. Beinhocker, J. Kaszowska-Mojsa, F. Lafond, P. Mealy, A. Pichler, and J. D. Farmer. Santa Fe, NM: SFI Press.

Boswijk, H. P., C. H. Hommes, and S. Manzan. 2007. "Behavioral Heterogeneity in Stock Prices." *Journal of Economic Dynamics & Control* 31:1938–1970. https://doi.org/10.1016/j.jedc.2007.01.001.

Brock, W., and S. Durlauf. 2006. "Multinomial Choice with Social Interactions." In *The Economy as an Evolving Complex System III,* edited by L. Blume and S. Durlauf, 175–206. Oxford, UK: Oxford University Press.

Brock, W. A. 1997. "Asset Price Behavior in Complex Environments." In *The Economy as an Evolving Complex System II,* edited by W. B. Arthur, S. N. Durlauf, and D. A. Lane, 385–423. Redwood City, CA: Addison-Wesley.

Brock, W. A., and S. R. Carpenter. 2010. "Interacting Regime Shifts in Ecosystems: Implication for Early Warnings." *Ecological Monographs* 80 (3): 353–367. https://doi.org/10.1890/09-1824.1.

Brock, W. A., W. D. Dechert, J. A. Scheinkman, and B. LeBaron. 1996. "A Test for Independence Based on the Correlation Dimension." *Econometric Reviews* 15 (3): 197–235. https://doi.org/10.1080/07474939608800353.

Brock, W. A., and C. H. Hommes. 1997. "A Rational Route to Randomness." *Econometrica* 65:1059–1095. https://doi.org/10.2307/2171879.

———. 1998. "Heterogeneous Beliefs and Routes to Chaos in a Simple Asset Pricing Model." *Journal of Economic Dynamics & Control* 22:1235–74. https://doi.org/10.1016/S0165-1889(98)00011-6.

Brock, W. A., D. A. Hsieh, and B. LeBaron. 1991. *Nonlinear Dynamics, Chaos, and Instability: Statistical Theory and Economic Evidence.* Cambridge, MA: MIT Press.

Brock, W. A., and C. L. Sayers. 1988. "Is the Business Cycle Characterized by Deterministic Chaos?" *Journal of Monetary Economics* 22:71–90. https://doi.org/10.1016/0304-3932(88)90170-5.

Carpenter, S. R., J. J. Cole, M. L. Pace, R. Batt, W. A. Brock, T. Cline, J. Coloso, *et al*. 2011. "Early Warnings of Regime Shifts: A Whole Ecosystem Experiment." *Science* 332 (6033): 1079–1082. https://doi.org/10.1126/science.1203672.

Crutchfield, J. P., J. D. Farmer, N. H. Packard, and R. S. Shaw. 1986. "Chaos: There is Order in Chaos: Randomness has an Underlying Geometric Form." *Scientific American* 254 (12): 46–57. https://www.jstor.org/stable/24976102.

Dawid, H., D. Delli Gatti, L. E. Fierro, and S. Poledna. 2026. "Implications of Behavioral Rules in Agent-Based Macroeconomics." In *The Economy as an Evolving Complex System IV,* edited by R. M. del Rio-Chanona, M. Pangallo, J. Bednar, E. D. Beinhocker, J. Kaszowska-Mojsa, F. Lafond, P. Mealy, A. Pichler, and J. D. Farmer. Santa Fe, NM: SFI Press.

Dechert, W. D. 1996. *Chaos Theory in Economics: Methods, Models, and Evidence.* Cheltenham, UK: Edward Elgar.

Duffy, J. 2016. "Macroeconomics: A Survey of Laboratory Research." In *The Handbook of Experimental Economics,* edited by J. H. Kagel and A. E. Roth, 2:1–90. Princeton, NJ: Princeton University Press.

Emenogu, U., C. H. Hommes, and M. Khan. 2021. *Detecting Exuberance in House Prices Across Canadian Cities.* Technical report Staff Analytical Note 2021-9. Bank of Canada.

Ghil, M., and V. Lucarini. 2020. "The Physics of Climate Variability and Climate Change." *Review of Modern Physics* 92 (3): 03001. https://doi.org/10.1103/RevModPhys.92.035002.

Grandmont, J.-M. 1987. *Nonlinear Economic Dynamics.* Boston, MA: Academic Press.

Grassberger, P., and I. Procaccia. 1983. "Measuring the Strangeness of Strange Attractors." *Physica* 90:189–208. https://doi.org/10.1016/0167-2789(83)90298-1.

Heemeijer, P., C. H. Hommes, J. Sonnemans, and J. Tuinstra. 2009. "Price Stability and Volatility in Markets with Positive and Negative Expectations Feedback: An Experimental Investigation." *Journal of Economic Dynamics and Control* 33:1052–1072. https://doi.org/10.1016/j.jedc.2008.09.009.

Hommes, C., S. Kozicki, S. Poledna, and Y. Zhang. 2026. "How an Agent-Based Model Can Support Monetary Policy in a Complex Evolving Economy." In *The Economy as an Evolving Complex System IV,* edited by R. M. del Rio-Chanona, M. Pangallo, J. Bednar, E. D. Beinhocker, J. Kaszowska-Mojsa, F. Lafond, P. Mealy, A. Pichler, and J. D. Farmer. Santa Fe, NM: SFI Press.

Hommes, C. H. 2013. *Behavioral Rationality and Heterogeneous Expectations in Complex Economic Systems.* Cambridge, UK: Cambridge University Press.

———. 2021. "Behavioral and Experimental Macroeconomics and Policy Analysis: A Complex Systems Approach." *Journal of Economic Literature* 59 (1): 149–219. https://doi.org/10.1257/jel.20191434.

Hommes, C. H., M. He, S. Poledna, M. Siqueira, and Y. Zhang. 2025. "CANVAS: A Canadian Behavioral Agent-Based Model for Monetary Policy." Forthcoming, *Journal of Economic Dynamics and Control,* https://doi.org/10.1016/j.jedc.2024.104986.

Hommes, C. H., and D. in 't Veld. 2017. "Booms, Busts and Behavioural Heterogeneity in Stock Prices." *Journal of Economic Dynamics & Control* 80:101–124. https://doi.org/10.1016/j.jedc.2017.05.006.

Hommes, C. H., A. Kopányi-Peuker, and J. Sonnemans. 2021. "Bubbles, Crashes and Information Contagion in Large-Group Asset Market Experiments." *Experimental Economics* 24:414–433. https://doi.org/10.1007/s10683-020-09664-w.

Hommes, C. H., J. Sonnemans, J. Tuinstra, and H. van de Velden. 2005. "Coordination of Expectations in Asset Pricing Experiments." *Review of Financial Studies* 18 (3): 955–980. https://www.jstor.org/stable/3598083.

Kuznetsov, Y. 1995. *Elements of Applied Bifurcation Theory.* Berlin, Germany: Springer-Verlag.

LeBaron, B., W. B. Arthur, and R. Palmer. 1999. "Time Series Properties of an Artificial Stock Market." *Journal of Economic Dynamics and Control* 23 (9--10): 1487–1516. https://doi.org/10.1016/S0165-1889(98)00081-5.

O'Brien, D. A., S. Deb, G. Gal, S. J. Thackeray, P. S. Dutta, S.-I. S. Matsuzaki, L. May, and C. F. Clements. 2023. "Early Warning Signals Have Limited Applicability to Empirical Lake Data." *Nature Communications* 14:7942. https://doi.org/10.1038/s41467-023-43744-8.

Scheffer, M., J. Bascompte, W. Brock, V. Brovkin, S. Carpenter, V. Dakos, H. Held, E. van Nes, M. Rietkerk, and G. Sugihara. 2009. "Early Warning Signals for Critical Transitions." *Nature* 461:53–59. https://doi.org/10.1038/nature08227.

Scheinkman, J. A., and B. LeBaron. 1989. "Nonlinear Dynamics and Stock Returns." *Journal of Business* 62:311–337. https://doi.org/www.jstor.org/stable/2353350.

Sugihara, G., and R. May. 1990. "Nonlinear Forecasting as a Way of Distinguishing Chaos from Measurement Error in Time Series." *Nature* 344:734–741. https://doi.org/10.1038/344734a0.

Takens, F. 1981. "Detecting Strange Attractors in Turbulence." In *Dynamical Systems and Turbulence,* edited by D. A. Rand and L. S. Young, 898:366–381. Lecture Notes in Mathematics. Berlin, Germany: Springer-Verlag.

Theiler, J., S. Eubank, A. Longtin, B. Galdrikian, and J. D. Farmer. 1992. "Testing for Nonlinearity in Time Series: The Method of Surrogate Data." *Physica D: Nonlinear Phenomena* 58:77–94. https://doi.org/10.1016/0167-2789(92)90102-S.

PART II

*Methods & Concepts
in Complexity Economics*

ECONOMIC COMPLEXITY ANALYSIS

Frank Neffke, Complexity Science Hub;
Angelica Sbardella, Enrico Fermi Research Center;
Ulrich Schetter, University of Pavia; and
Andrea Tacchella, Enrico Fermi Research Center

Abstract

Economic complexity analysis (ECA) is a newly emerging research program that aims to understand what determines the set of goods and services that a country can produce, and how this set changes over time. At its core, this research program assumes that production of a given good or service requires a combination of fine-grained and highly complementary capabilities. As a consequence, economic growth is driven by a process of diversification that is enabled by the acquisition of capabilities. This chapter traces the intellectual antecedents and origins of ECA and illustrates core tenets in a simple model of production that is analyzed using complex network theory. It then reviews current debates on core concepts in the field—in particular measures of relatedness and complexity of economic activities—and reflects on policy implications. We conclude by sketching a broad research agenda, identifying five key areas: (1) relaxing overly restrictive assumptions of current models; (2) better connecting ECA to debates in the wider field of economics; (3) exploring connections across scales, from countries to cities to firms; (4) addressing questions related to capability coordination; and (5) developing applications to important large-scale societal transitions, such as the green transition and the digitization of work.

Introduction

Why are some countries rich and others poor? Why do standards of living differ so much across cities in the same country? What determines whether these standards of living diverge or converge? These questions have inspired generations of economists. In essence, they ask what determines an economy's capacity to generate prosperity. The consensus since at least Moses Abramovitz (1986) is that the ultimate driver of prosperity is progress in "technology," that is, in the ways in which economies produce output. Initially, economists aimed to estimate a country's technological prowess from how efficiently it converts broad factors of production, such as capital, labor, and land, into aggregate output. However, by summing across many types of inputs and outputs, this "agglomerative approach" (Sbardella *et al.* 2018b; Balland *et al.* 2022) left much information on economies' production unused. Recently a new research program at the intersection of complexity science and economics has gained traction that aims to shed new light on the issue of economic development. This field uses detailed information on what economic entities produce, leveraging insights from complex network analysis. We will refer to this emerging field of research as economic complexity analysis (ECA).

ECA synthesizes insights from diverse intellectual traditions. It draws from the structuralist perspective, which emphasizes the importance of the composition of countries' activity baskets and views economic development as changes in how resources are allocated across sectors (Hirschman 1958; Prebisch 1962; Lin 2011). Echoing evolutionary economics and capability-based theories of the firm (Penrose 1959; Nelson and Winter 1982; Barney 1991; Dosi and Nelson 1994), ECA furthermore underscores the significance of productive capabilities—fine-grained inputs (broadly defined) to production—as the primary catalyst for transformative growth, where these capabilities range from specialized know-how to infrastructure and state capacities. In terms of complexity science, ECA incorporates principles from complexity economics, which emerged at the Santa Fe Institute in the late 1980s and 1990s as reviewed in earlier volumes of *The Economy as an Evolving Complex System*. This approach treats economic phenomena as complex systems, in which interconnected and heterogeneous actors evolve, learn, and adapt their strategies in ways that result in macro-level patterns of collective behavior (Arthur 2013). Exhaustive reviews of the literature on ECA can be found in Pierre-Alexandre Balland *et al.* (2022) and César Hidalgo (2023). Our aim here is not to repeat these efforts, but rather to describe the main features of ECA in a way that highlights current challenges, policy implications, and opportunities for future research.

Assumptions of ECA

ECA starts from a stylized depiction of the process by which economic entities, be they countries, cities, regions or firms, *make* things, where each product or service requires its own subset of a large but finite range of capabilities. This sets ECA apart from more traditional bodies of economic thought, which emphasize strategic behavior or focus on the puzzle of how to allocate scarce resources.

The capabilities that modern economies mobilize—ranging from specialized know-how to a variety of different public goods—are numerous. Moreover, capabilities are taken to be non-substitutable. Therefore, ECA assumes that, to a first approximation, economies can only produce the products and services for which they have all required capabilities.

Cast in the language of networks, this model of production can be described by a tripartite network that first connects economic entities to the capabilities they possess—shown in the left of figure 1 and captured by matrix C—and then connects these capabilities to the products or services that require them—shown in the center of figure 1 and captured by matrix P (Hidalgo and Hausmann 2009; Cristelli *et al.* 2013). This path also connects, albeit indirectly, countries to the products they can make, that is, the products for which they possess all required capabilities. This bipartite network is shown in figure 1 on the right and captured by matrix M. An important challenge in empirical applications is that, while an entity's economic activities, summarized in matrix M, are readily observable, the underlying capability structure in matrices C and P typically are not. We will get back to this point momentarily.

This tripartite model of production implies a highly nonlinear relationship between the capability endowments of economic entities and their capacity to make a product, that is, the structure of the resulting bipartite network that connects entities to the products they make. The reason is that the addition or removal of a single capability (that is, a link in the tripartite network) may trigger the appearance or disappearance of a link in the bipartite network. In other words, such changes result not just in a proportional increase or decrease in competitiveness, but in a complete gain or loss of the capacity to produce certain products. Moreover, the combinatorial nature of production—captured by the assumption that production combines product-specific sets of capabilities—implies that the added value of an extra capability increases superlinearly with the number of available capabilities (Hausmann and Hidalgo 2011; Fink and Reeves 2019; van Dam and Frenken 2022, and ch. 3 of the current volume). The reason is that each combination of capabilities potentially

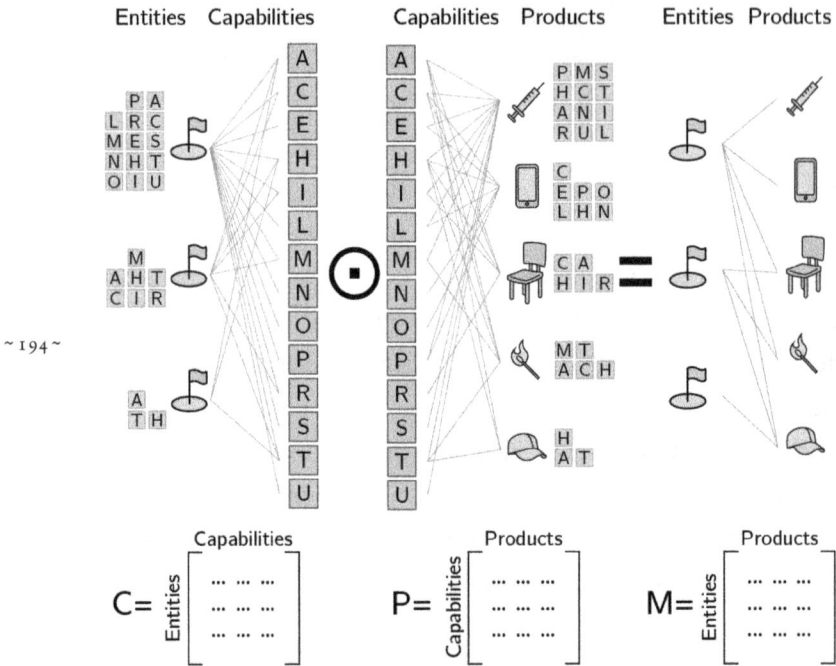

Figure 1. Model of production in ECA. Tripartite network composed of two bipartite networks, described by matrices P and C, that are connected through a generic operator that shows how capability endowments interact with capability requirements in economic production. The tripartite network connects capabilities to productive entities (such as countries, cities or firms) on the left and to products on the right. The operation yields a new network that shows which entities make which products, described by matrix M.

enables the production of a distinct product, and the greater the number of pre-existing capabilities, the more additional combinations become feasible when adding a capability. This introduces nonlinearities that lead to poverty traps and path dependence in economic development (Hidalgo *et al.* 2007; Hausmann and Hidalgo 2011; Pugliese *et al.* 2017; Diodato, Hausmann, and Schetter 2022).

An important finding in ECA is that many of the observed M matrices are *nested* (Hausmann and Hidalgo 2011; Tacchella *et al.* 2012; Mariani *et al.* 2019; Schetter 2024). Nestedness

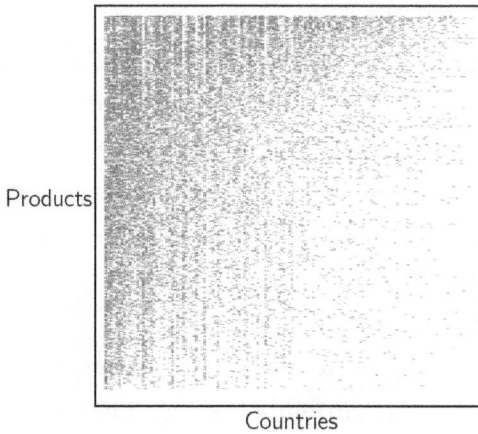

Products

Countries

Figure 2. Nested trade matrix. Matrix whose elements reflect the exports by countries (in rows) of products (in columns). The matrix is *nested* as shown by its triangular structure.

is a concept that originated in ecology, where it describes a situation in which generalist species, which are adaptable and able to draw from a large variety of food sources, interact with specialist species, which have a more limited diet and occupy narrower ecological niches. ECA carries this metaphor over to economic production, by highlighting that, empirically speaking, rare products (specialists) are often only exported by diversified countries (generalists), while diversified countries often export both rare and ubiquitous products. Visually, this results in the triangular shape (see fig. 2) that adjacency matrices of country–product networks assume when their rows and columns are appropriately sorted. We will return to this issue later on.

Empirical Implementation

A basic challenge of the ECA framework is that capabilities are hard to observe. As a consequence, the tripartite network of figure 1 is not immediately available to the researcher. What is readily observable is the bipartite network that connects economic entities to the products and services they produce. The core methodological insight of ECA is that, although the bipartite network in the far right of figure 1 cannot be easily used to uncover the exact capability endowments of an economic entity or the capability requirements of a product or service, it can be used to learn about the topology and complexity of capability endowments and requirements.

The idea of looking at the composition of output baskets to infer the topology of the capability base that underlies observed production patterns goes back to Ricardo Hausmann and Bailey Klinger (2006). Using international trade data, these authors infer the technological proximity between two exported products (that is, the degree to which they, presumably, share the same capability requirements) by counting how often two products are exported by the same countries. The idea of inferring the similarity between products in terms of their production requirements from coproduction patterns has a long history in management science, going back to at least the work of David Teece *et al.* (1994). These authors studied co-occurrences of industries in firms to estimate among which industries economies of scope exist. The rationale behind studying co-occurrences of products at the country level is similarly straightforward: If countries make the products for which they have all required capabilities, then (properly rescaled) co-occurrence counts should signal similarities in capability requirements. Subsequently, Hidalgo *et al.* (2007) cast this work in the language of network analysis, where products are connected if they are often co-exported. The resulting network, or *product space*, helps predict how countries diversify their export baskets: Countries are more likely to start producing new products that are closely related to—that is, are often co-exported with—their current exports than unrelated products.

This work has inspired a stream of research papers that apply this insight to predict how cities and regions diversify into new industries (Neffke, Henning, and Boschma 2011; Boschma, Minondo, and Navarro 2013; Essletzbichler 2017; Zhu, He, and Zhou 2017), technologies (Napolitano *et al.* 2018; Barbieri *et al.* 2022; Zhang and Rigby 2022), or academic research fields (Guevara *et al.* 2016; Patelli *et al.* 2017). These different types of activity can also interact. This is captured in multidimensional measures of relatedness that aim to map how, for instance, competitiveness in an academic research field facilitates innovation in technological areas, which in turn yields comparative advantage in exporting specific products (Pugliese *et al.* 2019a). By now, the prediction of related diversification has been replicated across so many datasets and domains that it has been dubbed the *principle of relatedness* (Hidalgo *et al.* 2018). Furthermore, the principle of relatedness also operates at levels other than countries or regions. For instance, individual careers (Gathmann and Schönberg 2010; Yildirim and Coscia 2014; Mealy, del Rio-Chanona, and Farmer 2018; Frank *et al.* 2024; Neffke, Nedelkoska, and Wiederhold 2024, see also ch. 24 in this volume) and firm diversification trajectories (Bryce and Winter 2009; Neffke and Henning 2013; Napolitano *et al.* 2018) also often follow a path of related diversification.

However, inferring the similarity of economic activities from colocation frequencies presents challenges, in part because the number of activities is much greater than the number of locations. As a consequence, most geographical areas host hundreds of activities that are often only indirectly related to one another. This has led to the development of statistical validation techniques for filtering network links (Saracco *et al.* 2015; Saracco *et al.* 2017; Cimini *et al.* 2019) or other types of similarity metrics (Neffke and Henning 2013; Li and Neffke 2024), as well as more sophisticated

~197~

machine-learning methodologies (Albora *et al.* 2023; Tacchella *et al.* 2023) and approaches that directly build on observable capabilities (Diodato, Hausmann, and Schetter 2022; Aufiero *et al.* 2024; Schetter *et al.* 2024).

The insight that the structure of output, that is, the portfolio of products, services, or industries in which an economic entity is active, can be used to estimate the complexity of the entity's underlying capability base goes back to Hausmann, Jason Hwang, and Dani Rodrik (2007). Once again relying on trade data, these authors propose that a country's productivity is reflected in its export basket: Countries that export high-income goods tend to be highly productive, or, "what you export matters." Hidalgo and Hausmann (2009) reformulate this work in terms of the network of the entity-product matrix, M. Starting from this matrix, they propose an iterative algorithm—the "method of reflections," equivalent to reciprocal averaging in correspondence analysis (Hill 1974; van Dam *et al.* 2021)—that assesses the complexity of countries using only information on the products they make and vice versa. The resulting metrics, the economic complexity index (ECI) for countries and the product complexity index (PCI) for economic activities, converge to the second eigenvectors of a rescaled version of M, multiplied with its transpose.

Hidalgo and Hausmann's (2009) work has generated substantial debate (e.g., Tacchella *et al.* 2012, Servedio *et al.* 2018, Mealy, Farmer, and Teytelboym 2019, Sciarra *et al.* 2020, van Dam *et al.* 2021, McNerney *et al.* 2021). A first challenge, by Andrea Tacchella *et al.* (2012), highlighted that the ECI/PCI pair does not always behave in the way the model of economic production of figure 1 would suggest. In particular, a country's ECI is, by construction, equal to the average PCI of the products it makes, and a product's PCI is the average ECI of the countries that make it. Consequently, a country's ECI may go up or down if the country adds a new product to its export basket, depending on whether the product is more or less complex than the existing products in this basket. However, a strict interpretation of the capability model of figure 1 prohibits this. After all, this model implies that adding extra products should never be associated with a decrease in the number of capabilities in a country. Similarly, the complexity of a product should not change if more complex economies start producing it, but should remain anchored by the least complex economy capable of doing so.

To remedy this, Tacchella *et al.* (2012) propose an alternative algorithm—the fitness complexity (FC) algorithm—that retains the monotonous relation between diversification and complexity and the nonlinearity required to bound a product's complexity to the least complex of its exporters. This leads to a new set of complexity or *fitness* measures. In terms of predictive validity, like a country's ECI, fitness is a strong predictor of the country's per capita gross domestic product (GDP) and GDP growth (Tacchella, Mazzilli, and Pietronero 2018). Interestingly, fitness also turns out to be closely connected to the concept, central to ECA, of nestedness. In fact, the FC algorithm ranks rows and columns of nested matrices in a close to optimal way to reveal the matrices' nested structure (Mariani *et al.* 2019; Mariani *et al.* 2024).[1]

Other authors have focused on the interpretation of the ECI. In particular, because the ECI and PCI metrics are eigenvectors of co-occurrence matrices, they also have more mundane interpretations. For instance, Penny Mealy, Doyne Farmer, and Alexander Teytelboym (2019) show that the ECI is equivalent to spectral clustering, dividing countries into two communities, based on the degree to which they produce similar products.

---

[1] As a consequence, the FC algorithm may find useful applications well beyond the field of economic complexity, from assessing the importance of species in mutualistic ecosystems (Domínguez-García and Muñoz 2015) to providing approximate solutions to optimal transport problems (Mazzilli *et al.* 2024).

Similarly, James McNerney *et al.* (2021) show that variants of the ECI can be derived from the assumption that the principle of relatedness accurately describes short-term dynamics of international exports. In this case, the principle of relatedness sets up a simple dynamical system where the first eigenvector closely resembles fitness and the second eigenvector resembles the PCI.

Ulrich Schetter (2022) shows that the ECI/PCI pair ranks countries and products in accordance with structural notions of complexity, provided that matrix M exhibits a *log-supermodular* structure, a general assumption that covers a range of special cases[2] and finds strong empirical support in international trade data. He embeds notions of complexity in a general equilibrium trade model to derive nonparametric productivity rankings that correlate highly with the ECI and PCI. Whereas Schetter (2022) provides a flexible micro-foundation for complexity *rankings*, Muhammed Yildirim (2021) builds on these insights by adding specific functional form assumptions that provide an approximate mapping to the underlying complexity *levels*.

The upshot of this debate is that the interpretation of the ECI/PCI pair depends on the assumptions one is willing to make about the data generating process behind the country–product matrix, M. Without any further assumptions, these variables simply recover communities of countries that produce similar products and of products that are produced by similar countries. If, instead, we are willing to assume that the principle of relatedness is a good description of short-run diversification dynamics, the PCI is an important axis of long-lived structural transformation, whereas the ECI approximately captures how far

---

[2]Examples range from ladders of specialization, to nested patterns of specialization, to the random capabilities model in Hausmann and Hidalgo (2011).

a country has moved along this axis.[3] Finally, if we are willing to assume log-supermodularity, the ECI ranks countries in terms of their underlying productive capabilities.

Economic Complexity Analysis and Policy

ECA is widely used in policymaking at the international, national, and subnational level to devise strategies for economic development, structural change, and innovation. ECA is used in policy frameworks for two main reasons: On the one hand, ECA provides novel perspectives on countries' productive capabilities and how they foster economic prosperity. This is most vividly illustrated by complexity metrics, which have been shown to correlate strongly not only with countries' current GDP per capita but also their future growth performance (Hidalgo and Hausmann 2009; Hausmann *et al.* 2014; Tacchella *et al.* 2012). Moreover, complex countries have been shown to be more inclusive (Hartmann *et al.* 2017; Sbardella, Pugliese, and Pietronero 2017; Hartmann and Pinheiro 2022; Barza *et al.* 2024), suggesting that complexity upgrading fosters more broadly shared prosperity.

On the other hand, the tools and methods developed in ECA and the fine-grained view on the economy they afford allow for context-specific, targeted strategies aimed at diversifying and upgrading economies. This starts with an assessment of an economy's current capabilities and its position in product and related spaces. By now, a range of publicly available tools allows us to readily assess this at the national and subnational level. Subsequent analysis allows us to identify products or services that may serve as stepping stones onto development ladders and

~201~

---

[3] Interestingly, another important axis of change closely resembles fitness, showing that in this interpretation, fitness and PCI are not in opposition but complement each other to describe the long-term patterns of change implied by the principle of relatedness.

strategies for amassing the capabilities to reach these stepping stones. The latter is facilitated by a recently proposed "genotypic" approach to the product space (Schetter *et al.* 2024), which takes the level of analysis from observed outcomes as summarized in matrix M to underlying capability structures as summarized in matrices C and P. The value of this approach is that it offers more actionable information on missing capabilities, but future work needs to explore this framework more carefully.

ECA has gained considerable traction among policymaking institutions, particularly those focused on investments in developing countries. These institutions often grapple with the challenge of prioritizing investments across diverse geographies and sectors. The ECA framework offers a valuable tool in this context, because it allows direct comparisons between highly diverse investment options across economies and sectors at a granular level of single products or technologies. It does so within a unified framework that is easily applicable at both the national and regional level and that provides a common language for evaluating otherwise incommensurable opportunities.

One of the key strengths of ECA in policy applications is its ability to quantify the likelihood that specific sectors or products can be developed based on an economy's existing competitiveness in related sectors. This stems from the principle of relatedness discussed earlier, which posits that a country is more likely to diversify into industries that require similar capabilities as the industries in which it is active already.

However, it is crucial to note that these quantitative predictions should not necessarily be interpreted as policy prescriptions or priorities. A common misunderstanding of ECA is that it suggests "picking winners." However, exclusively building on a country's existing comparative advantage by following related diversification trajectories may risk merely strengthening already competitive areas or provoking low-complexity lock-ins in regions that have so far failed to catch up with the technological frontier. Proper application of ECA in policymaking therefore often requires flanking it with more qualitative analysis. For instance, Li and Neffke (2024) suggest that ECA can be used to detect anomalies in a country's productive structure. Such anomalies suggest areas of the economy where deep dives would yield valuable information about binding constraints to development. That is, ECA can help identify activities that are surprisingly large or small and therewith point to hidden strengths or weaknesses of the economy as part of a broader diagnostic approach (Hausmann, Rodrik, and Velasco 2008).

Another common misconception is that ECA assumes that complexity itself should always be the main goal of economic development policy. However, policymakers may have different priorities. For instance, they may strive to reduce their economy's carbon footprint or want to create jobs for specific segments of the population. In this case, ECA can be used to assess impacts on complexity and productivity of development strategies guided by such priorities. It can also be used to map the associated development trajectories in spaces that reflect the economy's current capabilities and identify which capabilities would still need to be developed.

Therefore, policymakers and institutional investors, such as development banks, often use ECA as part of a broader decision-making process. They may choose to support projects that are closely related to existing activities, which are more likely to generate market-driven development. Alternatively, and perhaps more importantly, they might opt to invest in projects that are less related to the current productive structure, but offer higher potential for long-term diversification and productivity growth. This latter approach aims to fill capability gaps, enabling countries to access more complex and potentially more lucrative production opportunities in the future. For instance, the EU Commission (Pugliese and Tacchella 2020, 2021), the World Bank, and other development institutions frequently prioritize projects that are less likely to be developed by market forces alone to catalyze structural transformation that might not occur organically, pushing countries towards higher levels of economic complexity, fostering the acquisition of new capabilities, and facilitating entry into more sophisticated industries.

ECA also provides policymakers with a framework to understand the potential spillover effects of investments. By identifying the capability requirements of different industries and their connections in the product space, policymakers can better anticipate how support for one sector may create positive externalities for others (Diodato, Hausmann, and Schetter 2022). This system-level view can help in designing holistic development strategies that leverage synergies across different parts of the economy.

Finally, ECA can also be applied to development strategies for regions within countries. By analyzing the economic complexity and relatedness patterns at subnational levels, policymakers can tailor interventions to the capability endowments of different regions (Pugliese and Tacchella 2020, 2021; Sbardella *et al.* 2022). This can be particularly valuable in addressing regional inequalities and promoting balanced economic growth within a country.

In summary, while ECA offers powerful analytical tools for quantifying development potential, its true value in policymaking depends on how these insights are interpreted and applied. When used judiciously and in combination with other diagnostic tools, ECA can help guide policymakers in making strategic investments that not only build on existing strengths, but also pave the way for transformative economic development by systematically expanding a country's productive capabilities.

Future Directions for ECA

Despite the rapid expansion of the literature on ECA, many questions about its proposed process of economic development remain unresolved. In the model of production underlying ECA, economic growth can in principle come from three sources. First, economies may not yet have fully exploited the potential of their existing products. Second, economies may not yet have fully exploited the potential of their existing capabilities. Third, economies may acquire new capabilities, increasing the number and changing the nature of the products they can make. The first assumes that countries may produce a product at different levels of quality—or, alternatively, at different levels of efficiency (e.g., Schetter 2024). The second implies that countries do not make all products they possibly can. The third requires a theory of how capabilities diffuse or, if they are new to the world, how they are invented. A plausible hypothesis is that capabilities diffuse with the movement of people (e.g., migration) and teams of people (e.g., foreign direct investments [FDI]). This hypothesis finds support in a large literature on spillovers from highly skilled migrants (e.g., Hausmann and Nedelkoska 2018; Diodato, Hausmann, and Neffke 2023; Lissoni and Miguelez 2024; Bahar *et al.* 2024) and from FDI (e.g., Blomström and Kokko 1998; Javorcik 2004). Compared to these more traditional literatures, ECA emphasizes the *content* of the know-how that foreign migrants and firms possess, as well as the combinatorial possibilities the capabilities they bring with them unlock. Although such implications remain currently understudied, they suggest clear hypotheses that can be tested in future research.

Another fruitful direction scrutinizes the assumption in ECA that *having* capabilities is sufficient for *using* them. This assumption is problematic, because capabilities are likely to be distributed across individuals, firms, and locations. Mobilizing large sets of capabilities therefore requires mechanisms to coordinate them across people, teams, firms, and locations. In line with this, Matte Hartog *et al.* (2024) show that the advent of the industrial research lab—which offered new ways to coordinate the work of inventor teams—coincided with the take-off of teamwork, as well as with a strong increase in the capacity of teams to develop radical innovations. Similarly, in chapter 29 of this volume, Frenken and Neffke (2026) argue that institutions and multinational enterprises play key roles in coordinating capabilities in global value chains that are distributed across different locations (see also, Frenken, Neffke, and van Dam 2023).

Relatedly, ECA can be used to study how entities at different scales interact. For instance, Dario Laudati *et al.* (2023) show how nestedness breaks down when studying firm-level data. The reason is that the limited size of firms forces them to concentrate on specific subsets of products. However, after appropriately partitioning economic activities into communities of closely related industries or products, nestedness is recovered within these communities. That is, within a given community, the population of firms that are active there can be sorted according to the FC algorithm such that nested firm-activity submatrices emerge. Because economies are essentially collections of firms, this finding shows that nestedness at the macro level of entire economies builds on fine-grained nestedness patterns within firm populations. Building further on this insight may help us gain an understanding of how complexity is distributed and aggregated at different levels of analysis, from individuals, to firms, cities, countries, and global value chains.

The greater specialization at the firm *vis-à-vis* regional or national level is related to another issue often ignored in ECA: scarcity of capabilities. In essence, ECA typically assumes that capabilities are public goods. That is, capabilities are nonrivalrous and nonexcludable. As a consequence, once developed, capabilities can be freely used in the production of all products without impeding their use in other products. While this assumption simplifies the analysis, it is not obvious that it is always a reasonable approximation of reality. For example, one often-cited capability, specialized know-how, resides in the human capital of workers. Workers, however, can typically only be employed by one firm at a time. Scarcity of capabilities may have profound consequences for ECA. For instance, ECA proposes that the denser the web of related products a country is already making, the greater the prospects of starting to make a new product become. With scarce capabilities, this is no longer necessarily the case, and scarcity may thus help explain why the relationship between density and entry is not always monotonously increasing (Schetter *et al.* 2024). Analyzing more carefully the public-goods character of capabilities (or lack thereof) and its implications for ECA would be a fruitful avenue for future research.

At a more fundamental level, ECA differs from large parts of related literature in economics in two important respects: first, by emphasizing the granularity of capabilities that are needed as inputs for modern production and, second, by considering international specialization and structural change at the *extensive* margin, that is by analyzing at a highly disaggregated level the set of products that countries make and how this set evolves. Recently, a series of advances helped these literatures move closer together. This includes macroeconomic analyses of disaggregated economies, but with a focus on changes at the intensive margin of production (Baqaee and Farhi 2019, 2024); economic theories with a ladder of development at the extensive product margin (Lucas 1993; Foellmi and Zweimüller 2008; Sutton and Trefler 2016; Schetter 2024; Atkin, Costinot, and Fukui 2025; Diodato, Hausmann, and Schetter 2022); work that explains how a nested pattern of specialization can arise even if countries specialize according to their comparative advantage (Schetter 2024; Bruno *et al.* 2023) and macroeconomic implications of a nested pattern of specialization (Atkin, Costinot, and Fukui 2025; Gersbach, Schetter, and Schmassmann 2023); studies analyzing the macroeconomic implications of a greater division of labor in larger cities (Tian 2021); and papers that analyze related diversification of firms or countries (Napolitano *et al.* 2018; Pugliese *et al.* 2019b), using information about input-output relations (Boehm, Dhingra, and Morrow 2022) or occupational structures (Diodato, Hausmann, and Schetter 2022; Aufiero *et al.* 2024). Yet, there are many open questions at the intersection of economics and ECA that future research should address, concerning, for instance, the aforementioned topics of global value chains and of identifying fundamental drivers of economic growth.

Finally, ECA has proven to be highly effective in shedding light on important recent economic transformations, because it provides a comprehensive toolbox for analyzing the connections between the growth and spatial concentration of economic activities as well as the flows of knowledge, people, and resources among them. A particularly important area in this space applies elements of ECA to study the green transition. This research merges insights from sustainability studies and evolutionary economic geography with ECA to analyze the shift toward more sustainable and equitable socioeconomic systems.[4] For instance, connecting green products to the capability

---

[4]For a comprehensive review, see Caldarola *et al.* (2024).

bases of cities and countries allows us to evaluate the degree to which existing productive structures support new green specialization paths. Initial research suggests that green products and technologies (Barbieri, Marzucchi, and Rizzo 2020; Mealy and Teytelboym 2022) are comparatively complex and often intertwined with nongreen counterparts in product (Hamwey, Pacini, and Assunção 2013; Fankhauser *et al.* 2013; Fraccascia, Giannoccaro, and Albino 2018) and technology (Barbieri *et al.* 2022) spaces. In fact, nongreen and green innovation capabilities often complement one another (Montresor and Quatraro 2020; Perruchas, Consoli, and Barbieri 2020; Barbieri *et al.* 2022), such that the relatedness to preexisting nongreen knowledge bases often plays a key role in fostering new green technological advancements (Montresor and Quatraro 2020; Barbieri *et al.* 2022). Countries therefore typically adopt a dual strategy: diversifying into green technologies that align with their existing nongreen capabilities while specializing in mature green technologies where they have already accumulated expertise (Barbieri, Perruchas, and Consoli 2020; Perruchas, Consoli, and Barbieri 2020). This, however, may lead to significant gaps in green innovation capacity at subnational levels (Grashof and Basilico 2023; Napolitano *et al.* 2018; Sbardella *et al.* 2018a).

~ 209 ~

Currently, this literature is still in its infancy, and there are many ways to move this agenda forward. First, exploring the interplay between productive, technological, and scientific capabilities can help us understand transformation at the frontier of knowledge where much of these transitions are taking place (e.g., de Cunzo *et al.* 2022). Second, ensuring a just transition will require studying "left-behind regions" (Rodríguez-Pose *et al.* 2024) and the structural change and labor reallocation processes they face, as well as the distributional effects of the associated creation and destruction of jobs (see, e.g., Mealy, del

Rio-Chanona, and Farmer 2018; Vona, Marin, and Consoli 2019; Rughi, Staccioli, and Virgillito 2023, and ch. 23 in this volume). Third, green and digital technologies and products' reliance on critical minerals (Diemer *et al.* 2022; de Cunzo *et al.* 2023; International Energy Agency 2023) has led to concerns about supply-chain disruptions (European Commission 2023; Kowalski and Legendre 2023) for which ECA's product and technology spaces may help identify bottlenecks as well as workarounds.

The seamless connection between fine-grained and coarsened views on economic production that ECA provides, combined with the framework's focus on economic transformation, makes it an excellent starting point to study such transitions and their socioeconomic effects. Further developing the framework, both in terms of its theoretical foundations and its methodological toolbox, will therefore not only guide academic debates, but also inform economic policy on how to plot a course in periods that are characterized by large societal challenges, as well as substantial technological and structural change. ⚬

Acknowledgments

F. N. receives financial support from the Austrian Research Promotion Agency (FFG) in the framework of the project ESSENCSE (873927), within the funding program *Complexity Science*. A. S. and A. T. acknowledge financial support under the National Recovery and Resilience Plan (NRRP), Mission 4, Component 2, Investment 1.1, Call for tender No. 1409, published on 14 September 2022 by the Italian Ministry of University and Research (MUR), funded by the European Union – Next Generation EU – Project Title "Triple T – Tackling a Just Twin Transition: A Complexity Approach to the Geography of Capabilities, Labour Markets and Inequalities" – CUP F53D23010800001 - Grant Assignment Decree No. 1378, adopted on 1 September 2023 by the Italian Ministry of University and Research (MUR).

REFERENCES

Abramovitz, M. 1986. "Catching Up, Forging Ahead, and Falling Behind." *The Journal of Economic History* 46 (2): 385–406. https://www.jstor.org/stable/2122171.

Albora, G., L. Pietronero, A. Tacchella, and A. Zaccaria. 2023. "Product Progression: A Machine Learning Approach to Forecasting Industrial Upgrading." *Scientific Reports* 13 (1): 1481. https://doi.org/10.1038/s41598-023-28179-x.

Arthur, W. B. 2013. *Complexity and the Economy.* Oxford, UK: Oxford University Press.

Atkin, D., A. Costinot, and M. Fukui. 2025. "Globalization and the Ladder of Development: Pushed to the Top or Held at the Bottom?" *The Review of Economic Studies,* rdaf077. https://doi.org/10.1093/restud/rdaf077.

Aufiero, S., G. De Marzo, A. Sbardella, and A. Zaccaria. 2024. "Mapping Job Fitness and Skill Coherence into Wages: An Economic Complexity Analysis." *Scientific Reports* 14 (1): 11752. https://doi.org/10.1038/s41598-024-61448-x.

Bahar, D., A. Hauptmann, C. Özgüzel, and H. Rapoport. 2024. "Migration and Knowledge Diffusion: The Effect of Returning Refugees on Export Performance in the Former Yugoslavia." *Review of Economics and Statistics,* 1–18. https://doi.org/10.1162/rest_a_01165.

Balland, P.-A., T. Broekel, D. Diodato, E. Giuliani, R. Hausmann, N. O'Clery, and D. Rigby. 2022. "The New Paradigm of Economic Complexity." *Research Policy* 51 (3): 104450. https://doi.org/10.1016/j.respol.2021.104450.

Baqaee, D. R., and E. Farhi. 2019. "The Macroeconomic Impact of Microeconomic Shocks: Beyond Hulten's Theorem." *Econometrica* 87 (4): 1155–1203. https://doi.org/10.3982/ECTA15202.

Baqaee, D. R., and E. Farhi. 2024. "Networks, Barriers, and Trade." *Econometrica* 92 (2): 505–541. https://doi.org/10.3982/ECTA17513.

Barbieri, N., D. Consoli, L. Napolitano, F. Perruchas, E. Pugliese, and A. Sbardella. 2022. "Regional Technological Capabilities and Green Opportunities in Europe." *The Journal of Technology Transfer,* 1–30. https://doi.org/10.1007/s10961-022-09952-y.

Barbieri, N., A. Marzucchi, and U. Rizzo. 2020. "Knowledge Sources and Impacts on Subsequent Inventions: Do Green Technologies Differ from Non-Green Ones?" *Research Policy* 49 (2): 103901. https://doi.org/10.1016/j.respol.2019.103901.

Barbieri, N., F. Perruchas, and D. Consoli. 2020. "Specialization, Diversification, and Environmental Technology Life Cycle." *Economic Geography* 96 (2): 161–186. https://doi.org/10.1080/00130095.2020.1721279.

Barney, J. 1991. "Firm Resources and Sustained Competitive Advantage." *Journal of Management* 17 (1): 99–120. https://doi.org/10.1177/014920639101700108.

Barza, R., E. L. Glaeser, C. A. Hidalgo, and M. Viarengo. 2024. *Cities as Engines of Opportunities: Evidence from Brazil.* Technical report. National Bureau of Economic Research.

Blomström, M., and A. Kokko. 1998. "Multinational Corporations and Spillovers." *Journal of Economic Surveys* 12 (3): 247–277. https://doi.org/10.1111/1467-6419.00056.

Boehm, J., S. Dhingra, and J. Morrow. 2022. "The Comparative Advantage of Firms." *Journal of Political Economy* (Chicago, IL) 130 (12): 3025–3100.

Boschma, R., A. Minondo, and M. Navarro. 2013. "The Emergence of New Industries at the Regional Level in Spain: A Proximity Approach Based on Product Relatedness." *Economic Geography* 89 (1): 29–51. https://doi.org/10.1111/j.1944-8287.2012.01170.x.

Bruno, M., D. Mazzilli, A. Patelli, T. Squartini, and F. Saracco. 2023. "Inferring Comparative Advantage via Entropy Maximization." *Journal of Physics: Complexity* 4 (4): 045011. https://doi.org/10.1088/2632-072X/ad1411.

Bryce, D. J., and S. G. Winter. 2009. "A General Interindustry Relatedness Index." *Management Science* 55 (9): 1570–1585. https://doi.org/10.1287/mnsc.1090.1040.

Caldarola, B., D. Mazzilli, L. Napolitano, A. Patelli, and A. Sbardella. 2024. "Economic Complexity and the Sustainability Transition: A Review of Data, Methods, and Literature." *Journal of Physics: Complexity* 5 (2): 022001. https://doi.org/10.1088/2632-072X/ad4f3d.

Cimini, G., T. Squartini, F. Saracco, D. Garlaschelli, A. Gabrielli, and G. Caldarelli. 2019. "The Statistical Physics of Real-World Networks." *Nature Review Physics* 1:58–71. https://doi.org/10.1038/s42254-018-0002-6.

Cristelli, M., A. Gabrielli, A. Tacchella, G. Caldarelli, and L. Pietronero. 2013. "Measuring the Intangibles: A Metrics for the Economic Complexity of Countries and Products." *PloS One* 8 (8): e70726. https://doi.org/10.1371/journal.pone.0070726.

de Cunzo, F., D. Consoli, F. Perruchas, and A. Sbardella. 2023. "Mapping Critical Raw Materials in Green Technologies." *Papers in Evolutionary Economic Geography* 23.22. http://econ.geo.uu.nl/peeg/peeg2322.pdf.

de Cunzo, F., A. Petri, A. Zaccaria, and A. Sbardella. 2022. "The Trickle Down from Environmental Innovation to Productive Complexity." *Scientific Reports* 12 (1): 22141. https://doi.org/10.1038/s41598-022-25940-6.

Diemer, A., S. Iammarino, R. Perkins, and A. Gros. 2022. "Technology, Resources and Geography in a Paradigm Shift: The Case of Critical and Conflict Materials in ICTs." *Regional Studies* (July): 1–13. https://doi.org/10.1080/00343404.2022.2077326.

Diodato, D., R. Hausmann, and F. Neffke. 2023. "The Impact of Return Migration on Employment and Wages in Mexican Cities." *Journal of Urban Economics* 135:103557. https://doi.org/10.1016/j.jue.2023.103557.

Diodato, D., R. Hausmann, and U. Schetter. 2022. "A Simple Theory of Economic Development at the Extensive Industry Margin." *SSRN Electronic Journal,* https://doi.org/10.2139/ssrn.4227826.

Domínguez-García, V., and M. A. Muñoz. 2015. "Ranking Species in Mutualistic Networks." *Scientific Reports* 5 (1): 8182. https://doi.org/10.1038/srep08182.

Dosi, G., and R. R. Nelson. 1994. "An Introduction to Evolutionary Theories in Economics." *Journal of Evolutionary Economics* 4 (3): 153–172. https://doi.org/10.1007/BF01236366.

Essletzbichler, J. 2017. "Relatedness, Industrial Branching and Technological Cohesion in US Metropolitan Areas." In *Evolutionary Economic Geography,* edited by D. Kogler, 48–62. Milton Park, UK: Routledge.

European Commission. 2023. *Critical Raw Materials Act.* https : / / commission . europa . eu / strategy - and - policy / priorities - 2019 - 2024 / european - green - deal/green-deal-industrial-plan/european-critical-raw-materials-act_en.

Fankhauser, S., A. Bowen, R. Calel, A. Dechezleprêtre, D. Grover, J. Rydge, and M. Sato. 2013. "Who Will Win the Green Race? In Search of Environmental Competitiveness and Innovation." *Global Environmental Change* 23 (5): 902–913. https://doi.org/10.1016/j.gloenvcha.2013.05.007.

Fink, T. M. A., and M. Reeves. 2019. "How Much Can We Influence the Rate of Innovation?" *Science Advances* 5 (1): eaat6107. https : / / doi . org / 10 . 1126 / sciadv.aat6107.

Foellmi, R., and J. Zweimüller. 2008. "Structural Change, Engel's Consumption Cycles and Kaldor's Facts of Economic Growth." *Journal of Monetary Economics* 55 (7): 1317–1328. https : / / doi . org / 10 . 1016 / j . jmoneco . 2008 . 09.001.

Fraccascia, L., I. Giannoccaro, and V. Albino. 2018. "Green Product Development: What Does the Country Product Space Imply?" *Journal of Cleaner Production* 170:1076–1088.

Frank, M. R., E. Moro, T. South, A. Rutherford, A. Pentland, B. Taska, and I. Rahwan. 2024. "Network Constraints on Worker Mobility." *Nature Cities* 1 (1): 94–104. https://doi.org/10.1038/s44284-023-00009-1.

Frenken, K., and F. Neffke. 2026. "Economic Geography and Complexity Theory." In *The Economy as an Evolving Complex System IV,* edited by R. M. del Rio-Chanona, M. Pangallo, J. Bednar, E. D. Beinhocker, J. Kaszowska-Mojsa, F. Lafond, P. Mealy, A. Pichler, and J. D. Farmer. Santa Fe, NM: SFI Press.

Frenken, K., F. Neffke, and A. van Dam. 2023. "Capabilities, Institutions and Regional Economic Development: A Proposed Synthesis." *Cambridge Journal of Regions, Economy and Society* 16 (3): 405–416. https://doi.org/10.1093/cjres/rsad021.

Gathmann, C., and U. Schönberg. 2010. "How General is Human Capital? A Task-Based Approach." *Journal of Labor Economics* 28 (1): 1–49. https://doi.org/10.1086/649786.

Gersbach, H., U. Schetter, and S. Schmassmann. 2023. "From Local to Global: A Theory of Public Basic Research in a Globalized World." *European Economic Review* 160:104530. https://doi.org/10.1016/j.euroecorev.2023.104530.

Grashof, N., and S. Basilico. 2023. "The Dark Side of Green Innovation? Green Transition and Regional Inequality in Europe." *Papers in Evolutionary Economic Geography* 2314. http://econ.geo.uu.nl/peeg/peeg2314.pdf.

Guevara, M. R., D. Hartmann, M. Aristarán, M. Mendoza, and C. A. Hidalgo. 2016. "The Research Space: Using Career Paths to Predict the Evolution of the Research Output of Individuals, Institutions, and Nations." *Scientometrics* 109:1695–1709. https://doi.org/10.1007/s11192-016-2125-9.

Hamwey, R., H. Pacini, and L. Assunção. 2013. "Mapping Green Product Spaces of Nations." *The Journal of Environment & Development* 22 (2): 155–168. https://www.jstor.org/stable/26199441.

Hartmann, D., M. R. Guevara, C. Jara-Figueroa, M. Aristarán, and C. A. Hidalgo. 2017. "Linking Economic Complexity, Institutions, and Income Inequality." *World Development* 93:75–93. https://doi.org/10.1016/j.worlddev.2016.12.020.

Hartmann, D., and F. L. Pinheiro. 2022. *Economic Complexity and Inequality at the National and Regional Level.* arXiv preprint:2206.00818. https://doi.org/10.48550/arXiv.2206.00818.

Hartog, M., A. Gomez-Lievano, R. Hausmann, and F. Neffke. 2024. "Inventing Modern Invention: The Professionalization of Technological Progress in the US." *Papers in Evolutionary Economic Geography* 24.08.

Hausmann, R., and C. A. Hidalgo. 2011. "The Network Structure of Economic Output." *Journal of Economic Growth* 16:309–342. https://doi.org/10.1007/s10887-011-9071-4.

Hausmann, R., C. A. Hidalgo, S. Bustos, M. Coscia, S. Chung, J. Jimenez, A. Simoes, and M. A. Yildirim. 2014. *The Atlas of Economic Complexity: Mapping Paths to Prosperity.* The MIT Press, January. https://doi.org/10.7551/mitpress/9647.001.0001. eprint: https://direct.mit.edu/book-pdf/2269111/book_9780262317719.pdf.

Hausmann, R., J. Hwang, and D. Rodrik. 2007. "What You Export Matters." *Journal of Economic Growth* 12:1–25. https://doi.org/10.1007/s10887-006-9009-4.

Hausmann, R., and B. Klinger. 2006. *Structural Transformation and Patterns of Comparative Advantage in the Product Space.* Working Paper, KSG Working Paper; CID Working Paper No. 128 RWP06-041. Kennedy School of Government. https://ssrn.com/abstract=939646.

Hausmann, R., and L. Nedelkoska. 2018. "Welcome Home in a Crisis: Effects of Return Migration on the Non-Migrants' Wages and Employment." *European Economic Review* 101:101–132. https://doi.org/10.1016/j.euroecorev.2017.10.003.

Hausmann, R., D. Rodrik, and A. Velasco. 2008. "Growth Diagnostics." In *The Washington Consensus Reconsidered: Towards a New Global Governance,* edited by N. Serra and J. E. Stiglitz, 2008:324–355. New York, NY: Oxford University Press. https://doi.org/10.1093/acprof:oso/9780199534081.003.0015.

Hidalgo, C. A. 2023. "The Policy Implications of Economic Complexity." *Research Policy* 52 (9): 104863. https://doi.org/10.1016/j.respol.2023.104863.

Hidalgo, C. A., P.-A. Balland, R. Boschma, M. Delgado, M. Feldman, K. Frenken, E. Glaeser, C. He, D. F. Kogler, A. Morrison, *et al.* 2018. "The Principle of Relatedness." In *Unifying Themes in Complex Systems IX: Proceedings of the Ninth International Conference on Complex Systems 9,* edited by A. J. Morales, C. Gershenson, D. Braha, A. A. Minai, and Y. Bar-Yam, 451–457. Cham, Switzerland: Springer.

Hidalgo, C. A., and R. Hausmann. 2009. "The Building Blocks of Economic Complexity." *Proceedings of the National Academy of Sciences* 106 (26): 10570–10575. https://doi.org/10.1073/pnas.0900943106.

Hidalgo, C. A., B. Klinger, A.-L. Barabási, and R. Hausmann. 2007. "The Product Space Conditions the Development of Nations." *Science* 317 (5837): 482–487. https://doi.org/10.1126/science.1144581.

Hill, M. O. 1974. "Correspondence Analysis: A Neglected Multivariate Method." *Journal of the Royal Statistical Society: Series C (Applied Statistics)* 23 (3): 340–354. https://doi.org/10.2307/2347127.

Hirschman, A. O. 1958. *The Strategy of Economic Development.* New Haven, CT: Yale University Press.

International Energy Agency. 2023. *Energy Technology Perspectives 2023.* Technical report. Paris, France: International Energy Agency. https://www.iea.org/reports/energy-technology-perspectives-2023.

Javorcik, B. S. 2004. "Does Foreign Direct Investment Increase the Productivity of Domestic Firms? In Search of Spillovers Through Backward Linkages." *American Economic Review* 94 (3): 605–627. https://doi.org/10.1257/0002828041464605.

Kowalski, P., and C. Legendre. 2023. "Raw Materials Critical for the Green Transition: Production, International Trade and Export Restrictions." *OECD Trade Policy Papers* 269. https://doi.org/10.1787/c6bb598b-en.

Laudati, D., M. S. Mariani, L. Pietronero, and A. Zaccaria. 2023. "The Different Structure of Economic Ecosystems at the Scales of Companies and Countries." *Journal of Physics: Complexity* 4 (2): 025011. https://doi.org/10.1088/2632-072X/accb35.

Li, Y., and F. M. H. Neffke. 2024. "Evaluating the Principle of Relatedness: Estimation, Drivers and Implications for Policy." *Research Policy* 53 (3): 104952. https://doi.org/10.1016/j.respol.2024.104952.

Lin, J. Y. 2011. "New Structural Economics: A Framework for Rethinking Development and Policy." *The World Bank Research Observer* 26 (2): 193–221. https://doi.org/10.1596/978-0-8213-8955-3.

Lissoni, F., and E. Miguelez. 2024. "Migration and Innovation: Learning from Patent and Investor Data." *Journal of Economic Perspectives* 38 (1): 27–54. https://doi.org/10.1257/jep.38.1.27.

Lucas, R. E., Jr. 1993. "Making a Miracle." *Econometrica* 61 (2): 251–272. https://doi.org/10.2307/2951551.

Mariani, M. S., D. Mazzilli, A. Patelli, D. Sels, and F. Morone. 2024. "Ranking Species in Complex Ecosystems Through Nestedness Maximization." *Communications Physics* 7 (1): 102. https://doi.org/10.1038/s42005-024-01588-8.

Mariani, M. S., Z.-M. Ren, J. Bascompte, and C. J. Tessone. 2019. "Nestedness in Complex Networks: Observation, Emergence, and Implications." *Physics Reports* 813:1–90. https://doi.org/10.1016/j.physrep.2019.04.001.

Mazzilli, D., M. S. Mariani, F. Morone, and A. Patelli. 2024. "Equivalence Between the Fitness-Complexity and the Sinkhorn-Knopp Algorithms." *Journal of Physics: Complexity* 5 (1): 015010. https://doi.org/10.1088/2632-072X/ad2697.

McNerney, J., Y. Li, A. Gomez-Lievano, and F. Neffke. 2021. "Bridging the Short-Term and Long-Term Dynamics of Economic Structural Change." *Nature Communications* 16 (1). https://doi.org/10.1038/s41467-025-65043-0.

Mealy, P., R. M. del Rio-Chanona, and J. D. Farmer. 2018. "What You Do at Work Matters: New Lenses on Labour." *SSRN Electronic Journal*, https://doi.org/10.2139/ssrn.3143064.

Mealy, P., J. D. Farmer, and A. Teytelboym. 2019. "Interpreting Economic Complexity." *Science Advances* 5 (1): eaau1705. https://doi.org/10.1126/sciadv.aau1705.

Mealy, P., and A. Teytelboym. 2022. "Economic Complexity and the Green Economy." *Research Policy* 51 (8): 103948. https://doi.org/10.1016/j.respol.2020.103948.

Montresor, S., and F. Quatraro. 2020. "Green Technologies and Smart Specialisation Strategies: A European Patent-Based Analysis of the Intertwining of Technological Relatedness and Key Enabling Technologies." *Regional Studies* 54 (10): 1354–1365. https://doi.org/10.1080/00343404.2019.1648784.

Napolitano, L., E. Evangelou, E. Pugliese, P. Zeppini, and G. Room. 2018. "Technology Networks: The Autocatalytic Origins of Innovation." *Royal Society Open Science* 5 (6): 172445. https://doi.org/10.1098/rsos.172445.

Neffke, F., and M. Henning. 2013. "Skill Relatedness and Firm Diversification." *Strategic Management Journal* 34 (3): 297–316. https://doi.org/10.1002/smj.2014.

Neffke, F., M. Henning, and R. Boschma. 2011. "How Do Regions Diversify Over Time? Industry Relatedness and the Development of New Growth Paths in Regions." *Economic Geography* 87 (3): 237–265. https://doi.org/10.1111/j.1944-8287.2011.01121.x.

Neffke, F., L. Nedelkoska, and S. Wiederhold. 2024. "Skill Mismatch and the Costs of Job Displacement." *Research Policy* 53 (2): 104933. https://doi.org/10.1016/j.respol.2023.104933.

Nelson, R. R., and S. G. Winter. 1982. *An Evolutionary Theory of Economic Change.* Cambridge, MA: Belknap Press.

Patelli, A., G. Cimini, E. Pugliese, and A. Gabrielli. 2017. "The Scientific Influence of Nations on Global Scientific and Technological Development." *Journal of Informetrics* 11 (4): 1229–1237. https://doi.org/10.1016/j.joi.2017.10.005.

Penrose, E. T. 1959. *The Theory of the Growth of the Firm.* Oxford, UK: Oxford University Press.

Perruchas, F., D. Consoli, and N. Barbieri. 2020. "Specialisation, Diversification and The Ladder of Green Technology Development." *Research Policy* 49 (3): 103922. https://doi.org/10.1016/j.respol.2020.103922.

Prebisch, R. 1962. "The Economic Development of Latin America and its Principal Problems." *Economic Bulletin for Latin America.*

Pugliese, E., G. L. Chiarotti, A. Zaccaria, and L. Pietronero. 2017. "Complex Economies Have a Lateral Escape from the Poverty Trap." *PloS One* 12 (1): e0168540. https://doi.org/10.1371/journal.pone.0168540.

Pugliese, E., G. Cimini, A. Patelli, A. Zaccaria, L. Pietronero, and A. Gabrielli. 2019a. "Unfolding the Innovation System for the Development of Countries: Coevolution of Science, Technology and Production." *Scientific Reports* 9 (1): 16440. https://doi.org//10.1038/s41598-019-52767-5.

Pugliese, E., L. Napolitano, M. Chinazzi, and G. Chiarotti. 2019b. *The Emergence of Innovation Complexity at Different Geographical and Technological Scales.* arXiv preprint:1909.05604. https://doi.org/10.48550/arXiv.1909.05604.

Pugliese, E., and A. Tacchella. 2020. *Economic Complexity for Competitiveness and Innovation: A Novel Bottom-Up Strategy Linking Global and Regional Capacities.* Technical report. Joint Research Centre (Seville site). https://publications.jrc.ec.europa.eu/repository/handle/JRC122086.

———. 2021. "Economic Complexity Analytics: Country Factsheets." *Publications Office of the European Union* 10 (KJ-NA-30711-EN-N): 368138. https://doi.org/10.2760/368138.

Rodríguez-Pose, A., F. Bartalucci, N. Lozano-Gracia, and M. Dávalos. 2024. "Overcoming Left-Behindedness: Moving beyond the Efficiency versus Equity Debate in Territorial Development." *Regional Science Policy & Practice* 16 (12): 100144. https://doi.org/10.1016/j.rspp.2024.100144.

Rughi, T., J. Staccioli, and M. E. Virgillito. 2023. "Climate Change and Labour-Saving Technologies: the Twin Transition via Patent Texts." *SSRN Electronic Journal,* https://doi.org/10.2139/ssrn.4407851.

Saracco, F., R. Di Clemente, A. Gabrielli, and T. Squartini. 2015. "Randomizing Bipartite Networks: The Case of the World Trade Web." *Scientific Reports* 5:10595. https://doi.org/10.1038/srep10595.

Saracco, F., M. J. Straka, R. Di Clemente, A. Gabrielli, G. Caldarelli, and T. Squartini. 2017. "Inferring Monopartite Projections of Bipartite Networks: An Entropy-Based Approach." *New Journal of Physics* 19:053022. https://doi.org/10.1088/1367-2630/aa6b38.

Sbardella, A., N. Barbieri, D. Consoli, L. Napolitano, F. Perruchas, and E. Pugliese. 2022. *The Regional Green Potential of the European Innovation System.* Technical report JRC124696. Joint Research Centre (Seville site). https://publications.jrc.ec.europa.eu/repository/handle/JRC124696.

Sbardella, A., F. Perruchas, L. Napolitano, N. Barbieri, and D. Consoli. 2018a. "Green Technology Fitness." *Entropy* 20 (10): 776. https://doi.org/10.3390/e20100776.

Sbardella, A., E. Pugliese, and L. Pietronero. 2017. "Economic Development and Wage Inequality: A Complex System Analysis." *PloS one* 12 (9): e0182774. https://doi.org/10.1371/journal.pone.0182774.

Sbardella, A., E. Pugliese, A. Zaccaria, and P. Scaramozzino. 2018b. "The Role of Complex Analysis in Modelling Economic Growth." *Entropy* 20 (11): 883. https://doi.org/10.3390/e20110883.

Schetter, U. 2022. "A Measure of Countries' Distance to Frontier Based on Comparative Advantage." *CID Research Fellow and Graduate Student Working Paper,* no. 135, https://doi.org/10.2139/ssrn.4227848.

———. 2024. "Quality Differentiation, Comparative Advantage, and International Specialization Across Products." *European Economic Review* 170:104869. https://doi.org/10.1016/j.euroecorev.2024.104869.

Schetter, U., D. Diodato, E. Protzer, F. Neffke, and R. Hausmann. 2024. "From Products to Capabilities: Constructing a Genotypic Product Space." *CEPR Discussion Paper No. 19369,* https://cepr.org/publications/dp19369.

Sciarra, C., G. Chiarotti, L. Ridolfi, and F. Laio. 2020. "Reconciling Contrasting Views on Economic Complexity." *Nature Communications* 11 (1): 3352. https://doi.org/10.1038/s41467-020-16992-1.

Servedio, V. D. P., P. Buttà, D. Mazzilli, A. Tacchella, and L. Pietronero. 2018. "A New and Stable Estimation Method of Country Economic Fitness and Product Complexity." *Entropy* 20 (10): 783. https://doi.org/10.3390/e20100783.

Sutton, J., and D. Trefler. 2016. "Capabilities, Wealth, and Trade." *Journal of Political Economy* 124 (3): 826–878. https://doi.org/10.1086/686034.

Tacchella, A., M. Cristelli, G. Caldarelli, A. Gabrielli, and L. Pietronero. 2012. "A New Metrics for Countries' Fitness and Products' Complexity." *Scientific Reports* 2 (1): 723. https://doi.org/10.1038/srep00723.

Tacchella, A., D. Mazzilli, and L. Pietronero. 2018. "A Dynamical Systems Approach to Gross Domestic Product Forecasting." *Nature Physics* 14 (8): 861–865. https://doi.org/10.1038/s41567-018-0204-y.

Tacchella, A., A. Zaccaria, M. Miccheli, and L. Pietronero. 2023. "Relatedness in the Era of Machine Learning." *Chaos, Solitons & Fractals* 176:114071. https://doi.org/10.1016/j.chaos.2023.114071.

Teece, D. J., R. Rumelt, G. Dosi, and S. Winter. 1994. "Understanding Corporate Coherence: Theory and Evidence." *Journal of Economic Behavior & Organization* 23 (1): 1–30. https://doi.org/10.1016/0167-2681(94)90094-9.

Tian, L. 2021. *Division of Labor and Productivity Advantage of Cities: Theory and Evidence from Brazil.* Technical report. CEPR Discussion Paper No. DP16590. Centre for Economic Policy Research. https://ssrn.com/abstract=3960170.

van Dam, A., M. Dekker, I. Morales-Castilla, M. Á. Rodríguez, D. Wichmann, and M. Baudena. 2021. "Correspondence Analysis, Spectral Clustering and Graph Embedding: Applications to Ecology and Economic Complexity." *Scientific Reports* 11 (1): 8926. https://doi.org/10.1038/s41598-021-87971-9.

van Dam, A., and K. Frenken. 2022. "Variety, Complexity, and Economic Development." *Research Policy* 51 (8): 103949. https://doi.org/10.1016/j.respol.2020.103949.

Vona, F., G. Marin, and D. Consoli. 2019. "Measures, Drivers and Effects of Green Employment: Evidence from US Local Labor Markets, 2006–2014." *Journal of Economic Geography* 19 (5): 1021–1048. https://doi.org/10.1093/jeg/lby038.

Yildirim, M. 2021. "Sorting, Matching and Economic Complexity." *CID Faculty Working Paper Series* 392. https://nrs.harvard.edu/URN-3:HUL.INSTREPOS:37369328.

Yildirim, M. A., and M. Coscia. 2014. "Using Random Walks to Generate Associations Between Objects." *PloS One* 9 (8): e104813. https://doi.org/10.1371/journal.pone.0104813.

Zhang, Y., and D. L. Rigby. 2022. "Do Capabilities Reside in Firms or in Regions? Analysis of Related Diversification in Chinese Knowledge Production." *Economic Geography* 98 (1): 1–24. https://doi.org/10.1080/00130095.2021.1977115.

Zhu, S., C. He, and Y. Zhou. 2017. "How to Jump Further and Catch Up? Path-Breaking in an Uneven Industry Space." *Journal of Economic Geography* 17 (3): 521–545. https://doi.org/10.1093/jeg/lbw047.

BACK TO THE FUTURE: AGENT-BASED MODELING AND DYNAMIC MICROSIMULATION

Matteo Richiardi, University of Essex; and
Justin van de Ven, University of Essex

Abstract

We look at the commonalities and differences between agent-based and dynamic microsimulation analytical approaches. Starting from a shared history, we discuss how the two literatures quickly diverged. Our discussion concludes with evidence of some recent convergence between agent-based and dynamic microsimulation methods and emerging opportunities for mutual reinforcement of the two methodologies.

Introduction

Agent-based models (ABMs) have become standard tools for research in complexity economics. At the same time, increasing availability of survey microdata and computational power has prompted interest in the related field of dynamic microsimulation modeling (DMM), particularly in relation to economics, healthcare, urban planning, and environmental studies.[1] Nevertheless, practitioners of both fields are often only vaguely aware of related work conducted in the other field. Furthermore, when work beyond a practitioner's specific field is discussed, it is sometimes regarded with interest, sometimes

---

[1] We use the initialism ABM for agent-based modeling and ABMs for agent-based models. Similarly, DMM refers to the methodology (dynamic microsimulation modelling) and DMMs to the applications (dynamic microsimulation models).

with suspicion, often with a focus on perceived methodological deficiencies. In this chapter, we focus on the relationship between agent-based and microsimulation approaches, point to the foundational dynamic microsimulation work of the 1960s as a precursor to agent-based models, and suggest that mutual knowledge and convergence have the potential to strengthen the complexity-economics approach. More specifically, we advance five claims:

Claim 1: Mathematically, the two methodologies are the same.

Claim 2: Historically, they have evolved separately, even though some of the first examples of agent-based models in the 1960s and 1970s, now largely forgotten, were developed as "microsimulations."

Claim 3: The microsimulation literature offers important insights for advancing agent-based studies, particularly in relation to:

  i. introducing greater realism through empirical descriptions of agent-specific heterogeneity founded on survey micro-data;

  ii. stylized approaches for obtaining realistic descriptions of the effects of complex real-world systems, including (e.g.) tax and benefit entitlements;

  iii. approaches to improve alignment between models and survey observations; and

  iv. consideration of statistical uncertainty associated with model projections.

Claim 4: The agent-based literature, in turn, offers important insights for advancing microsimulation studies, particularly in relation to:

i. agent interactions and decision-making; and

ii. macro to micro interactions.

Claim 5: The time is ripe for a convergence between agent-based and microsimulation literatures, with some novel developments such as data-assimilation techniques and integration of multiple data sources into realistic synthetic populations being at the forefront of research in both communities.

The second section presents a stylized description of ABMs and DMMs, forming the basis for claim 1. The third section investigates the historical development of the two methodologies, leading to claim 2. The fourth section discusses ongoing trends that outline a convergence between ABMs and DMMs and elaborates on the opportunities that this convergence opens (claims 3–5). The fifth section concludes.

Modeling Approaches

Readers might be more familiar with ABMs than DMMs. As is well known, ABMs are composed of many entities, or agents, acting and interacting with each other and the environment (Axtell and Farmer 2025). DMMs are very similar.[2] The field of microsimulation originates from work conducted by Guy Orcutt in the late 1950s (Orcutt 1957, 1960, 1962). The 1950s

---

[2]For reviews of dynamic microsimulation models, see O'Donoghue (2001), Li and O'Donoghue (2013), O'Donoghue (2014), and O'Donoghue and Dekkers (2018). A burgeoning area of application is health modeling (Schofield *et al.* 2018). Dynamic microsimulation models study changes in populations of agents over time. By contrast, static models (e.g., tax-benefit models) look at the short- to medium-term effects of policy changes—or more generally changes in the environment—and are of less interest here.

saw intense research interest in the development of econometric models of the macroeconomy, following Jan Tinbergen's foundational work (Solow 2004). Orcutt was motivated by the observation that nonlinearities in individual behavior and the effects of policy on microunits complicate macroeconomic projections framed on "representative" microagents; in this case valid macroeconomic forecasts require taking into account the distribution of microagents.

Orcutt's revolutionary contribution was his advocacy for a new type of modeling approach that uses as inputs representative distributions of individuals, households, or firms and puts emphasis on their heterogeneous decision-making, as observed in the real world. In so doing, not only can economy-wide averages be correctly computed, but their entire distribution can be analyzed.

In Orcutt's words, "this new type of model consists of various sorts of interacting units which receive inputs and generate outputs. The outputs of each unit are, in part, functionally related to prior events and, in part, the result of a series of random drawings from discrete probability distributions" (Orcutt *et al.* 1961). Anders Klevmarken (2022) puts it in a way that will be familiar to any agent-based modeler: "In micro simulation modeling there is no need to make assumptions about the average economic man. Although unpractical, we can in principle model every man."

Two key methodological stances are common to ABMs and DMMs: a focus on the role of individual specific heterogeneity underlying dynamic projections, and use of recursion in the computational approach to model solution. Other features of ABMs and DMMs generally emphasized in the literature, as a focus on interaction for ABMs and on policies for DMMs, are to be understood more as distinctive flavors, not altering the underlying common analytical architecture.

Building on Jakob Grazzini and Matteo Richiardi (2015), we can formalize ABMs as composed of many entities, or agents, acting and interacting with each other and the environment. Agents can be of different types $j = 1, ..., J$ (e.g., workers, firms, banks), where each agent i at time t is described by a set of state variables $x_{i,t}^j$. Then the evolution over (discrete) time of the state variables of agent i can be described by the law of motion:

~227~

$$x_{i,t+1}^j = f_j(x_{i,t}^j, X_{-i,t}^j, X_t^{-j}, \xi_{i,t}^j, \theta_t, P_t) \qquad (1)$$

where $X_{-i,t}^j$ refers to the state of agents other than i of the type j at time t, X_t^{-j} refers to the state of all agents of types other than j, ξ are random draws that represent innovations beyond the explicit scope of the model, θ are (possibly time-variant) behavioral parameters, and P are (possibly time-variant) environmental parameters, including public policies. The phase line f_j can be time-dependent, although this is rarely the case. The same structure for the data generating process holds for DMMs. The emphasis, however, is different. In DMMs, there are generally fewer agent types (j), and their interaction ($X_{-i,t}^j$, X_t^{-j}) is more limited than in an agent-based setting.[3] Furthermore, the number of parameters governing agents' behavior (θ) is often larger in DMMs, and policies (P) are spelled out in greater detail. The initial configuration of states (D_0^j, D_0^{-j}) are also often synthetically generated in ABMs, whereas they are typically drawn from survey data for DMMs.

---

[3] For instance, in Patryk Bronka et al.'s (2025) state-of-the-art SimPaths modeling framework, individuals form partnerships based on assortative mating (i.e., the probability of forming a partnership depends on the joint distribution of the characteristics of each pair of possible partners). Once partnered, an individual's trajectories depend on the partner's characteristics, and couples make joint decisions over some domains (e.g., labor supply), alongside individual decisions of the partners over other domains.

Another frequently cited distinction between ABMs and DMMs is that the ABMs typically feature two-way interactions between the macro environment and micro agents, while DMMs focus more on the micro-to-macro direction of causality.[4] The micro–macro interplay defines the property of *reflexivity*, where individual actions change the environment in which the same individuals operate. Complex reflexive systems lie at the far end of the complexity spectrum (Beinhocker 2013), but they are indeed common when it comes to social interaction. However, not all ABMs target complex reflexive systems: In many cases, they are motivated "simply" by a desire to understand the aggregate implications of given micro behavior, or to uncover what micro behavior can possibly generate given macro phenomena of interest. An example is crowd dynamics, where the environment in which agents operate remains the same.[5] Similarly, while many DMMs do not exhibit reflexivity, some do—see the next section. There is nothing *a priori* that limits ABMs to complex reflexive systems and DMMs to noncomplex, nonreflexive ones. Drawing a distinction between the two approaches based on the complexity of the system under investigation is akin to confusing the proverbial hammer with the nail. Another false dichotomy refers to the partial versus complete nature of the models. For example, a dynamic microsimulation might include detailed household behavior but abstract from the production side of the economy. In this context, the model might project the supply of labor and demand for consumption goods, treating wages and prices as exogenous.[6] This focus on subsystems is commonly referred to as a

---

[4]Equation 1 allows the evolution of each agent to depend on the state of all other agents, hence on the aggregate state of the system.

[5]See, for instance, Makinoshima and Oishi (2022).

[6]In DMMs of households, wages are typically related to individual characteristics such as age and education; they can also follow some exogenously given macro trend (for instance, related to productivity growth).

partial equilibrium in the economics literature. In contrast, many ABMs are specifically designed to explore the dynamic feedback between different subsystems; in the above example this would imply accounting for the dynamic interactions between agents in the domestic and production sectors that underly market prices. The economics literature commonly refers to this type of analysis as a *general equilibrium*; given the focus of ABMs on nonequilibrium dynamics (see chs. 4 and 5), we prefer the term *system closure*.[7]

A common source of confusion is the term *nonequilibrium* itself, which in agent-based modeling refers to temporal dynamics of the evolving economic environment (outside a fixed-point solution), rather than the absence of market-clearing regularities at any point in time. This is ultimately related to the nature of ABMs as dynamic systems, something that equation (1) is designed to reflect. In microsimulation, there is less interest in the long-run properties of models, given the strict mapping between the initial conditions of the system and the populations of interest: The temporal horizon is usually restricted to the time span of interest for policy purposes. This may be taken to reflect the empirical, policy-driven orientation of DMMs, relative to the more theoretical orientation of ABMs. The differences above have implications in terms of the estimation strategy. DMMs are typically estimated piecewise on microdata, one component of $x^{j}_{i,t+1}$ at a time, controlling for the lagged values of the other state variables of each individual agent. By contrast, ABMs are often calibrated using various sources of micro- and macrodata, as in Sebastian Poledna

---

The wage *structure*, however—that is, the premium that each individual characteristic commands—remains constant.

[7] A systematic application of the system closure approach requires *stock-flow* consistency, the requirement that all the monetary and physical flows in a model are properly accounted for by both originators and recipients (see, for instance, Caiani *et al.* 2016).

et al. (2023), or estimated via indirect inference on macrodata (e.g., Platt 2020; Dyer *et al.* 2022).[8] The difference in methods employed for estimation, however, reflects not a fundamental discriminant between the two approaches but rather specific constraints coming from the data, given specific applications. In particular, recourse to indirect inference is necessary when some of the variables are not observable. Typically, in DMM all the state variables characterizing individuals have a counterpart in survey data. Hence, standard estimation techniques can be used. In contrast, ABMs often include details that cannot be observed, such as the topology of interaction networks. The associated parameters must then be deduced by means of indirect inference. However, when datasets are available with the relevant information, they are used for direct parameterization.[9]

Historical Developments

A classic example of the ABM literature is Thomas Schelling's (1971) segregation model. This toy model has two populations of agents distinguished by one characteristic: ethnicity. Individuals are characterized by two interdependent state variables: their location and their ethnicity. Individuals possess one behavioral parameter: their tolerance for living as an ethnic minority in their neighborhood. Individuals make one choice each period: to stay or change location. This highly stylized model, characterized by limited heterogeneity and interaction brought about only by a

---

[8] Recent work employing a transformer-based, machine-learning architecture to predict sequences of life events (Savcisens *et al.* 2023) is promising but seems to rely on "almost ideal" large-scale administrative data.

[9] An example is Pangallo *et al.* (2024), where they use census, survey, and mobility data to parameterize a granular model of the interplay between economic and health factors during the COVID-19 pandemic in the New York metropolitan area. van de Ven (2017) is, on the other hand, an instance of a microsimulation model partially parameterized by indirect inference (method of simulated moments).

shared space, allows for interesting dynamics that lead to high levels of ethnic segregation, even when the tolerance for living in a minority is relatively high.[10]

It is interesting to contrast the Schelling segregation model with one of the longest-running and actively used DMMs: MOSART (Andreassen *et al*. 2020). MOSART is a life-course model based on administrative data for the entire Norwegian population, which projects birth, death, migration, marriage, divorce, educational activities, labor-force participation, retirement, income, and wealth based on estimated transition probabilities. The model is used by Statistics Norway and the Norwegian government for projections and policy analyses related to the pension system. MOSART features heterogeneity across a wide range of dimensions but allows for limited interaction between agents and features no emergent properties.[11]

While the pairwise comparison presented above captures the essence of traditional differences between the ABM and DMM literatures, such generalities break down with respect to the scope of variation described by the contemporary literature. For example, the ABM literature includes increasing examples of large-scale, data-driven models, as in Adrian Carro *et al*.'s (2023) model of the UK housing market and Poledna *et al*.'s (2023) model of the Austrian economy. These models include diverse behavioral and policy parameters and are extensively calibrated and validated.

A cursory observer might now conclude that ABMs have expanded to *encompass* DMMs, presenting comparable functionality

---

[10]"At home . . . I made a 16 × 16 checkerboard, located zincs and coppers at random with about a fifth of the spaces blank, got my twelve-year-old to sit across from me at the coffee table, and moved discontented zincs and coppers to where their demands for like or unlike neighbors were met. The dynamics were sufficiently intriguing to keep my twelve-year-old engaged" (Schelling 2006).

[11]*Emergent properties* refer to phenomena that result from interactions between agents.

(rich heterogeneity, realistic behaviors, and policy details), with the addition of coherent assumptions concerning system closure. However, what is perhaps less well known is that early examples from the DMM literature feature the same emphasis on empirically founded interactions between agents and agent types underlying macrophenomena as the most recent macroeconomic ABMs. Two striking examples in this regard are Barbara Bergmann's US Transactions Model (Bergmann 1974; Bennett and Bergmann 1986) and Gunnar Eliasson's MOSES (Model of the Swedish Economic System) (Eliasson 1976, 1977).

In Bergmann's model, workers, firms,[12] banks, financial intermediaries, the government, and the central bank interact to determine aggregate and distributional outcomes. The model is simulated on a weekly basis. Each period firms make production plans based on past sales and inventory position; firms attempt to adjust the size of their workforce; wages are set and the government adjusts public employment; production occurs; firms adjust prices; firms compute profits, pay taxes, and buy inputs for the next period; workers receive wages, government transfers, and property income; workers pay taxes and make payments on outstanding loans; households decide how much to consume and save, choosing among different consumption goods and adjust their portfolios of assets; firms invest; the government purchases public procurement from firms; firms make decisions on seeking outside financing; the government issues public debt; banks and financial intermediaries buy or sell private and public bonds; the monetary authority buys or sells government bonds; and interest rates are set. In the labor market, firms offer jobs to particular workers, some of which are accepted; some vacancies remain unfilled, with the vacancy rate affecting the wage-setting mechanism. The model could easily be described as an agent-based

---

[12]Distinguished in 12 sectors.

macro model *ante litteram*. However, Bergmann consistently referred to her work as a "microsimulation."

Relatedly, Bergmann also developed "toy" simulation models that she used to investigate specific theoretical mechanisms of interest. These models are broadly comparable to Schelling's, and when considered as such represent early examples of agent-based analyses that appear in high-grade economic journals (e.g., Bergmann 1990). Importantly, Bergmann considered her article as providing "an introduction to microsimulation."

Meanwhile, Eliasson's MOSES (Eliasson 1976, 1977) was rooted in the Wicksellian/Stockholm School of *ex ante* plans and *ex post* outcomes (Jonung 1991), an intellectual tradition that agent-based modelers and the complexity-economics literature should perhaps rediscover. MOSES featured a one-to-one mapping between the universe of Swedish firms and their simulated counterparts, endowed with real balance sheets (some synthetic firms were also introduced with calibrated balance sheets to match sector totals). In the model, operating on a quarterly basis, firms make production plans, invest, hire workers in advance, and set prices. They then review their choices based on realized profits, leading to endogenous growth dynamics. A monetary system, later integrated with a stock and financial derivatives market, and a rudimentary venture capital market completed the description of the economy.

Echoing Bergmann, Eliasson described his model as a "microsimulation," although in later work he referred to it as an "agent-based macroeconomic model" (Eliasson 2018). Indeed, Bergmann and Eliasson played an important role, together with Orcutt, in developing the microsimulation approach to economics. In the introduction to a jointly edited volume, they discuss the role of microsimulation in "integrating theory and measurement": "Aspects of a theory that do not pass the test against observation are not allowed to survive in a true scientific environment. In a complex world one should consider it natural to live with many conflicting interpretations of economic reality; but in a world of scientific progress the interpretations should change, old erroneous doctrine should be unloaded and new theory allowed to enter. . . . We believe that micro simulation opens up new possibilities for estimation and analysis based on direct access to the wealth of data that exists at the micro level" (Bergmann, Eliasson, and Orcutt 1980, 11). This same research program resonates well with many modern macro ABMs.

The above discussion suggests that microsimulation should be considered not only a closely related field to agent-based modeling but one of its precursors—arguably its oldest—alongside evolutionary economics (Nelson and Winter 1982) and complexity economics (Anderson, Arrow, and Pines 1988).[13]

Notwithstanding their common roots, the DMM and ABM literatures have evolved along different trajectories. In this regard, it is of note that the Bergmann, Eliasson, and Orcutt (1980) volume cited above discusses three state-of-the-art DMMs of the time: Bergmann's US Transactions Model, Eliasson's MOSES model, and Orcutt's Urban Institute–Yale model. In contrast to the extensive agent interactions of the Bergmann and Eliasson models, Orcutt's model focuses primarily on behavior in the household sector, including an auxiliary module to capture macro trends that might affect households' decisions. History reveals that it is Orcutt's model that won the day, inspiring the majority of subsequent developmental work conducted within the DMM literature.[14]

With hindsight, Bergmann's and Eliasson's models were

---

[13] See Richiardi (2018) for a discussion.
[14] Orcutt's model evolved into DYNASIM (Orcutt, Caldwell, and Wertheimer 1976), which remains under development at the Urban Institute (Favreault, Smith, and Johnson 2015).

ahead of their time. It is notable that it took Bergmann more than fifteen years to complete her model, and Eliasson managed to complete his thanks to support from IBM, which offered "unlimited programming and computer support for two years" (Eliasson 2018, 11). Unsurprisingly, these examples were not considered a viable research avenue by the contemporary scientific community. Simplification of the Bergmann–Eliasson examples was necessary in context of prevailing data, developmental, and computational limitations. In the case of the evolving DMM literature, empirical relevance was prioritized at the expense of the diversity of considered agent types and their interactions. Meanwhile, the ABM literature resolved the tradeoff in the opposite direction, slowly adding realistic features to the first, simple cellular automata models (e.g., John Conway's Game of Life). Differences between the DMM and ABM literatures that now exist can consequently be understood as the product of differences in researcher priorities, given the prevailing analytical tradeoffs. In this regard it is notable that recent advances in computer power have supported the emergence of data-driven, empirically calibrated ABMs (Dawid and Delli Gatti 2018).[15] This same increase in computational power may support a shift in DMMs back toward their origin. Seen from this perspective, Bergmann's and Eliasson's models can be understood as being illustrative of an approach, thereby providing a glimpse of a brighter modeling future that will take fifty years to develop. The reconvergence of ABM and DMM literatures presents exciting research opportunities, to which we now turn.

---

[15]Interestingly, the ASPEN model of the US economy (Basu, Pryor, and Quint 1998), heavily built on the Orcutt–Bergmann experience, was branded an "agent-based microeconomic simulation model," a sign that agent-based modeling was replacing microsimulation as the new game in town.

Convergence

As discussed previously, there is evidence of increasing realism in the empirical descriptions of agent-specific heterogeneity in recent ABM literature, which tends to reduce the disparity with DMMs. This shift involves estimation of transitional relationships that are well understood in the existing econometric literature, facilitated by publicly available data sources and widely available statistical software. There are, nevertheless, several ways in which the development of associated ABMs could benefit by drawing on the contemporary DMM literature. The developmental overhead associated with implementing increasing statistical detail in an ABM can be mitigated by one of the generic software packages that have been developed within the DMM literature. Several now exist (OpenM++, JAS-mine, AnyLogic, LIAM2, Modgen, GENESIS), each of which generally includes libraries that support common model-building tasks.[16]

Another approach is to integrate results from specialist model structures. Consider, for example, attempts to obtain a realistic reflection of tax and benefit payments, which may reasonably be posited to influence labor decisions. Most advanced countries have tax and benefit systems that are nontrivially complex and subject to constant revision. Capturing a realistic reflection of this dynamic complexity represents a substantial modeling challenge. It is consequently notable that there are well-maintained static tax-benefit calculators that focus exclusively on encoding the variation of policy through time (e.g., TaxBEN, developed and maintained by the Organisation for Economic Cooperation and Development, OECD, and EUROMOD, originally developed

---

[16]OpenM++: https://openmpp.org; JAS-mine: Richiardi and Richardson (2017); AnyLogic: https://www.anylogic.com; LIAM2: de Menten *et al.* (2014); Modgen: https://www.statcan.gc.ca/en/microsimulation/modgen/modgen; GENESIS: Gillman (2017).

at the University of Essex and currently maintained by the Joint Research Council).[17] Methods have been developed in the DMM literature to integrate results generated by these purpose-specific applications into more general DMMs/ABMs (e.g., van de Ven, Bronka, and Richiardi 2025).

Integrating increased realism into a dynamic microsimulation by adding new agent-specific heterogeneity introduces a range of associated issues of concern. Many of these have been addressed in the DMM literature, presenting off-the-shelf solutions for related development in ABMs. Consider, for example, the problem of empirically estimating model parameters. Ideally, all model parameters would be estimated endogenous to a model's structure. In the case of microsimulation models, however, this is often not computationally feasible. Hence, the equations governing temporal dynamics of individual specific characteristics are often estimated on external survey data, exogenous to a model's structure. Although this approach mitigates the computational burden of parameter estimation, it increases the likelihood of model mis-specification. Hence, even when multiple processes are estimated using a common data source, their combination within a DMM or ABM context can result in unrealistic projections.

Alignment methods have been devised within the DMM literature to mitigate model mis-specification attributable to exogenous estimation of parameters. These methods are typically divided into two broad categories, distinguishing methods that seek to (implicitly or explicitly) adjust model parameters from those that seek to adjust *ex post* simulation projections given model parameters (e.g., Baekgaard 2002; Li and O'Donoghue 2014). An example of the former approach

~237~

---

is logit scaling, which seeks to identify adjusted probabilities that match a model to external targets (e.g., Stephensen 2016). An example of the latter approach is resampling, where the set of random draws used to project uncertain outcomes through time are redrawn until a target is obtained. Most of the generic software packages referred to above also facilitate implementation of alignment methods (e.g., Richiardi and Richardson 2017, sec. 4.4).

Another issue associated with attempts to reflect greater statistical detail in ABMs concerns accompanying measures of uncertainty in projections. Coming from a more theoretical interest, ABMs have traditionally emphasized exploration of the possible behaviors of the models under alternative parameterizations, by means of sensitivity analysis of model outputs to model parameters (e.g., Lee *et al.* 2015). Empirically relevant ABMs, on the other hand, share with DMMs a focus on narrower parameterizations—those supported by the data. Uncertainty regarding a model's projections arise for a variety of reasons (Bilcke *et al.* 2011; Creedy, Kalb, and Kew 2007), including the representativeness of input data, a model's structure and parameterization, and the random draws underlying a dynamic projection. Some of these are difficult or impossible to quantify (e.g., model structure), while others might reasonably be abstracted from (e.g., input data). Uncertainty stemming from model parameters and simulated random events, however, qualifies for neither of these potential exemptions. State-of-the-art methods for uncertainty quantification in DMMs include bootstrapping of model parameters based on estimated variance–covariance matrixes (e.g., Bronka *et al.* 2025).

There are also methods and practices from the ABM literature that can be usefully imported into DMM, as models incorporate more agent types and increased interaction. In particular, use of simple heuristics in the decision-making of at least some of the agents will help keep models manageable. Expectation formation is another area where DMM can learn from ABM, either in a repeated-decision context (Mu, Zheng, and Trott 2022) or in a social-learning environment (Nowak, Matthews, and Parker 2017).

Recent developments highlight areas of common interest. Data-assimilation techniques, initially developed in meteorology and the earth sciences, have started to be used in ABM (e.g., Ward, Evans, and Malleson 2016). Data assimilation aims at a reparameterization of the models as new data becomes available and provides a more general framework for "learning" from data than alignment. In particular, alignment emphasizes hitting the (macro) targets more or less exactly, while data assimilation weights the new information against previous information. Data assimilation is more relevant for high-frequency applications—such as finance or traffic models—but the increasing availability of big data updated almost in real time will spread its use to other areas where DMM has been more active, such as household behavior.

~239~

The use of synthetic population is another area of convergence. Because ABMs evolved mostly with a theoretical rather than an applied interest, they tended to be based on artificial and rather abstract simulated populations (as in the Schelling segregation model). However, an increased attention to empirical relevance means that the synthetic populations of ABMs are becoming more sophisticated (e.g., Dyer *et al.* 2024). By contrast, as discussed above, DMMs usually build their simulated populations from survey data. However, as DMMs grow in scope, they need to integrate information from multiple sources (such as wealth and asset holdings, and consumption or mobility patterns, all not generally available in household surveys) and disaggregate information at more fine-grained geographical levels. This is done through probabilistic (regression-based or hot-deck) imputation and use of spatial microsimulation techniques (e.g., Wu *et al.* 2022). The usefulness of building simulations on synthetic populations is also connected to the ability to share the data when privacy concerns can be overcome. This is increasingly relevant in a scientific environment emphasizing open-source development, which is common to ABM and DMM.

Realism in model structure and parameterization also calls for validation of model outputs against real data, a critical issue for both ABMs and DMMs. Approaches to model validation are quite well-established in the literature (see, e.g., National Research Council 2012, and Alarid-Escudero, Gulati, and Rutter 2020), although no universal prescription for determining when a model passes the validity test exists.

Conclusions

The real world is a complex system. But modeling complex systems does not necessarily produce realistic results. Although a few visionaries had imagined microsimulation models that are at the same time complex and realistic and built working prototypes at a time when PCs did not even exist, until recently computational and data limitations have precluded a wide adoption of this research program. Since the 1970s, ABMs and DMMs have taken the challenge of modeling a complex world from two different perspectives: ABMs emphasizing complexity, to the detriment of realism, and DMMs emphasizing realism, to the detriment of complexity. Advances in computational power and a wider availability of good-quality microdata has changed this landscape, prompting ABMs to increase in realism and DMMs to increase in complexity. A convergence in the two approaches is therefore under way, with the potential to finally realize Bergmann and Eliasson's vision. Rarely has the phrase "back to the future" been more apt to describe the emergence of a new research paradigm. 🖝

REFERENCES

Alarid-Escudero, F., R. Gulati, and C. M. Rutter. 2020. "Validation of Microsimulation Models Used for Population Health Policy." In *Complex Systems and Population Health,* edited by Y. Apostolopoulos, K. Hassmiller Lich, and M. K. Lemke. Oxford, UK: Oxford University Press.

Anderson, P. W., K. Arrow, and D. Pines. 1988. *The Economy As An Evolving Complex System.* 1st ed. Boca Raton, FL: CRC Press.

Andreassen, L., D. Fredriksen, H. M. Gjefsen, E. Halvorsen, and N. M. Stølen. 2020. "The Dynamic Cross-Sectional Microsimulation Model MOSART." *International Journal of Microsimulation* 13 (1): 92–113. https://doi.org/10.34196/IJM.00214.

Axtell, R., and J. D. Farmer. 2025. "Agent-Based Modeling in Economics and Finance: Past, Present, and Future." *Journal of Economic Literature* 63 (1): 197–287. https://doi.org/10.1257/jel.20221319.

Baekgaard, H. 2002. *Micro–Macro Linkage and the Alignment of Transition Processes: Some Issues, Techniques and Examples.* Technical paper 25. National Centre for Social and Economic Modelling (NATSEM).

Basu, N., R. Pryor, and T. Quint. 1998. "ASPEN: A Microsimulation Model of the Economy." *Computational Economics* 12:223–241. https://doi.org/10.1023/A:1008691115079.

Beinhocker, E. D. 2013. "Reflexivity, Complexity, and the Nature of Social Science." *Journal of Economic Methodology* 20 (4): 330–342. https://doi.org/10.1080/1350178X.2013.859403.

Bennett, R. L., and B. R. Bergmann. 1986. *A Microsimulated Transactions Model of the United States Economy.* Baltimore, MD: John Hopkins University Press.

Bergmann, B. R. 1974. "A Microsimulation of the Macroeconomy with Explicitly Represented Money Flows." In *Annals of Economic and Social Measurement*, 3:475–489. 3. National Bureau of Economic Research. http://www.nber.org/chapters/c10173.

———. 1990. "Micro-to-Macro Simulation: A Primer With a Labor Market Example." *The Journal of Economic Perspectives* 4 (1): 99–116. https://doi.org/10.1257/jep.4.1.99.

Bergmann, B. R., G. Eliasson, and G. Orcutt, eds. 1980. *Micro Simulation – Models, Methods and Applications. Proceedings of a Symposium in Stockholm, Sept. 19–22, 1977*. Stockholm, Sweden: The Industrial Institute for Economic and Social Research.

Bilcke, J., P. Beutels, M. Brisson, and M. Jit. 2011. "Accounting for Methodological, Structural, and Parameter Uncertainty in Decision-Analytic Models: A Practical Guide." *Medical Decision Making* 31 (4): 675–692. https://doi.org/10.1177/0272989X11409240.

Brock, W. A., and C. H. Hommes. 1997. "A Rational Route to Randomness." *Econometrica* 65 (5): 1059–1096. https://doi.org/10.2307/2171879.

Bronka, P., J. van de Ven, D. Kopasker, S. V. Katikireddi, and M. Richiardi. 2025. "SimPaths: An Open-Source Microsimulation Model for Life Course Analysis." *International Journal of Microsimulation* 18 (1): 95–133. https://doi.org/10.34196/ijm.00318.

Caiani, A., A. Godin, E. Caverzasi, M. Gallegati, S. Kinsella, and J. E. Stiglitz. 2016. "Agent Based-Stock Flow Consistent Macroeconomics: Towards a Benchmark Model." *Journal of Economic Dynamics and Control* 69:375–408. https://doi.org/10.1016/j.jedc.2016.06.001.

Carro, A., M. Hinterschweiger, A. Uluc, and J. D. Farmer. 2023. "Heterogeneous Effects and Spillovers of Macroprudential Policy in an Agent-Based Model of the UK Housing Market." *Industrial and Corporate Change* 32 (2): 386–432. https://doi.org/10.1093/icc/dtac030.

Creedy, J., G. Kalb, and H. Kew. 2007. "Confidence Intervals for Policy Reforms in Behavioural Tax Microsimulation Modelling." *Bulletin of Economic Research* 59 (1): 37–65. https://doi.org/10.1111/j.0307-3378.2007.00250.x.

Dawid, H., and D. Delli Gatti. 2018. "Agent-Based Macroeconomics." Chap. 2 in *Handbook of Computational Economics,* edited by C. Hommes and B. LeBaron, 4:63–156. Amsterdam, Netherlands: Elsevier. https://doi.org/10.1016/bs. hescom.2018.02.006.

de Menten, G., G. Dekkers, G. Bryon, P. Liégeois, and C. O'Donoghue. 2014. "LIAM2: A New Open Source Development Tool for Discrete-Time Dynamic Microsimulation Models." *Journal of Artificial Societies and Social Simulation* 17. https://doi.org/10.18564/jasss.2574.

Dyer, J., P. Cannon, J. D. Farmer, and S. M. Schmon. 2022. *Black-Box Bayesian Inference for Economic Agent-Based Models,* Working Paper No. 2022-05. INET Oxford. https://ideas.repec.org/p/amz/wpaper/2022-05.html.

Dyer, J., A. Quera-Bofarull, N. Bishop, J. D. Farmer, A. Calinescu, and M. Wooldridge. 2024. "Population Synthesis as Scenario Generation for Simulation-Based Planning under Uncertainty." In *23rd International Conference on Autonomous Agents and Multiagent Systems (AAMAS 2024),* 490–498. International Foundation for Autonomous Agents and Multiagent Systems.

Eliasson, G. 1976. *A Micro-Macro Interactive Simulation Model of the Swedish Economy.* Technical report, Economic Research Report B15. Reprinted in International Journal of Microsimulation 17(24), 60–128 (2024). Federation of Swedish Industries, December. https://doi.org/10.34196/ijm.00292.

———. 1977. "Competition and Market Processes in a Simulation Model of the Swedish Economy." *The American Economic Review* 67:277–281. https://www.jstor.org/stable/1815916.

———. 2018. "Why Complex, Data Demanding and Difficult to Estimate Agent Based Models? Lessons from a Decades Long Research Program." *International Journal of Microsimulation* 11 (1): 4–60. https://doi.org/10.34196/IJM.00173.

Favreault, M. M., K. E. Smith, and R. W. Johnson. 2015. *The Dynamic Simulation of Income Model (DYNASIM).* Research Report. Urban Institute. https://www. urban . org / research / publication / dynamic - simulation - income - model - dynasim-overview.

Gillman, M. S. 2017. "GENESIS - The GENEric SImulation System for Modelling State Transitions." *Journal of Open Research Software* 5 (1): 24. https://doi.org/10. 5334/jors.179.

Grazzini, J., and M. Richiardi. 2015. "Estimation of Ergodic Agent-Based Models by Simulated Minimum Distance." *Journal of Economic Dynamics and Control* 51:148–165. https://doi.org/10.1016/j.jedc.2014.10.006.

Jonung, L. 1991. *The Stockholm School of Economics Revisited*. Historical Perspectives on Modern Economics. Cambridge, UK: Cambridge University Press.

Klevmarken, A. 2022. "Microsimulation. A Tool for Economic Analysis." *International Journal of Microsimulation* 15 (1): 6–14. https://doi.org/10.34196/IJM.00246.

Lee, J.-S., T. Filatova, A. Ligmann-Zielinska, B. Hassani-Mahmooei, F. Stonedahl, I. Lorscheid, A. Voinov, J. G. Polhill, Z. Sun, and D. C. Parker. 2015. "The Complexities of Agent-Based Modeling Output Analysis." *Journal of Artificial Societies and Social Simulation* 18 (4): 4. https://doi.org/10.18564/jasss.2897.

Li, J., and C. O'Donoghue. 2013. "A Survey of Dynamic Microsimulation models: Uses, Model Structure and Methodology." *International Journal of Microsimulation* 6 (2): 3–55. https://doi.org/10.34196/IJM.00082.

———. 2014. "Evaluating Binary Alignment Methods in Microsimulation Models." *Journal of Artificial Societies and Social Simulation* 17 (1): 15. https://doi.org/10.18564/jasss.2334.

Li, J., C. O'Donoghue, and G. Dekkers. 2014. "Dynamic Microsimulation." In *Handbook of Microsimulation Modelling*, edited by C. O'Donoghue, 293:305–343. Contributions to Economic Analysis. Leeds, UK: Emerald Group Publishing Limited. https://doi.org/10.1108/S0573-855520140000293009.

Makinoshima, F., and Y. Oishi. 2022. "Crowd Flow Forecasting via Agent-Based Simulations with Sequential Latent Parameter Estimation from Aggregate Observation." *Scientific Reports* 12:11168. https://doi.org/10.1038/s41598-022-14646-4.

Mu, T., S. Zheng, and A. Trott. 2022. "Modeling Bounded Rationality in Multi-Agent Simulations Using Rationally Inattentive Reinforcement Learning." *Transactions on Machine Learning Research* 12. https://openreview.net/forum?id=DY1pMrmDkm.

National Research Council. 2012. *Assessing the Reliability of Complex Models: Mathematical and Statistical Foundations of Verification, Validation, and Uncertainty Quantification*. Washington, DC: The National Academies Press.

Nelson, R. R., and S. G. Winter. 1982. *An Evolutionary Theory of Economic Change.* Cambridge, MA: Harvard University Press.

Nowak, S. A., L. J. Matthews, and A. M. Parker. 2017. *A General Agent-Based Model of Social Learning.* Technical report. Santa Monica, CA: RAND Corporation. https://www.rand.org/pubs/research_reports/RR1768.html.

O'Donoghue, C. 2001. "Dynamic Microsimulation: A Methodological Survey." *Brazilian Electronic Journal of Economics* 4:77.

———, ed. 2014. *Handbook of Microsimulation Modelling.* Vol. 293. Contributions to Economic Analysis. Leeds, UK: Emerald Group Publishing Limited.

O'Donoghue, C., and G. Dekkers. 2018. "Increasing the Impact of Dynamic Microsimulation Modelling." *International Journal of Microsimulation* 11 (1): 61–96. https://doi.org/10.34196/IJM.00174.

OECD. 2022. *TaxBEN: The OECD Tax-Benefit Simulation Model.* Technical report. Paris, France: OECD.

Orcutt, G. H. 1957. "A New Type of Socio-Economic System." *The Review of Economics and Statistics* 39:116–123. https://doi.org/10.2307/1928528.

———. 1960. "Simulation of Economic Systems." *American Economic Review* 50 (5): 894–907. https://www.jstor.org/stable/1810951.

———. 1962. "Microanalytic Models of the United States Economy: Need and Development." *American Economic Review* 52 (2): 229–240. https://www.jstor.org/stable/1910887.

Orcutt, G. H., S. Caldwell, and R. Wertheimer. 1976. *Policy Exploration Through Microanalytic Simulation.* Washington, DC: Urban Institute.

Orcutt, G. H., M. Greenberger, J. Korbel, and A. M. Rivlin. 1961. *Microanalysis of Socioeconomic Systems. A Simulation Study.* New York, NY: Harper & Brothers.

Pangallo, M., A. Aleta, R. M. del Rio-Chanona, D. Pichler A. Martín-Corral, M. Chinazzi, F. Lafond, M. Ajelli, *et al.* 2024. "The Unequal Effects of the Health–Economy Trade-off During the COVID-19 Pandemic." *Nature Human Behavior* 8:264–275. https://doi.org/10.1038/s41562-023-01747-x.

Platt, D. 2020. "A Comparison of Economic Agent-Based Model Calibration Methods." *Journal of Economic Dynamics and Control* 113:103859. https://doi.org/10.1016/j.jedc.2020.103859.

Poledna, S., M. G. Miess, C. Hommes, and K. Rabitsch. 2023. "Economic Forecasting with an Agent-Based Model." *European Economic Review* 151:104306. https://doi.org/10.1016/j.euroecorev.2022.104306.

Richiardi, M. 2018. "Agent-Based Computational Economics: What, Why, When." In *Agent-Based Models in Economics,* edited by D. Delli Gatti, G. Fagiolo, M. Gallegati, M. Richiardi, and A. Russo. Cambridge, UK: Cambridge University Press.

Richiardi, M., and R. E. Richardson. 2017. "JAS-mine: A New Platform for Microsimulation and Agent-Based Modelling." *International Journal of Microsimulation* 10:106–134. https://doi.org/10.34196/IJM.00151.

Savcisens, G., T. Eliassi-Rad, L. K. Hansen, L. H. Mortensen, L. Lilleholt, A. Rogers, I. Zettler, and S. Lehmann. 2023. "Using Sequences of Life-Events to Predict Human Lives." *Nature Computational Science* 4:43–56. https://doi.org/10.1038/s43588-023-00573-5.

Schelling, T. C. 1971. "Dynamic Models of Segregation." *The Journal of Mathematical Sociology* 1 (2): 143–186. https://doi.org/10.1080/0022250x.1971.9989794.

———. 2006. "Some Fun, Thirty-Five Years Ago." Chap. 37 in *Handbook of Computational Economics,* 1st ed., edited by L. Tesfatsion and K. L. Judd, 2:1639–1644. Amsterdam, Netherlands: Elsevier.

Schofield, D. J., M. J. B. Zeppel, O. Tan, S. Lymer, M. M. Cunich, and R. N. Shrestha. 2018. "A Brief, Global History of Microsimulation Models in Health: Past Applications, Lessons Learned and Future Directions." *International Journal of Microsimulation* 11 (1): 97–142. https://doi.org/10.34196/IJM.00175.

Solow, R. M. 2004. "Progress in Economics Since Tinbergen." *De Economist* 152:159–160. https://doi.org/10.1023/B:ECOT.0000023252.86552.5b.

Stephensen, P. 2016. "Logit Scaling: A General Method for Alignment in Microsimulation Models." *International Journal of Microsimulation* 9:89–101. https://doi.org/10.34196/IJM.00144.

van de Ven, J. 2017. "Parameterising a Detailed Dynamic Programming Model of Savings and Labour Supply using Cross-Sectional Data." *International Journal of Microsimulation* 10 (1): 135–166. https://doi.org/10.34196/IJM.00152.

van de Ven, J., P. Bronka, and M. Richiardi. 2025. "Dynamic Simulation of Taxes and Welfare Benefits by Database Imputation." *International Journal of Microsimulation* 18 (1): 124–155. https://doi.org/10.34196/ijm.00326.

Ward, J. A., A. J. Evans, and N. S. Malleson. 2016. "Dynamic Calibration of Agent-Based Models Using Data Assimilation." *Royal Society Open Science* 3:150703. https://doi.org/10.1098/rsos.150703.

Wu, G., A. Heppenstall, P. Meier, R. Purshouse, and N. Lomax. 2022. "A Synthetic Population Dataset for Estimating Small Area Health and Socio-Economic Outcomes in Great Britain." *Scientific Data* 9 (1): 19. https://doi.org/10.1038/s41597-022-01124-9.

THE SELF-ORGANIZED CRITICALITY PARADIGM IN ECONOMICS AND FINANCE

*Jean-Philippe Bouchaud, Capital Fund Management
and Académie des Sciences*

Abstract

"Self-organized criticality" (SOC) is the mechanism by which complex systems are spontaneously driven towards, or even across, a critical point at the edge between stability and chaos. These special points are characterized by fat-tailed fluctuations and long-memory correlations. Such a scenario can explain why insignificant perturbations may generate large disruptions, through the propagation of "avalanches" across the system. This short review discusses how SOC could offer a plausible solution to the excess volatility puzzle in financial markets and the analogue "small shocks, large business cycle puzzle" for the economy at large, as initially surmised by Per Bak et al. (1993) and, in a different language, by Hyman Minsky. I argue that in general the quest for efficiency and the necessity of resilience may be mutually incompatible and require specific policy considerations.

1 Introduction

Many systems made up of a large number of interacting items appear to be "marginally stable," that is, close to an incipient instability, or tipping point. The scenario according to which these systems are spontaneously driven towards such a fragile state is called "self-organized criticality" (SOC), and was proposed by Per Bak (2013) as a generic mechanism that allows one to explain large endogenous fluctuations in complex systems, such as natural systems (avalanches, earthquakes, floods, solar flares, turbulence, etc.; see Bak, Tang, and Wiesenfeld 1987; Sornette and Sornette 1989; Frisch 1995; Sethna, Dahmen, and Myers 2001; Sachs *et al.* 2012; Watkins *et al.* 2016), ecological systems (mass extinctions; May 1972; Bak and Paczuski 1997), biological neural networks (epilepsy; de Arcangelis, Perrone-Capano, and Herrmann 2006; Chialvo 2010; Osorio *et al.* 2010; Kinouchi, Pazzini, and Copelli 2020), bird flocks and fish schools (collective motion; Bialek *et al.* 2014), sociotechnical systems (black-outs, traffic jams, failure cascades, etc.; Dekker and Panja 2021; Laval 2023; Moran *et al.* 2024).

Financial markets and global economies are equally prone to such wild fluctuations. History is strewn with bubbles and crashes, booms and busts, crises and upheavals of all sorts. Understanding the origin of these dramatic events (and the remote possibility of curbing them) is arguably one of the most important problems in economic theory. Are these events exogenous (due to a shock from outside the system) or endogenous (generated by internal feedback loops) (Sornette 2006)? As John Cochrane (1994) quipped, "What shocks are responsible for economic fluctuations? Despite at least two hundred years in which economists have observed fluctuations in economic activity, we still are not sure."

This did not escape Per Bak's sagacity. Together with Kan Chen, José Scheinkman, and Michael Woodford (Bak *et al.* 1993; Scheinkman and Woodford 1994), he was quick to transpose ideas born in the context of statistical physics to economic situations, with the suggestion that the substantial, unexplained year-on-year gross domestic product (GDP) fluctuations of large developed economies—the so-called "small shocks, large business cycle puzzle" (Bernanke, Gertler, and Gilchrist 1996)—could in fact result from the *inherent fragility* of economic systems that prevents fluctuations from vanishing, even for large system sizes.

The financial markets analogue of excess GDP fluctuations is the equally well-known "excess volatility puzzle" elicited by Robert Shiller (1981, 1987) and by Stephen LeRoy and Richard Porter (1981).[1] Asset prices frequently undergo large jumps for no particular reason, when financial economics asserts that only unexpected news can move prices (Cutler, Poterba, and Summers 1988; Joulin *et al.* 2008; Marcaccioli, Bouchaud, and Benzaquen 2022). Volatility is an intermittent, scale-invariant process that resembles the velocity field in turbulent flows Frisch 1995; Muzy and Bacry 2000. Small, seemingly innocuous perturbations can end up sending the market in shambles—like Black Monday in October 1987 or the infamous Flash Crash of May 6, 2010. But such extreme events are not isolated outliers. Just as earthquake severity can span orders of magnitude, from hardly detectable tremors to devastating calamities, the probability distribution of price returns exhibits a power-law tail typical of complex systems sitting in the vicinity of a critical point (see, e.g., Cont 2001; Bouchaud and Potters 2003; Gabaix 2009). Again, it is tempting to invoke a kind of spontaneous coordination of competing market participants right at the border of chaos.

The aim of this (short) review is to discuss whether such a paradigm makes sense in the context of economic systems and financial markets. Indeed, as the French saying goes, *comparaison n'est pas raison*. It may well be that all these analogies are misleading, and that the true underlying mechanisms leading to excess volatility should be found elsewhere. For one thing, some would still argue that volatility is not excessive at all, but merely reflects the existence of genuine exogenous shocks of large

---

[1] For a nice summary of the literature on this point, see LeRoy (2008).

amplitude, perhaps invisible to the economist, but adequately processed by smart market participants and economic agents. Needless to say, such an "invisible hand" explanation often sounds preposterous, but reflects that it is hard to accept (and quite disturbing indeed) that large events can occur without "large causes"—whereas it is the major epistemological lesson of complexity science in general, and SOC in particular, that tiny perturbations can induce full crises. A complex system can actually be defined as a system where small perturbations *can—* but not necessarily *do—*trigger incommensurate effects, (see, e.g., Parisi 2007, 1999; Sethna, Dahmen, and Myers 2001; Bouchaud 2021).

Other explanations have been proposed, which are more plausible than phantom shocks. For example, Xavier Gabaix's (2011) "granularity hypothesis" (GH) is an attempt to rationalize the fact that idiosyncratic shocks may survive at the aggregate level, because some firms have a disproportionate size and cannot be averaged out (see also Moran, Secchi, and Bouchaud 2024, and Moran et al., ch. 25 of this volume). This is the case when firm sizes S are power-law distributed, with a probability distribution function tail decaying slow enough, for example as S^{-2} as is indeed the case empirically (Axtell 2001). Similarly, according to Gabaix *et al.* (2003), large price jumps in financial markets would result from orders sent by large investors or funds, the size of which is also known to be heavy-tailed (Gabaix *et al.* 2006), but probably not power-law tailed (see Schwarzkopf and Farmer 2010).

Although superficially very different, SOC and GH share common ingredients. Whereas the granularity hypothesis ascribes the persistence of aggregate fluctuations to large, stable entities (firms, investors), the self-organized criticality scenario relies on large, fleeting "coalitions" that are dynamically generated by contagion across the system. In other words, the survival of idiosyncratic shocks at the aggregate level is a result of their propagation over large scales—a phenomenon made possible by *fragility*, that is, the proximity of a critical point. While large coalitions or clusters are temporary in the SOC scenario, they are persistent within GH.

Yet another mechanism, which I will discuss further in sections 3.4 and 4.3, is that the equilibrium state of the system is in fact dynamically unstable, leading to a quasi-periodic or chaotic evolution even in the absence of any exogenous shock. Excess volatility is then purely self-induced. As illustrated by the model of section 4.3, one does not necessarily expect to see power-law statistics or long-memory effects when such systems are left to their own device. Interestingly, however, attempts to stabilize the system (through, for example, regulation, monetary policy, or learning) might yet again drive the system close to a critical point, reinstalling, in a sense, the SOC scenario.

Finally, let me mention two related but different concepts that I will not develop further here: highly optimized tolerance (HOT) (Carlson and Doyle 1999) and self-organized bistability (SOB) (di Santo *et al.* 2016), which have been argued to be relevant in different situations. Such scenarios might also be interesting to consider in the context of economic or financial systems (see, e.g., Harras, Tessone, and Sornette 2012, for similar ideas, and sec. 6).

I first explain in section 2 what is special about critical (marginally unstable) points. I then motivate the concept of self-organized criticality in section 3 and give a few examples where such a concept naturally applies. I next turn to possible applications of SOC in economical (sec. 4) and financial (sec. 5) contexts. Finally, I discuss in the concluding section 6 the policy implications of the possible fragility of socio-economic systems. I argue in particular that efficiency and resilience are often incompatible, and that operators, regulators, and policymakers should take stock of the unintended consequences that can appear when *resilience* (i.e., tolerance to tail events) is not explicitly included in the welfare function.

2 Stability and Criticality

2.1 A TRIVIAL EXAMPLE

Consider a simple dynamical equation for a quantity $x(t)$ that describes deviations from equilibrium, for example the difference between the actual production of a firm and the long term equilibrium (optimal) value. I write:

$$\frac{\mathrm{d}x}{\mathrm{d}t} = -\kappa x(t) + \eta(t), \qquad (1)$$

where κ is the anchoring parameter, which has dimensions of inverse-time, and $\eta(t)$ is a Gaussian white noise describing unanticipated exogenous shocks, with $\mathbb{E}[\eta(t)] = 0$ and $\mathbb{E}[\eta(t)\eta(t')] = \sigma^2\delta(t - t')$. This equation defines an Ornstein–Uhlenbeck process.

   In the absence of noise, $\sigma = 0$ and $x(t) = x_0 \exp(-\kappa t)$, that is, the system relaxes to equilibrium in a time $t_{\mathrm{eq.}} \sim \kappa^{-1}$. In the presence of noise, $x(t)$ randomly fluctuates around zero, with a stationary correlation function given by

$$\lim_{\kappa t \gg 1} \mathbb{E}[x(t + \tau)x(t)] = \frac{\sigma^2}{2\kappa}e^{-\kappa\tau}, \qquad (2)$$

with in particular $\mathbb{E}[x^2] = \sigma^2/2\kappa \propto t_{\mathrm{eq.}}$. The important message of this ultra-simple model is that as equilibrium becomes unstable, i.e. $\kappa \to 0$, the variance of fluctuations and the relaxation time both diverge at the same rate, as κ^{-1}.

   A slightly less trivial example is the multidimensional generalization of equation (1), namely:

$$\frac{\mathrm{d}x_i}{\mathrm{d}t} = -\sum_{ij} \mathbb{K}_{ij}x_j(t) + \eta_i(t), \qquad i, j = 1, \dots, N \quad (3)$$

where $\mathbb{K}$ is the so-called stability matrix and $\eta_i(t)$ is a multidimensional white noise. Now, equilibrium stability depends on the eigenvalues of matrix $\mathbb{K}$, which are in general complex. If all eigenvalues have negative real parts, equilibrium is stable. Let us denote by $-\kappa^\star$ the real part of the eigenvalue closest to zero. Then one finds that generically, when $\kappa^\star \to 0$, the variance of the fluctuations of any component $\mathbb{E}[x_i^2]$ diverges as $(\kappa^\star)^{-1}$, as does the relaxation time of the system.

Hence, the main message is that in the limit of marginal stability $\kappa^\star \to 0$, the system both amplifies exogenous shocks and becomes auto-correlated over very long time scales.

2.2 CRITICAL BRANCHING

A less trivial yet classic example, is the critical branching transition. The model describes a very large class of situations: sand pile avalanches, brain activity, epidemic propagation, default/bankruptcy waves, word of mouth, etc. In the language of sand piles, one assumes that one rolling grain can dislodge a certain number n of other grains that start rolling and may themselves dislodge more grains downhill. Suppose that n is an independent and identically distributed (IID) random variable with distribution $\rho(n; R_0)$ with a finite second moment and $\mathbb{E}[n] = R_0$.

When $R_0 < 1$, a first unstable grain dislodges on average a finite number of other grains. More precisely, the average size of avalanches is given by $(1 - R_0)^{-1}$. The probability of very large avalanches is exponentially small. However, as R_0 approaches 1, one single grain can lead to large disruptions. The probability of triggering an avalanche involving $S \gg 1$ grains is given by Harris (1963):

$$P(S) \propto S^{-3/2} \exp(-\varepsilon^2 S), \qquad \varepsilon = 1 - R_0 \to 0. \quad (4)$$

In other words, when $R_0 = 1$ the distribution of avalanche sizes is a scale-free, power-law distribution $S^{-3/2}$, with infinite mean.[2] Still, the probability ϕ that the avalanche continues forever is zero.

When $R_0 > 1$, this probability becomes non-zero (and grows as $\phi \propto R_0 - 1$ when R_0 is close to unity). With probability ϕ, the size of the avalanche is formally infinite, and in practice only limited by the size of the sand pile. With probability $1 - \phi$, the avalanche stops and its size is again distributed as in equation (4).

The value $R_0 = 1$ is therefore a *critical point*, separating a stable regime with avalanches of finite size from an unstable regime where landslides occur endogenously, triggered by the initial motion of a single grain. As for the trivial example of the previous section, such an instability is accompanied by a diverging time scale. Indeed, the duration τ of an avalanche is related to its size as $S \propto \tau^2$ (Harris 1963). Correspondingly, durations are distributed again as a power-law ($\propto \tau^{-2}$) for $R_0 = 1$.

Note that we have called R_0 the mean number of offspring $\mathbb{E}[n]$ by analogy with the notation used for the basic reproduction number in epidemics, with a rather obvious mapping. When infected individuals transfer the virus to $R_0 > 1$ (on average) other individuals, there is a nonzero probability ϕ for a full-blown pandemic to spread across the population.

3 Self-Organized Criticality

3.1 MOTIVATION

The previous example shows that, close to criticality, a system can reveal a very interesting type of phenomenology: when $R_0 \lesssim 1$, small shocks *can* generate large perturbations, but do not

---

[2] Note that when the distribution of offspring $\rho(n; R_0)$ decays as $n^{-1-\alpha}$ with $1 < \alpha < 2$, the critical avalanche size distribution has a power-law tail given by $S^{-1-1/\alpha}$. Note that for $\alpha = 2$ one recovers the classic $S^{-3/2}$ behavior.

necessarily do so. In fact, the critical power-law distribution $S^{-3/2}$ means that most "avalanches" are of small size, although *some* can be very large. In other words, the system looks stable, but occasionally goes haywire with no apparent cause.

However, this only occurs for a very special value of R_0. In generic cases, the system is either stable, or completely unstable. Why would the proximity of $R_0 = 1$ play a role in practice, except if the system is fine-tuned "by hand" close to criticality? Why would criticality be relevant in practice, in particular in the context of economics and finance?

Self-organized criticality relies on the following seminal idea (Bak, Tang, and Wiesenfeld 1987): model parameters themselves are dynamical variables and evolve in such a way that the system *spontaneously evolves towards the critical point*, or at least visits its neighborhood frequently enough (for a precursor of this general idea, see sec. 8 of Keeler and Farmer 1986, and for reviews, see Bak 2013; Jensen 1998; Watkins *et al.* 2016). Take for example the case of a highly transmissible virus, with a natural value of R_0 above unity. As the disease spreads, containment measures are adopted (masks, social distancing, etc.) and the risk of getting sick is taken seriously by the population. This leads to a significant reduction of R_0, perhaps even below unity. Now, the epidemic slows down and people become more complacent, which causes R_0 to rise again. It is not absurd to imagine that the process will indeed self-organize around a point where the spread of the disease is curbed at a minimal amount of individual constraints. This is achieved precisely at the critical value $R_0 = 1$.

3.2 SWEEPING THROUGH AN INSTABILITY

Consider now the sand pile example initially put forth by Bak, Tang, and Wiesenfeld (1987). Start with a flat layer of sand and slowly add grains from a point source above it. A conical pile will form, with a progressively steeper slope. As the slope increases, so does the probability R_0 that a rolling grain destabilizes more grains increases. We will thus describe the dynamics of the pile in terms of R_0 rather than its slope.

As long as $R_0 < 1$, the resulting avalanches are small and do not change the overall shape of the pile. But as soon as $R_0 > 1$, there is a possibility of a landslide that collapses the pile and reduces its slope. The dynamical evolution of the probability to find a pile with "slope" R_0 is then described as[3]

$$\frac{\partial P(R_0, t)}{\partial t} = -\mu \frac{\partial P(R_0, t)}{\partial R_0} - \gamma (R_0 - 1)^+ P(R_0, t), \quad (5)$$

where $P(R_0, t)$ is the probability to find a slope characterized by a certain value of R_0, μ describes the rate at which R_0 increases due to the addition of new grains, and γ the rate at which single grains start rolling down the slope, and $(R_0 - 1)^+$ the probability that a single grain does trigger a landslide. For simplicity, one assumes that after the landslide, the pile restarts at $R_0 = 0$.

The stationary state $Q_{\text{st.}}(R_0) = P(R_0, t \to \infty)$ of this process reads:

$$Q_{\text{st.}}(R_0) = Z^{-1} \qquad\qquad\qquad (0 \le R_0 \le 1),$$

$$= Z^{-1} \exp\left(-\frac{\gamma}{2\mu}(R_0 - 1)^2\right) \qquad (R_0 \ge 1), \quad (6)$$

with $Z = 1 + \sqrt{\pi\mu/2\gamma}$ such that $Q_{\text{st.}}(R_0)$ is normalized to one. This equation shows that there is a finite probability density for the slope to be exactly critical. This is enough to ensure that one will observe a power-law distribution of avalanche sizes (Sornette 1994), which can be computed from

$$P(S) = \int_0^\infty dR_0 \, Q_{\text{st.}}(R_0) P(S),$$

---

[3] Throughout the paper I use the standard notation $(x)^+ = \max(x, 0)$.

where $P(S)$ is given by equation (4). It is not very difficult to check that $\mathcal{P}(S)$ behaves at large S as S^{-2} without the exponential truncation term present in equation (4), that is, as a pure power-law with exponent -2 instead of $-3/2$. Note however that $\mathcal{P}(S)$ only captures avalanches of finite size (i.e., much smaller than the total size of the system). One should also consider system-wide avalanches which relax the system all the way down to $R_0 = 0$. These contribute to an additional hump for very large S of the order of the system size—events that Didier Sornette and Guy Ouillon (2012) call "dragon-kings" (see fig. 1).

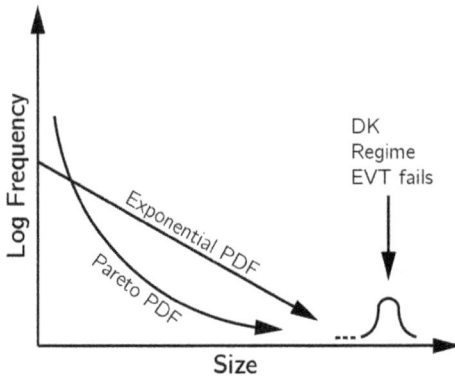

Figure 1. Schematic illustration of the concept of dragon-kings (DK). Plot of the distribution of avalanche sizes S (say) that follow a Pareto law for sizes S much less than the maximum possible size S_{max}, followed by an anomalous hump around S_{max} (corresponding to system-wide avalanches). The mechanism generating those dragon-kings is, in many cases, the excursion of the system in an unstable region, whereas the power-law regime comes from the vicinity of the critical point. *From https://encyclopedia.pub/entry/30734.*

The above example illustrates one possible scenario through which intermittent dynamics, with a power-law distribution of event sizes, can occur (see also Keeler and Farmer 1986 for an early discussion). The critical point is regularly visited by the natural dynamics of the system, as it travels from the stable phase to the unstable phase, the instability driving the system back to the stable phase. As discussed below, such a scenario is also reminiscent of the Minsky cycle in financial markets.

3.3 DYNAMICAL CONVERGENCE TOWARD CRITICALITY

A different scenario is when the natural dynamics stops precisely at the critical point. For example, gradient descent applied to complex *constrained satisfaction problems* generically stops when a local minimum of the loss function is reached. In many cases, however, such minima are marginally stable, that is, the Hessian of the loss function has all its eigenvalues non negative, but the smallest one is very close to zero (see, e.g., Cavagna, Giardina, and Parisi 1998; Müller and Wyart 2015; Franz *et al.* 2017). This is precisely the situation described by equation (3), where stability matrix $\mathbb{K}$ is the Hessian matrix. Therefore, small shocks get amplified by the proximity of an instability and small static perturbations can disproportionately modify the equilibrium state (see also Carlson and Doyle 1999 for related ideas). In fact, such perturbations can even change the sign of the smallest eigenvalue, forcing the system to rearrange and find another (marginally stable) equilibrium, often very different from the original (unperturbed) one.

As an illustration, let us consider the so-called generalized Lotka–Volterra model, describing the population dynamics of many interacting species (Bunin 2017). We denote $x_i(t)$ the number of living individuals of species $i = 1, \ldots, N$ and assume that these numbers evolve as

$$\frac{\mathrm{d}x_i}{\mathrm{d}t} = x_i \left(\mu_i + \sum_{j=1}^{N} \mathbb{A}_{ij} x_j \right), \qquad (7)$$

where $\mu_i > 0$ is the "fitness" of species i (i.e., its growth rate in the absence of interaction with other species, including itself) and the matrix $\mathbb{A}$ captures beneficial ($\mathbb{A}_{ij} > 0$) or detrimental ($\mathbb{A}_{ij} < 0$) interactions with other species (for example, $\mathbb{A}_{ij}\mathbb{A}_{ji} < 0$ describes a predator–prey situation).

The ecological equilibria corresponding to equation (7) must be such that either $x_i^\star = 0$ (extinct species) or $\mu_i + \sum_{j=1}^{N} \mathbb{A}_{ij} x_j^\star = 0$. Naively solving the second equation yields, in a vector notation,

$$\vec{x}^\star = -\mathbb{A}^{-1}\vec{\mu}. \tag{8}$$

However, the resulting vector $\vec{x}^\star$ will in general contain negative entries, which of course does not make sense since the number of living individuals must be positive or zero. Feasible equilibria are thus such that only a subset of species survives, such that equation (8) restricted to that subset returns a vector $\vec{x}^\star$ with only positive entries. As first anticipated by the late Lord Robert May (1972), not all ecologies can be stable.

What about the dynamics generated by equation (7)? In the special case where the matrix $\mathbb{A}$ is symmetric, one can show that generically the system evolves towards a marginally stable state, that is, enough species disappear for a feasible equilibrium to exist, but the sensitivity matrix defined as

$$\frac{\partial x_i^\star}{\partial \mu_j} = -(\mathbb{A}^{-1})_{ij},$$

has a diverging eigenvalue $\lambda^\star$ in the large N limit (Biroli, Bunin, and Cammarota 2018). This means that any small change in the fitness of one species can have dramatic consequences on the whole system—in the present case, mass extinctions (Bak and Paczuski 1997). Similarly, the stability matrix $\mathbb{K}$, given by (cf. eq. 3)

$$\mathbb{K}_{ij} = -x_i^\star \mathbb{A}_{ij},$$

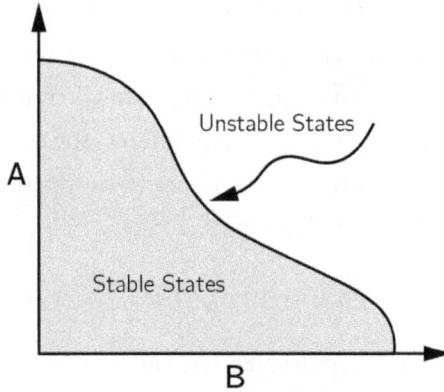

Figure 2. Schematic explanation of the prevalence of marginally stable states in complex systems. One starts from an unstable point (white region) and follows the arrow. The system stops as soon as it becomes stable, that is, at the boundary between stability and instability. *From Müller and Wyart (2015).*

has an eigenvalue $\kappa^\star \propto \lambda^{\star-1}$ very close to zero, meaning again that one is in a situation where small exogenous shocks are amplified by marginal stability (see, e.g., Stone 2018). The generalized Lotka–Volterra equation therefore provides a non-trivial realization of equation (3) with $\kappa^\star \to 0$, as a natural consequence of the dynamics of the system that converges to a marginally stable equilibrium (see fig. 2). I will discuss in section 4.2 a direct analogy between Lotka–Volterra dynamics and production networks.

There are many other concrete examples where such a scenario is at play (see, e.g., Müller and Wyart 2015; Aspelmeier and Moore 2019; Patil and Bouchaud 2024), in particular in the context of complex learning games (Garnier-Brun, Benzaquen, and Bouchaud 2024). As mentioned above, most complex optimization systems are in a sense fragile, as the solution to the optimization problem is highly sensitive to the precise value of the parameters of the specific instance one wants to solve, like the $\mathbb{A}_{ij}$ entries in the Lotka–Volterra model. Small changes of these parameters can completely upend the structure of the optimal state, and trigger large-scale rearrangements (see, e.g., Fisher and Huse 1991; Aspelmeier 2008; Krząkała and Bouchaud 2005) and, in the context of optimal portfolio construction (Garnier-Brun *et al.* 2021).

3.4 STABILIZING AN UNSTABLE EQUILIBRIUM

Another interesting generic scenario for self-organized critical-ity was proposed by Felix Patzelt and Klaus Pawelzik (2011). The narrative is that of the balancing stick problem (Cabrera and Milton 2004, 2012). Imagine trying to keep a long stick vertical on the tip of your finger. The angle with the vertical is noted $\theta(t)$. The vertical position $\theta = 0$ is an unstable equilib-rium, so the dynamics for small θ reads

$$\frac{d\theta}{dt} = G(t)\theta(t) + \eta(t),$$

where $G(t) > 0$ is the amplification rate, possibly time-dependent, and $\eta(t)$ is a white noise describing unexpected perturbations (e.g., wind). Because $G(t) > 0$, the dynamics is unstable and without any control, the stick just falls off.

You will try to keep the stick vertical by moving your hand in response to the motion of the stick. However, the observation of $\theta(t)$ is usually both noisy (measurement noise) and time-delayed. The observer can therefore only compute the *control policy* $F(t, \{\theta(t' \leq t - \tau)\})$ with a certain lag τ. The equation of motion then becomes

$$\frac{d\theta}{dt} = G(t)\theta(t) + \eta(t) - F\left(t, \{\theta(t' \leq t - \tau)\}\right).$$

The optimal control policy is computed in Patzelt and Pawelzik (2011). You can indeed stabilize the position of the stick around $\theta = 0$, but the time series of $\theta(t)$ shows very interesting features (intermittent dynamics, power-law distributions), which are indeed observed in real balancing stick experiments (Cabrera and Milton 2004, 2012) and reminiscent of returns in financial markets (Cont 2001; Bouchaud and Potters 2003; Patzelt and Pawelzik 2013).

What is the underlying mechanism leading to such special dynamical features? The basic idea here is that the closer $\theta(t)$ is to zero, the more uncertain is the most likely estimate of the amplification rate $\hat{G}(t) = \dot{\theta}/\theta$ and the larger the error in the control policy $F(.)$.

In other words, the better one is able to stabilize the system, the more difficult it becomes to predict its future evolution! This in turn generates occasional large errors, leading to large swings that are later corrected as the control policy becomes efficient again (Patzelt and Pawelzik 2011). This rather beautiful scenario is quite universal. For example, heartbeat regulation is a result of two conflicting forces: excitation vs. inhibition (sympathetic vs. parasympathetic nervous system), or brain activity regulation (Lombardi, Herrmann, and Arcangelis 2017). Such ideas are probably also relevant in economics and finance (see secs. 5.2, 5.4 below). In fact, the title of the present subsection is tantalizingly close to that of Hyman Minsky's foundational book *Stabilizing an Unstable Economy* (2008).

4 Self-Organized Criticality in Economics

As mentioned in the introduction, Bak, Chen, Scheinkman and Woodford first attempt to apply SOC to economic systems in 1993. Although their paper is brimming with exciting ideas, the actual model the authors put forth to describe spontaneous large-scale output fluctuations appears somewhat ad hoc. The firm input–output network is modeled as a square lattice:

downstream firms buying exactly zero or one unit of good from exactly two upstream firms. At each time step, firms can only produce two units of goods or none at all, depending on their inventory and the orders received from their downstream clients. These orders in turn depend on how much these clients need to produce. Demand fluctuations for final goods (i.e., the top row of the lattice) are exogenous, but propagate upstream as "avalanches" with a power-law distribution of sizes, like grains on the slope of a sand pile. Hence, again, small local shocks can lead to large-scale disruptions of the supply chain. The problem with such a highly stylized model is that it is not clear how generic the results are. In fact, the somewhat constrained production rules have been chosen such that the system is critical—but, somewhat paradoxically in view of the title of the paper, it is not clear how supply chains in general would dynamically "self-organize" into criticality.

In revisiting the issue and discussing more recent proposals that lead to self-organized criticality in economic systems, based on arguably less artificial assumptions, I hope to put the SOC scenario on firmer theoretical ground and revive the interest of economists. It is indeed fair to say that, despite the high profile of its authors, Bak *et al.*'s (1993) paper did not reach the level of attention that (in my view) it deserved[4]— probably because of its purely conceptual nature, and the lack of a precise, actionable proposal to fit on data.

---

[4]At the time of writing, the joint number of citations of Bak *et al.* (1993) and Scheinkman–Woodford (1994) is ~ 800 according to Google Scholar, compared for example to 2,050 for Gabaix's 2011 "granularity" paper, or 2,700 for Acemoglu *et al.* (2012), on the network origin of aggregate fluctuations.

4.1 MACROECONOMIC FLUCTUATIONS AND "TIMELINESS CRITICALITY"

As a possibly more realistic incarnation of the scenario proposed by Bak *et al.*, consider the propagation of production delays along the supply chain. The idea is that when input goods are not timely, production stops—unless "buffers" (i.e., inventories) are present. The same logic holds in other sociotechnical systems, such as train or plane networks (Dekker and Panja 2021). Indeed, the next train can only leave on time if the previous train has arrived and the driver and train managers are available. In this case, time buffers are embedded within the time table, allowing for enough time between different scheduled events, such that delays can be (to some extent) absorbed.

Recently, José Moran *et al.* (2024) proposed a minimal, stylized model of delay propagation that reveals the existence of a critical point as the size of mitigating buffers is reduced. In a nutshell, the model assumes that the delay $\tau_i(n)$ accumulated on node i of the network at the nth iteration step evolves as

$$\tau_i(n+1) = \left[\max_{j \in \partial_i}(\tau_j(n)) - B\right]^+ + \epsilon_i(n+1), \qquad (9)$$

where ∂_i denotes the set of tasks that need to be completed for task i to start, B is the time buffer and $\epsilon_i(n+1)$ is a positive noise describing idiosyncratic events that delay the start of task i. The interpretation of the first term in the right hand side of equation (9) is quite transparent: Provided the worst delay is less than the buffer, there is no impact of previously accumulated delay on the start of the new task.

Above a certain critical buffer size B_c, delays are found to self-heal, and large-scale system-wide delays are avoided. On the contrary, when temporal margins are not wide enough, delays accumulate without bounds, creating system-wide disruptions. Close to the critical point, delay "avalanches" of all sizes are observed (Moran *et al.* 2024).

Now, there exist a variety of incentives for operators, often reinforced by competitive pressures, to increase time efficiencies (i.e., reduce B) in order to achieve superior operational results. For example, train operators may have the goal to maximize the number of passengers to be transported by the network. This reduction will have a minor impact on the global functioning of the system provided B is larger than B_c—only small, occasional disruptions ensue. Operators then feel legitimate in reducing B further, until a major breakdown occurs. Taking stock of the recent events, regulators usually step in and insist that reasonable safeguards are put in place. Such a scenario is very close to a sand pile sweeping through its critical angle (see sec. 3.2) or the dynamics of disease propagation, with a reproduction number hovering around $R_0 = 1$ (sec. 3.1).

The model can be naturally extended to the propagation of supply fluctuations in production networks, where firms reduce costs by keeping the inventory of production inputs at a minimal level. Any delay in such a "just-in-time" supply chain can then propagate down the production network with major disruptions, as exemplified by the Suez Canal obstruction in 2021. Inventories are conceptualized as buffers that allow firms to keep producing in the absence of inputs for some amount of time. Low levels of inventories can thus cause strong output fluctuations at the aggregate level (see, e.g., Colon and Ghil 2017). Myopic cost-cutting measures at the local scale may unwittingly push the whole system into criticality. Such a mechanism provides a clear justification for the seminal idea of Bak *et al.*'s (1993; 1994) and is perhaps key to explaining why large economies are so much more volatile than expected based on economic equilibrium models.

It is of course tempting to think of other situations in similar terms. For example, bank regulators impose a certain level of capital "cushions" (the analogue of buffers) that bank should keep aside to avoid the propagation of liquidity crises. This obviously comes at a cost, since a certain amount of capital cannot be put to work in profitable (but risky) investments. In relatively stable economic environments, bankers will sooner or later argue that safeguards need to be reduced...until the all but inevitable "Minsky moment" (Minsky 2015) when the default of a single financial institution avalanches through the whole sector. Network models have indeed been proposed to explain the system-wide breakdowns of the banking sector in 2008 (Gai and Kapadia 2010; Haldane and May 2011; Squartini, van Lelyveld, and Garlaschelli 2013; Caccioli, Barucca, and Kobayashi 2018).

~267~

4.2 THE ECOLOGY OF PRODUCTION NETWORKS (STATICS)

The importance of network effects for the propagation of shocks across production chains was also considered in the context of *economic equilibrium* models in Daron Acemoglu *et al.* (2012) and Vasco Carvalho and Alireza Tahbaz-Salehi (2019), elaborating on the classic paper of John Long, Jr. and Charles Plosser (1983). Assuming a Cobb–Douglas production function for all firms (see eq. 10 below), one can show that idiosyncratic productivity shocks get amplified at the aggregate level, through the existence of client-supplier links between firms. In fact, if the input–output network is such that the degree distribution is broad enough—that is, if some firms act as hubs in the supply network—then the volatility of aggregate production decreases slowly (or even not at all) with the number of firms (Acemoglu *et al.* 2012), possibly explaining why the activity of large economies fluctuates so much.

However, if one drills down further into the mathematics of the model, one realizes that these anomalous fluctuations are in fact due to Gabaix's (2011) "granularity" mechanism, namely, when the topology of the network induces a broad distribution of firm sizes (measured as total sales). The idiosyncratic shocks hitting those large firms then overwhelmingly contribute to GDP fluctuations. A statistical model explaining how the production network might self-organize to generate such a power-law distribution of firm sizes is discussed in Enghin Atalay *et al.* (2011).

There are, however, important issues with equilibrium models. First and foremost, it sounds more plausible to think that the large aggregate disruptions we are trying to decipher are *disequilibrium* effects. Indeed, the assumption that during such periods all markets instantaneously clear (no stock-outs, no inventories) and firms muddle through without defaulting is clearly untenable. Second, even if equilibrium is assumed, the choice of a Cobb–Douglas production function is extremely special, as it allows a feasible equilibrium (where all firms produce a positive amount of goods and sell them at positive prices) to exist for arbitrary input–output networks and any value of firm productivities. This is a result of the relatively high amount of substitutability of input goods implied by Cobb–Douglas. As soon as elasticity of substitution σ is lower, some firms may have to disappear for a feasible equilibrium to exist—much like what happens in the context of the Lotka–Volterra description of ecological communities discussed in section 3.3. Hence, bankruptcies and bankruptcy waves must be considered within such generalized models, which means that a fully out-of-equilibrium, dynamical description of shock propagation is necessary to describe the resulting aggregate fluctuations.

For definiteness, let us consider the constant elasticity of substitution (CES) family of production functions. Calling π_i the production of firm i, one writes

$$\pi_i = z_i \left(\sum_j a_{ij} \left(\frac{J_{ij}}{Q_{ij}} \right)^{\frac{1}{q}} \right)^{-q} \quad \text{with} \quad \sum_j a_{ij} = 1, \quad (10)$$

where z_i is the firm productivity, Q_{ij} is the number of goods firm i buys from firm j and $J_{ij}, a_{ij} \geq 0$ are weight parameters.

~269~

Parameter q captures the substitutability of input goods, and is related to the standard elasticity of substitution through $\sigma = q/(q+1)$. The Cobb–Douglas function $\pi_i = z_i \prod_j (Q_{ij}/J_{ij})^{a_{ij}}$ corresponds to $q \to \infty$, $\sigma = 1$ and is often used to describe the average aggregate production of economic sectors (Long, Jr. and Plosser 1983), or of the economy as a whole. When $q \to 0^+$, $\sigma \to 0$, on the other hand, no substitutes are available, and equation (10) reduces to the classical Leontief production function: $\pi_i = z_i \min_j (Q_{ij}/J_{ij})$. It models a situation where redundancy is costly. Firms therefore choose their suppliers with parsimony and cannot "rewire" (i.e., find alternative suppliers) on short time scales in the real economy.

In the case of a general CES production function, the competitive, market-clearing equilibrium equation for prices reads (Moran and Bouchaud 2019):

$$(z_i p_i)^\varsigma - \sum_{j \neq 0} a_{ij}^{q\varsigma} (J_{ij} p_j)^\varsigma = a_{i0}^{q\varsigma} (J_{i0} w_i)^\varsigma \quad (>0), \qquad \varsigma := \frac{1}{1+q}, \tag{11}$$

where good 0 is labor and w_i is the wage of firm i. A similar equation can be written for equilibrium productions π_i.

Now, in order for the equilibrium to make sense, the solutions to equation (11) must be such that $p_i > 0$, $\forall i$; that is, that equilibrium prices and quantities must all be strictly positive—otherwise some firms are for all purposes bankrupt.

As first noted by David Hawkins and Herbert Simon (1949), this is *not* automatic and requires matrix $\mathbb{M}$, defined by $\mathbb{M}_{ij} = z_i^\zeta \delta_{ij} - a_{ij}^{q\zeta} J_{ij}^\zeta$ to be a so-called "M-matrix,"[5] that is, such that *all its eigenvalues have nonnegative real parts.* Therefore some conditions on productivities z_i and linkages J_{ij}, a_{ij} must be fulfilled for the economy to work. Rather interestingly, equation (11) is identical, *mutatis mutandis*, to the equation determining the equilibrium size of species in a generalized Lotka–Volterra model discussed in section 3.3.[6]

Expanding equation (11) to first order in ζ, one notices that since $\sum_{j>0} a_{ij} < 1$, the Perron–Frobenius theorem ensures that $q \to \infty$ Cobb–Douglas networked economies (such as those considered in Long, Jr. and Plosser 1983; Acemoglu *et al.* 2012) *always* have a feasible equilibrium where all firms can survive, for any network and any productivities. Therefore, the type of crisis that takes place for $q < +\infty$, where bankruptcies occur, has no counterpart in a Cobb–Douglas economy.

In all other cases, when the smallest real part of the eigenvalues of matrix $\mathbb{M}$ reaches zero, the system approaches a critical state, and bankruptcies must occur for the economy to remain viable. If the system self-stabilizes around such a critical point, then small productivity shocks can cause bankruptcy avalanches which are found to be scale-free when the production network is sufficiently heterogeneous (Moran and Bouchaud 2019).

What drives the smallest eigenvalue of $\mathbb{M}$ to zero? Several

---

[5]Note that if $\mathbb{M}$ is an M-matrix, $\mathbb{M}^\top$ is also an M-matrix. An interesting property of an M-matrix is that all the elements of its inverse are non negative.

[6]There is a long tradition of using the *two-species* classical Lotka–Volterra equation in economics, initiated by Richard Goodwin (see, e.g., Flaschel 2010). However the dynamical version of the *multi-species* Lotka–Volterra model has not been studied in the context of firm growth, and might be a fruitful path to follow, using in particular the recent results of Herskovic *et al.* (2020) and Mazzarisi and Smerlak (2024).

types of evolutionary forces act to that effect. One is simply the creation of new firms, that connect to the preexisting network at constant average productivity $\bar{z}$. As shown in Moran and Bouchaud (2019), this can only *decrease* the smallest eigenvalue of the matrix $\mathbb{M}$ that describes the preexisting firms. One concludes that a growing economy can only become more unstable with time, unless productivity increases.[7] This argument is actually closely related to Robert May's (1972) original argument about the stability of large ecologies.

~271~

In fact, one can show that as the number of links to the most connected node of the network increases, the smallest eigenvalue of $\mathbb{M}$ decreases (Castellano and Pastor-Satorras 2017), until the instability threshold is reached. In this case, the fragility of the network comes from the most central hubs, a scenario akin to, but different from, the one of Acemoglu *et al.* (2012). This effect might be amplified if firms systematically favor links toward hubs, leading to a "scale-free" free input–output network (Atalay *et al.* 2011). Interestingly, a stability-constrained growth mechanism for networks, whereby a node is freely added to the network if it does not destabilize the system but induces rewirings in the network until stability is found again if it does, has been found to generate such self-organized scale-free networks (Perotti *et al.* 2009).

The second evolutionary effect is the complexification of the production process, that is, technology progress means that a wider array of products are needed as inputs. If the average productivity $\bar{z}$ remains the same while the average connectivity of the network increases, the system eventually reaches the instability point. Hence, again, productivity must increase at some minimum rate for the economy to remain stable.

---

[7]For a related story, see Patil and Bouchaud (2024), where the incipient instability generates emergent wealth inequalities.

But since increasing productivity is costly, one can postulate that the average productivity $\bar{z}$ will tend to hover around the minimal viable threshold, and sometimes lag behind, leading to occasional endogenous crises. Similarly, firms tend to optimize their portfolios of suppliers, thereby reducing their redundancy but, by the same token, reducing the effective substitution effects captured by the CES parameter q. As q decreases, the economy will again inevitably become unstable.

Finally, competitive equilibrium corresponds to zero profit. Now, in more realistic situations, firms attempt to realize positive profits and distribute dividends. In the present framework, this amounts (for $q = 0$) to a reduction of effective productivity $z_i \rightarrow z_i(1 - \varphi_i)$, where φ_i is the fraction of the total sales firm i targets as profit. As firms attempt to increase their profits, the average effective productivity goes down, until the marginal stability point is reached and a crisis ensues. After the crisis, economic actors revert to more reasonable levels of markups, which makes the economy viable again—until the next crisis: We are again back to a Minsky cycle of sorts.

One may therefore conjecture that evolutionary and behavioral forces repeatedly drive the economy close to marginal stability, as argued by Bak *et al.* (Bak *et al.* 1993; Scheinkman and Woodford 1994) but here within the context of a more traditional economic equilibrium model. It is now the very structure of the supply network and the nonsubstitutability of input goods that determines the proximity of the critical point.

4.3 THE ECOLOGY OF PRODUCTION NETWORKS (DYNAMICS)

Still, the above story is all within the realm of economic equilibrium, where profits are maximized, markets clear and all produced goods are consumed. As exogenous shocks buffet the economy, a new equilibrium is assumed to be almost immediately reached, corresponding to the new values of productivities. As already discussed above, this is a heroic assumption that disregards the dynamics of, for example, inventories, stockouts and bankruptcies. At the very least, one should explain the dynamical process through which equilibrium is reached, and how long one should wait before it is reached—if it is ever reached (on this point, see the insightful introductory chapter of Fisher 1989 and refs. therein, and Pangallo 2025, ch. 5 of this volume).[8]

~273~

There is, however, no consensus on how to model such dynamics. Theo Dessertaine *et al.* (2022) proposed a simplified, linear approach to model the dynamics of small deviations away from equilibrium, where supply–demand and profit imbalances are used as restoring forces towards equilibrium. The resulting equations are of the general form equation (3), with $x_i = \{p_i, \pi_i\}$ and $N \rightarrow 2N$ and a $2N \times 2N$ stability matrix $\mathbb{K}$. Interestingly, the smallest negative eigenvalue $-\kappa^\star$ of $\mathbb{K}$ goes to zero precisely when the smallest positive eigenvalue of $\mathbb{M}$ goes to zero, that is, when the supply network becomes critical in the sense defined in the previous subsection (Dessertaine *et al.* 2022). As discussed in section 2.1, as the system reaches criticality, convergence towards equilibrium becomes infinitely slow and small exogenous shocks are disproportionately amplified. Hence, as for the generalized Lotka–Volterra dynamics, the system becomes infinitely *fragile* as it reaches the critical point, just before the network has to rearrange. In the present context, it means that

---

[8] Although it lays out a research program that is in my view of fundamental importance, Franklin Fisher's 1989 book *Disequilibrium Foundations of Equilibrium Economics* only has 882 citations to date. This shows that mainstream economics is still not really interested in these topics, as testified by "state-of-the-art" macroeconomics models that completely disregard out-of-equilibrium effects (see, e.g., Kaplan, Moll, and Violante 2018; Baqaee and Farhi 2019, 2020). For a related discussion, see Farmer (2024).

some firms have to disappear and/or the supply network has to rewire for the economy to remain functional.[9]

However, the above "naive" linear implementation of the out-of-equilibrium dynamics is theoretically inconsistent, as some incontrovertible constraints are violated (for example, consumption cannot exceed amount of available goods). The fully consistent, agent-based description proposed in Dessertaine *et al.* (2022) goes beyond the scope of the present paper. The results are summarized in the form of the phase diagram shown in figure 3. Dessertaine *et al.* (2022) find that, depending on the parameters of the system (strength of the reaction in the face of imbalances, perishability of goods, . . .) the economy can either (a) completely collapse; (b) reach the competitive equilibrium after some time, which can become very large; or else (c) enter a state of perpetual disequilibrium, with purely endogenous fluctuations that can be periodic or completely chaotic, corresponding to a coordination breakdown of firms along the supply chain (Sterman 1989).[10] In this last regime, economic equilibrium still exists but is linearly unstable, and is dynamically out-of-reach. This regime is interesting because it offers yet another explanation for "small shocks, large business cycles": The economy might be permanently in a turbulent state, far from equilibrium, without necessarily being close to a critical point.

Dessertaine *et al.*'s model therefore suggests *two* distinct out-of-equilibrium routes to excess volatility: (1) purely endogenous cycles, resulting from over-reactions and non-linearities; or (2) self-organized criticality—that is, persistence and amplification of exogenous shocks, governed by the

---

[9] On the slow, complex dynamics of rewiring, see, e.g., Colon and Bouchaud (2022).

[10] There is actually another possibility (d) where some deflationary equilibrium is reached; see Dessertaine *et al.* (2022).

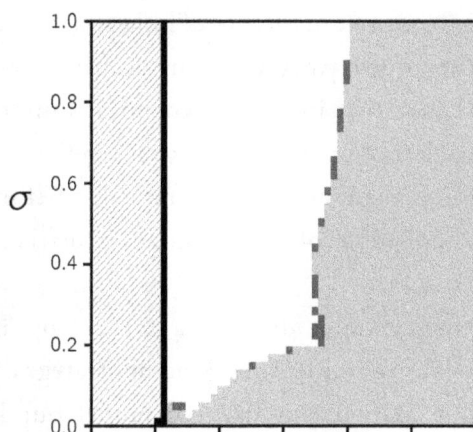

Figure 3. Phase diagram of the agent-based firm network model of Dessertaine *et al.* (2022). x−axis: strength of forces counteracting supply/demand and profit imbalances; y−axis: goods perishability. Leftmost region (a, diagonal gray lines): the economy collapses; middle region (b, white): the economy reaches equilibrium relatively quickly; right region (c, solid gray): the economy is in perpetual disequilibrium, with purely endogenous fluctuations. The black vertical sliver (d) corresponds to a deflationary equilibrium, see footnote 10 and Dessertaine *et al.* (2022).

proximity of a boundary in parameter space where the competitive equilibrium becomes unstable. While scenario (1) may appear at first sight to be more generic, the economic forces discussed above could make (2) plausible as well. Specific empirical work is needed to distinguish between these two scenarios, which lead to rather different predictions for the structure and dynamics of fluctuations. The SOC scenario suggests in particular that fluctuations should be correlated over long time scales, corresponding to the propagation of "avalanches" over large portions of the production network.

4.4 INFLATION "AVALANCHES"

In a very interesting recent paper, Makoto Nirei and José Scheinkman (2024) have suggested that repricing events (i.e., firms increasing their price in the face of inflation) are clustered in time—in other words that inflation occurs in waves, or avalanches. The idea that inflation dynamics is intermittent actually dates back to Robert Engle (1982), and it motivated the introduction of the famous ARCH model,[11] before it was generalized and applied to volatility in financial markets (Engle 2001, and see sec. 5.1).

Nirei and Scheinkman develop a rather sophisticated equilibrium model to describe the repricing strategy of profit-optimizing firms. Here I simplify the model but keep its essential ingredients. I assume that the inflation rate index is $I(t) > 0$. In a mean-field spirit, I do not model the details of the client–supplier network but assume that firms are only sensitive to the overall inflation index. The *real* (i.e., deflated) log-price p_i of firm i then evolves, in the absence of repricing, as

$$\frac{\mathrm{d}p_i}{\mathrm{d}t} = -I(t).$$

At one point, firms have to reprice up otherwise production costs will exceed their expected income. I assume that firm i chooses to do so at a Poisson rate γ if $p_- < p_i < p_+$, in which case the new price is $p_i = p_+ > 0$. But if p_i reaches p_-, expected future losses are deemed too high and the firm reprices immediately at p_+. The evolution of the distribution of prices $P(p, t)$ is then given by (compare with eq. 5)

$$\frac{\partial P(p, t)}{\partial t} = I(t)\frac{\partial P(p, t)}{\partial p} - \gamma P(p, t), \qquad p \in [p_-, p_+] \quad (12)$$

with an equation describing the "reinjection" process at p_+,

$$I(t)P(p_+, t) = \gamma + I(t)P(p_-, t), \qquad (13)$$

---

[11]See equation (17) below. The 1982 Engle paper has, according to Google scholar, 34,000 citations as of today!

which expresses that all firms that reprice, either spontaneously or constrained to do so, contribute to the probability flux at p_+, equal to $I(t)P(p_+, t)$.

Now, the crux of the matter is that a firm that reprices up contributes to the average inflation seen by its customers, and therefore pushes other firms towards repricing. This is the basic mechanism for repricing "avalanches." Let us thus write, again in a mean-field spirit, the effect of repricing on inflation as

~ 277 ~

$$I(t) = I_0 + J\left[\gamma(p_+ - \overline{p}(t)) + I(t)P(p_-, t)(p_+ - p_-)\right],$$
$$\overline{p}(t) = \int_{p_-}^{p_+} \mathrm{d}p\, p P(p, t). \tag{14}$$

where I_0 is the primary good inflation (say energy) and J measures an average client-supplier coupling strength between firms in the production network.[12] The two terms inside the square brackets correspond to spontaneous repricing (at rate γ) and forced repricing (firms with $p_i \approx p_-$).

I now look for a stationary state for the system, i.e. $P(p, t) = P_{st.}(p)$ and $I(t) = I_{st.}$. From equation (12) one derives

$$P_{st.}(p) = Z^{-1} \exp\left[\frac{\gamma p}{I_{st.}}\right], \quad Z = \frac{I_{st.}}{\gamma}\left(\exp\left[\frac{\gamma p_+}{I_{st.}}\right] - \exp\left[\frac{\gamma p_-}{I_{st.}}\right]\right)$$

where Z is such that $\int_{p_-}^{p_+} \mathrm{d}p\, P_{st.}(p) = 1$. One can check that equation (13) is automatically satisfied, as it should be. In the limit $\gamma(p_+ - p_-) \ll I_{st.}$, the average inflation rate is then given by

$$I_{st.} \approx \frac{I_0}{1 - J}, \qquad (J < 1). \tag{15}$$

Now, what happens when a single firm with price close to p_- reprices? By doing so, it increases inflation index I by an

---

[12]Parameter J is the direct analogue of parameter θ defined in Nirei and Scheinkman (2024) and Leal *et al.* (2021).

amount $J(p_+ - p_-)$, such that other firms close to p_- tip over and also reprice. The branching ratio R_0 (average number of firms tipping over) is thus given by

$$R_0 = \int_{p_-}^{p_- + J(p_+ - p_-)} \mathrm{d}p \, P_{\text{st.}}(p) \approx J \qquad (16)$$

where the last equality holds in the same limit $\gamma(p_+ - p_-) \ll I_{\text{st.}}$ considered above.

Hence, the above repricing model becomes exactly equivalent to the branching process described in section 2. The model therefore becomes critical when $J \to 1$, as repricing avalanches of diverging size appear, leading to runaway inflation, see equation (15). When $J \to 1$, the distribution of the size S of repricing avalanches is given by equation (4), and more generally by a generalized Poisson distribution (see Nirei and Scheinkman 2024). Quite remarkably, a calibration of the model on empirical data suggests that J is quite close to unity ($J \sim 0.88$–0.95 for a variety of countries) (Leal *et al.* 2021), suggesting again that the economy is in a strongly coupled regime, not far from a critical point. Interestingly, inflation volatility is also predicted to increase with inflation itself (as $I_{\text{st.}}^{3/2}$), as empirically observed (Nirei and Scheinkman 2024).

Quite remarkably, exactly the same mechanism was proposed in Gualdi *et al.* 2015a to explain the endogenous bankruptcy waves observed in the stylized macroeconomic agent-based model Mark0 proposed in Gualdi *et al.* (2015b). In that model, firms default when their debt-to-sales ratio exceeds a certain threshold, while newborn firms appear with a positive initial balance sheet at some Poisson rate. Much like for the above inflation story, the default of a single firm deteriorates the economic environment of other firms, pushing some of them beyond their bankruptcy threshold. The mathematical encryption of this mechanism is very similar to equation (12) (see Gualdi *et al.* 2015a), and is related to the propagation of neuronal activity in the brain (de Arcangelis, Perrone-Capano, and Herrmann 2006; Lombardi *et al.* 2021), which is also argued to be close to critical (Chialvo 2010; Kinouchi, Pazzini, and Copelli 2020), and for which the regime $J > 1$ corresponds to epilepsy crises.

In Gualdi *et al.* (2015a), special attention was devoted to the case $J > 1$, where no stationary state can be reached. Instead, an oscillating "business-cycle" dynamics appears, where the feedback mechanism described above leads to the synchronization of bankruptcies and spikes of unemployment (see also Sharma *et al.* 2021, for similar effects). One can expect similar inflation cycles to set in for $J > 1$ within the context of the Nirei–Scheinkman model, although the introduction of monetary policy would probably significantly affect the results. Still, the question of why J appears to be empirically so close to unity (i.e., why is the system close to critical) is an open question that certainly requires more thought. In that respect, the analogy between economic activity and the critical brain activity, with its occasional seizures (de Arcangelis, Perrone-Capano, and Herrmann 2006; Chialvo 2010; Osorio *et al.* 2010; Kinouchi, Pazzini, and Copelli 2020), is quite enticing—after all, economic agents are, like neurons, processing information, computing prices and productions, and sending signals to other agents.

5   Self-Organized Criticality in Finance

Financial price series exhibit several "anomalous" statistical properties reminiscent of critical systems. For example, as it is well known, the probability distribution function of daily returns r has fat, power-law tails both for positive and negative returns, decaying as $|r|^{-1-\alpha}$ with α in the range $(3,5)$ for almost all traded instruments Cont 2001; Bouchaud and Potters 2003; Gabaix 2009 (including rather exotic ones like cryptocurrencies, credit default swaps or implied volatility; see Bouchaud 2011; Bouchaud and Challet 2017).

The second intriguing feature is *long memory*: Realized volatility, market activity (measured as transaction volumes, number of price changes, or order book events), and sign of market orders all show a time auto-correlation function that can be fitted by a slow power law. In the case of market activity, one can measure such long-memory effects from minutes to years. This is captured in different ways by, for example, rough/multifractal volatility models (Muzy and Bacry 2000; Muzy, Baïle, and Bacry 2013; Bayer *et al.* 2023) or by Hawkes processes (Hardiman and Bouchaud 2014; Bacry, Mastromatteo, and Muzy 2015) which all rely on scale-free, slowly decaying kernels that propagate the impact of shocks forward in time. As discussed in section 2, the appearance of slow time scales (here much slower than the time between two transactions) is often a signature of critical behavior.

Perhaps the most striking observation suggesting some form of *market fragility* is the fact that most major price jumps do not seem to be related to anything particular happening in the world, at least seen from major news feeds like Bloomberg (see, e.g., Cutler, Poterba, and Summers 1988; Fair 2002; Joulin *et al.* 2008; Marcaccioli, Bouchaud, and Benzaquen 2022; Aubrun *et al.* 2024).[13] Of course, some news does make prices jump, but most of the time jumps seem to be a

---

[13] Note that private information should not make prices jump. Traders who want to benefit from private information should execute their orders in a stealthy manner, as clearly illustrated by Albert Kyle's (1985) model. By making the price jump, they would immediately lose their informational advantage.

result of endogenous feedback loops. Seemingly innocuous shocks get amplified, resulting in micro-, meso-, or even macro-liquidity crises (see sec. 5.2 below). Hence, markets appear to operate "at the edge of chaos," which would explain why they tend to overreact to small perturbations. The precise mechanism(s) driving markets close to a critical point are, however, not known—or rather several such mechanisms have been proposed (as reviewed below), but none have been accepted on the basis of clear, falsifiable predictions.

~ 281 ~

There are actually other, noncritical explanations for power laws and long memory. For example, Gabaix *et al.* (2003) argue that the power-law tail of the return distribution is inherited from the power-law distribution of assets under management.[14] Large investment firms trade large quantities and should induce large price moves. This idea, however, does not seem to fly: Large order sizes are sliced and diced in such a way to be commensurate to the trading volume (see footnote 13). Hence those large orders do not create large returns, but rather long-range correlations in the sign of market orders (see Lillo, Mike, and Farmer 2005; Bouchaud, Farmer, and Lillo 2009; Bouchaud *et al.* 2018; Sato and Kanazawa 2023). Furthermore, large price moves seem to be related to sudden liquidity shortages, not to outsized incoming volumes (see Farmer and Lillo 2004; Gillemot, Farmer, and Lillo 2006; Joulin *et al.* 2008; Fosset, Bouchaud, and Benzaquen 2020).

Long memory effects, on the other hand, could be attributed not to the proximity of a critical point but more trivially to the existence of several natural time scales in human activity, roughly distributed on a log scale (hours, days, weeks,

---

[14]The fact that the distribution of assets is a power law could itself be a signature of criticality. However, simple plausible models based on random multiplicative growth can easily explain such a behavior; see the discussion in Bouchaud and Mézard (2000) and Schwarzkopf and Farmer (2010).

months, quarters, years) (Bouchaud 2001; Bochud and Challet 2007). If different market participants are somehow tuned into these human frequencies, multi-time scales should naturally transpire in market activity and volatility (Zumbach and Lynch 2001). Although such effects might indeed contribute to long memory (there is, for example, a clear periodicity in activity/volatility due to quarterly earnings announcements), excess volatility and volatility clustering seem to be two sides of the same coin, related to self-excitation effects inherent to financial markets, as I discuss now.

5.1 ARCH AND HAWKES PROCESSES

That endogenous feedback loops may be responsible for the excess volatility of financial markets has been suggested many times in the past (see, e.g., Bouchaud 2011; Fosset, Bouchaud, and Benzaquen 2020; Wehrli and Sornette 2022). In fact, the classic ARCH framework mentioned in section 4.4 is arguably the simplest way to describe such a feedback loop from past realized volatility to future volatility. The model posits that daily returns can be written as $r_t = \sigma_t \xi_t$ where ξ_t is an IID $N(0,1)$ Gaussian noise and σ_t the time dependent volatility. One then postulates the following feedback equation:

$$\sigma_t^2 = \sigma_0^2 + \sum_{t'<t} \Phi(t-t')\, r_{t'}^2, \qquad (17)$$

where $\Phi(.)$ is a kernel describing how the perceived volatility $r_{t'}^2$ at time $t' < t$ impacts volatility at time t. Averaging equation (17) over time and assuming stationarity, one finds that the average square volatility $\sigma_\infty^2 = \mathbb{E}_t[\sigma_t^2]$ is given by

$$\sigma_\infty^2 = \frac{\sigma_0^2}{1-g}; \qquad g := \sum_{\tau=1}^{\infty} \Phi(\tau) < 1, \qquad (18)$$

showing that feedback, measured by parameter g, leads to excess volatility. When $g \nearrow 1$, self-referential effects become so strong that volatility explodes (compare with eq. 15). $g = 1$ again appears as a critical point, in fact accompanied by slowing down, as in section 2. More precisely, the autocorrelation function of σ_t^2 decays over a time scale much longer than the decay time of the kernel $\Phi(\tau)$ when $g \nearrow 1$.

Hawkes processes provide an alternative description for self-exciting feedback loops. A Hawkes process describes Poisson events whose intensity increases with the past activity of the process itself (for a review in the context of financial markets, see Bacry, Mastromatteo, and Muzy 2015). Such an increase is mediated by a kernel $\Phi(\tau)$, as for ARCH processes, but now in continuous time. When $g := \int d\tau \, \Phi(\tau)$ becomes larger than unity, the intensity of the process diverges. When g is precisely equal to 1, the process is critical and only exists for long-memory (power-law) kernels (Brémaud and Massoulié 2001).

Both ARCH and Hawkes models have been calibrated on data. In all cases, one finds very high values $g \gtrsim 0.8$ for the integrated feedback parameter (see, e.g., Chicheportiche and Bouchaud 2014; Hardiman, Bercot, and Bouchaud 2013; Filimonov *et al.* 2014; Hardiman and Bouchaud 2014; Blanc, Donier, and Bouchaud 2017; Wheatley, Wehrli, and Sornette 2019; Wehrli and Sornette 2022), with a power-law decaying kernel $\Phi(\tau)$. This is in line with Shiller's excess volatility puzzle (Shiller 1981, 1987; LeRoy and Porter 1981): Volatility in financial markets is at least five times larger than what it "should" be in the absence of feedback—and, as noted in the previous section, many jumps appear without obvious causes.

But yet again, one might ask: Why is the feedback kernel a power law and the feedback parameter so high? What are the driving forces explaining why the system appears to be not far from criticality, as we saw in the case of inflation as well?

5.2 CRITICAL LIQUIDITY PROVISION

One plausible explanation for the fragility of financial markets is that at the very heart of all markets lies a fundamental competitive tension: Buyers want to buy low and sellers want to sell high. The role of *market-makers* (MM), also known as liquidity providers, is to resolve (part of) this tension. MM buy from sellers and sell later to buyers (or vice versa), allowing markets to solve the immediacy problem and ensure, most of the time, "fair and orderly trading." MM make a living from the bid–ask spread s_t—buying slightly below current midprice p_t and selling slightly above p_t, while trying to keep their inventory as close to zero as possible.

MM, however, bear the risk of adverse selection, which results from the fact they must post binding quotes (bid or ask prices), which can be "picked off" by more informed traders who see an opportunity to buy low or to sell high. In other words, if the MM sells (or buys) at a price $p_t \pm s_t/2$ that turns out be lower (higher) than the future price at which they will need to unwind their position, they will suffer a loss that can be very large. More formally, let us define the average adverse price move as

$$\Delta := \mathbb{E}\left[\epsilon_t \cdot (p_\infty - p_t)\right], \qquad (19)$$

where ϵ_t is the sign of the incoming trade and p_∞ is the price in a sufficiently distant future (for more precise statements, see Wyart *et al.* 2008; Bouchaud *et al.* 2018, ch. 18). Market-making is thus only profitable if the bid–ask spread exceeds, on average, adverse selection, that is, if $\mathbb{E}[s_t]/2 > \Delta$.

Unfortunately for MM, the distribution of price changes after a trade is very broad, with a heavy tail in the direction of the trade. Whereas most trades contain relatively little information and are therefore innocuous for MM, some rare trades are followed by extreme adverse moves, wiping out the small profits accumulated on "normal" transactions. In other words, the distribution of MM profits is strongly negatively skewed (see Bouchaud *et al.* 2018, ch. 1).

Because liquidity providers compete, the spread is compressed to values close to break even, that is, $s \approx 2\Delta$. But since the risk of large losses is huge, MM are quick to increase the spread and reduce the amount of liquidity that they offer for purchase or sale when volatility ticks up.

This simple argument illustrates why liquidity is fragile and can disappear quickly.[15] In fact, the classic Glosten–Milgrom model (1985) predicts that high volatilities can lead to liquidity crises, in the sense that no value of the spread can allow MM to break even. Hence, there is a universal built-in destabilizing feedback loop, intrinsic to the way markets operate but independent of their specific microstructure: *More volatility leads to higher spreads and less liquidity, which itself leads to higher volatilities.* Such a mechanism is discussed and empirically validated in Fosset, Bouchaud, and Benzaquen (2020), Dall'Amico *et al.* (2019), and Fosset, Bouchaud, and Benzaquen (2022), and sounds like a plausible explanation for the large fraction of endogenous price jumps mentioned previously (Joulin *et al.* 2008; Marcaccioli, Bouchaud, and Benzaquen 2022; Aubrun *et al.* 2024).

Markets thus appear to be maintained in a subtle equilibrium between liquidity takers (the excitatory force) and liquidity providers (the inhibitory force; see Bouchaud *et al.* 2003), much as for the balancing-stick problem of section 3.4, or in the context of heartbeats or neuronal activity (Cabrera and Milton 2004, 2012; Patzelt and Pawelzik 2011; Lombardi, Herrmann, and Arcangelis 2017). In such a story, criticality in financial markets is enforced by competition that drives the system in the vicinity of the break-

---

[15] As the saying goes: *Liquidity is a coward. It is never there when it is needed.*

even point $s = 2\Delta$, such that neither liquidity takers nor liquidity providers are systematically favored see (see Wyart *et al.* 2008; Bouchaud *et al.* 2018, ch. 17). Much like in section 4.1, efficiency (i.e., small spreads offered to liquidity takers) comes with instability. It would be rewarding to come up with a stylized model, in the spirit of section 4.1, that would neatly encapsulate the above scenario and allow one to compute, for example, the distribution of the size of endogenous jumps, that is, when liquidity spontaneously disappears through feedback effects.

5.3 COMPETITIVE MARKET ECOLOGY

Another path to understand the apparent criticality of financial markets is to think in terms of an ecological, competitive equilibrium between different investment strategies (Farmer 2002). Thomas Lux and Michele Marchesi's (1999) paper is the first to explicitly exploit this idea and insist on the (self-organized) criticality of the resulting artificial market (see also Scholl, Calinescu, and Farmer 2021, for closely related ideas).

Following Carl Chiarella (1992), investors are split into three broad classes: value investors (arbitraging the difference between market price and their estimate of value), trend followers (buying when the price went up, and selling when it went down), and noise traders (with essentially random strategies). Lux and Marchesi add the possibility for investors to switch between trend following and value depending on the relative pay-off of the two strategies (see also Palmer *et al.* 1994; Brock and Hommes 1998). Quite interestingly, these ingredients appear to be enough to generate fat-tailed returns and long-range volatility correlations, which come from the fact that agents can stick to one kind of strategy for very long times—see Giardina and Bouchaud (2003) for more details on this point. The coexistence of trend followers and value traders is further documented empirically in Bouchaud *et al.* (2017), Majewski, Ciliberti, and Bouchaud (2020), Schmitt and Westerhoff (2021), and Lux (2021) and references therein.

There are several variations on the same theme. For example, the minority game is a multi-agent stylized model of market ecology amenable to analytic calculations (Challet, Marsili, and Zhang 2004), which reveals a *genuine phase transition* between "no-arbitrage" and exploitable regularities. As the number of agents M playing the game increases, profit opportunities wither and actually vanish for a critical value M_c, beyond which the average profit of agents becomes negative due to transaction costs.

The attractive nature of the critical point is then quite clear (Challet and Marsili 2003; Giardina and Bouchaud 2003; Alfi, Pietronero, and Zaccaria 2009) and somewhat akin to the classic Grossman–Stiglitz (1980) paradox: Either $M < M_c$ and new investors are lured in by profit opportunities, such that $M \nearrow$; or $M > M_c$, and current investors then make losses and progressively leave the market, until $M \approx M_c$. As markets become more complex and unpredictable, they also become more fragile, as discussed in section 3.4.

This generic path to criticality is actually quite enticing and lends credence to the SOC scenario for financial markets (see Alfi, Pietronero, and Zaccaria 2009 for a review, and Farmer 2002; Amir *et al.* 2005; Caccioli, Marsili, and Vivo 2009; Patzelt and Pawelzik 2011; Marsili 2014 for related ideas), in particular the difficulty for prices to reach equilibrium when incentives provided by mispricing tend to zero (Grossman and Stiglitz 1980; Cherkashin, Farmer, and Lloyd 2009; Scholl, Calinescu, and Farmer 2021).

Note finally that, long ago, J. Doyne Farmer (2002) derived a generalized Lotka–Volterra description of the financial market ecology, along the lines of equation (7). As market participants trade and impact prices, other participants either benefit or suffer from those price changes. This could provide another justification of market criticality: Investment strategies feed into one another in mutually beneficial or predator–prey relationships, coevolving into a marginally stable equilibrium, as in section 3.3.

5.4 CONTAGION AND FINANCIAL STABILITY

Finally, contagion mechanisms are evidently at play in financial markets, much like in the banking sector (Haldane and May 2011; Gai and Kapadia 2010; Caccioli, Barucca, and Kobayashi 2018; see also Caccioli et al., ch. 16 of this volume).

As I just mentioned, the trades of one investor impacts the holdings of all other market participants. When two investors (say, Alice and Bob) have *overlapping portfolios*, that is, similar positions in the market—the deleveraging of Alice's portfolio will lead to a depreciation of the marked-to-market value of Bob's portfolio (Thurner, Farmer, and Geanakoplos 2012; Caccioli, Bouchaud, and Farmer 2012; Cont and Wagalath 2013; Caccioli *et al.* 2015). If the drop in value is too large, Bob may be tempted (or compelled by his broker) to deleverage his own portfolio, further lowering the value of Alice's holdings, and perhaps that of other investors as well. Such a contagion mechanism can set off "deleveraging spirals" (Brunnermeier and Pedersen 2009), avalanches of trades in the same direction, collectively detrimental to all investors holding similar positions. Such a mechanism has been argued to be at the origin of the infamous "quant crunch" of 2007 (Khandani and Lo 2007), and was shown in Thurner, Farmer, and Geanakoplos (2012) to generate a fat left-tail in the distribution of equity returns.

The basic model for such a phenomenon is again the branching process of section 2.2. The contagion parameter R_0 is then a measure of the similarity between portfolios and of the funds' sensitivity to losses. If leverage constraints are binding, for example, R_0 will be higher. It is not clear, however, why in this story R_0 should be particularly close to the critical value $R_0 = 1$. But one could easily argue with Minsky that as complacency increases, portfolio risk also increases, making them more exposed to downside events—thereby realizing the growing sand-pile scenario of section 3.2.

5.5 DISCUSSION

The previous sections presented rather disparate scenarios for critical-like behavior in financial markets. It is not clear which one is to be favored at this stage. Different mechanisms might actually be needed to understand the dynamics of markets over different time scales.

My personal view is that the critical liquidity scenario of section 5.2 is the one explaining the fat-tailed distribution of intra-day and daily returns. The basic feedback loop between increased volatility and decreased liquidity appears to be universal across both markets and epochs, and has the potential to destabilize markets in the absence of news (Dall'Amico *et al.* 2019; Fosset, Bouchaud, and Benzaquen 2020). As argued in Bouchaud *et al.* (2018, chs. 2 and 20), excess volatility generated at high frequency then naturally cascades to lower frequencies due to the random-walk nature of prices.

Other mechanisms might, however, be at play in the emergence of long memory in activity fluctuations that extend up to months or even years (on this particular point, see also Muzy, Baïle, and Bacry 2013). The competitive market ecology picture of section 5.3 could be relevant to understand such lower-frequency phenomena. Similarly, contagion effects—as discussed in section 5.4—most likely play an important role in explaining extreme tail events such as major crashes (portfolio insurance in 1987, overlapping portfolios during the 2007 quant crunch, global deleveraging in 2008, etc.).

Still, both the very relevance of the SOC scenario and the relative importance of its possible incarnations in the dynamics of financial markets are, in my opinion, very much open questions. Their resolution might have substantial impact on how markets should be organized and regulated, such as to actually ensure "fair and orderly trading."

6   Conclusion: Efficiency vs. Resilience

Elucidating the very mechanisms leading to instabilities, failures and system-wide crises in socioeconomic and financial systems is crucial to finding remedies and mitigation measures and proposing adequate regulations. Indeed, as argued throughout this paper and in Moran *et al.* (2023), the quest for "efficiency" and "optimality" and the necessity of *resilience*—that is, of robustness against small perturbations, may be mutually incompatible, a point forcefully made in Carlson and Doyle (1999).

The self-organized criticality scenario goes even further, as it suggests that optimization necessarily leads in many complex systems to a marginally stable point that is particularly *fragile*—that is, at the edge of the proverbial cliff. We have seen, for example, how "just-in-time" policies low inventories, or over-optimized timetables necessarily push the system towards a functioning point where disruption avalanches of all sizes can appear (secs. 4.1, 4.2). I have argued that liquidity in financial markets is fragile because competition between market-makers drives their profits close to zero, making them hyper-sensitive to volatility blips (sec. 5.2). I have discussed how competitive ecologies (of living systems, of firms, of economic agents, of investment strategies) tend to co-evolve towards marginally stable, fragile equilibrium points where just enough species have to disappear for others to survive (secs. 3.3, 4.2, 4.3, 5.3). Finally, I have emphasized how stabilizing unstable systems may be inherently difficult, and may generate anomalously large fluctuations when stabilization occasionally falters (secs. 3.4, 5.2). More generally, systems driven by antagonist forces (excitatory and inhibitory) tend to operate close to criticality. The point is that fragility and flexibility/adaptability are often two sides of the same coin.

Several plausible mechanisms could thus bring financial markets and economies as a whole close to, or even through, a critical point where the system loses stability. At this juncture, however, we have many more narratives than hard empirical evidence for such an interpretation of the pervasive "excess volatility" or "small shocks, large business cycle" effects routinely observed in these systems. A renewed data-driven effort, like in Leal *et al.* (2021), is certainly needed to convince the economics community of the relevance of *fragility* to understand excess fluctuations, without having to invoke major exogenous shocks (which are often difficult to identify and substantiate).

Similarly, we also sometimes lack clear mechanisms to explain why the system is attracted to the critical point, for example in the case of inflation avalanches (see sec. 4.4, Nirei and Scheinkman 2024). I have actually mentioned several times that excess fluctuations could be due to other mechanisms, *not* related to the immediate proximity of a critical point. The system might be such that its dynamics cycles through an instability, like a slowly growing sand pile that periodically collapses, or, in a financial context, a growing complacency toward risk that leads to market crashes—the famous Minsky (2015) cycle. Another possibility is that the system is linearly unstable but nonlinearly stable, chaotically evolving in a region of parameter space and generating *purely endogenous* fluctuations, as we discussed in the context of firm networks in section 4.3. Yet another path is provided by the role of noise in bistable systems (di Santo *et al.* 2016; Harras, Tessone, and Sornette 2012).

In any case, the main policy consequence of fragility in socioeconomic systems is that any welfare function that system operators, policymakers, or regulators seek to optimize should contain a measure of the robustness of the solution to small perturbations, to the uncertainty about parameters value, and to extreme events. Adding such a resilience penalty will for sure increase costs and degrade strict economic performance, but it will keep the solution at a safe distance away from the cliff edge. As argued by Nassim Nicholas Taleb (2014), and also using a different language in Hynes *et al.* (2022), good policies should ideally lead to "anti-fragile" systems, that is, systems that spontaneously improve when buffeted by large shocks. ↲

Acknowledgments

I wish to thank Doyne Farmer for asking me to write this piece and allowing me to put my ideas together on this exciting topic. His comments on my initial draft have been very useful. My understanding of the subject owes a lot to many conversations with him over the years, and also with many friends and collaborators: F. Aguirre-Lopez, C. Aubrun, E. Bacry, M. Benzaquen, G. Biroli, J. Bonart, G. Bunin, F. Caccioli, D. Challet, C. Colon, Th. Dessertaine, J. Donier, A. Fosset, X. Gabaix, J. Garnier-Brun, J. Gatheral, I. Giardina, M. Gould, S. Gualdi, S. Hardiman, K. Kanazawa, A. Kirman, P. Le Doussal, A. Majewski, R. Marcaccioli, M. Marsili, I. Mastromatteo, M. Mézard, J. Moran, R. Morel, J.-F. Muzy, D. Panja, N. Patil, F. Patzelt, F. Pijpers, M. Potters, M. Rosenbaum, J. Scheinkman, A. Secchi, J. Sethna, M. Smerlak, D. Sornette, N. Taleb, M. Tarzia, U. Weitzel, M. Wyart, and F. Zamponi.

REFERENCES

Acemoglu, D., V. M. Carvalho, A. Ozdaglar, and A. Tahbaz-Salehi. 2012. "The Network Origins of Aggregate Fluctuations." *Econometrica* 80 (5): 1977–2016. https://doi.org/10.3982/ECTA9623.

Alfi, V., L. Pietronero, and A. Zaccaria. 2009. "Self-Organization for the Stylized Facts and Finite-Size Effects in a Financial-Market Model." *Europhysics Letters* 86 (5): 58003. https://doi.org/10.1209/0295-5075/86/58003.

Amir, R., I. Evstigneev, T. Hens, and K. Schenk-Hoppé. 2005. "Market Selection and Survival of Investment Strategies." *Journal of Mathematical Economics* 41 (1-2): 105–122. https://EconPapers.repec.org/RePEc:eee:mateco:v:41:y:2005:i:1-2:p:105-122.

Aspelmeier, T. 2008. "Bond Chaos in the Sherrington-Kirkpatrick Model." *Journal of Physics A: Mathematical and Theoretical* 41 (20): 205005. https://doi.org/10.1088/1751-8113/41/20/205005.

Aspelmeier, T., and M. A. Moore. 2019. "Realizable Solutions of the Thouless-Anderson-Palmer Equations." *Physical Review E* 100 (3): 032127. https://doi.org/10.1103/PhysRevE.100.032127.

Atalay, E., A. Hortaçsu, J. Roberts, and C. Syverson. 2011. "Network Structure of Production." *Proceedings of the National Academy of Sciences* 108 (13): 5199–5202. https://doi.org/10.1073/pnas.1015564108.

Aubrun, C., R. Morel, M. Benzaquen, and J.-P. Bouchaud. 2024. *Riding Wavelets: A Method to Discover New Classes of Price Jumps.* ArXiv preprint:2404.16467. https://doi.org/10.48550/arXiv.2404.16467.

Axtell, R. L. 2001. "Zipf Distribution of US Firm Sizes." *Science* 293 (5536): 1818–1820. https://doi.org/10.1126/science.1062081.

Bacry, E., I. Mastromatteo, and J.-F. Muzy. 2015. "Hawkes Processes in Finance." *Market Microstructure and Liquidity* 1 (01): 1550005. https://doi.org/10.1142/S2382626615500057.

Bak, P. 2013. *How Nature Works: The Science of Self-Organized Criticality.* New York, NY: Copernicus. https://doi.org/10.1007/978-1-4757-5426-1.

Bak, P., K. Chen, J. Scheinkman, and M. Woodford. 1993. "Aggregate Fluctuations from Independent Sectoral Shocks: Self-Organized Criticality in a Model of Production and Inventory Dynamics." *Ricerche Economiche* 47 (1): 3–30. https://doi.org/10.1016/0035-5054(93)90023-V.

Bak, P., and M. Paczuski. 1997. "Mass Extinctions vs. Uniformitarianism in Biological Evolution." In *Physics of Biological Systems: From Molecules to Species,* edited by H. Flyvbjerg, J. Hertz, M. H. Jensen, O. G. Mouritsen, and K. Sneppen, 341–356. Berlin, Germany: Springer.

Bak, P., C. Tang, and K. Wiesenfeld. 1987. "Self-Organized Criticality: An Explanation of the 1/f Noise." *Physical Review Letters* 59 (4): 381–384. https://doi.org/10.1103/PhysRevLett.59.381.

Baqaee, D. R., and E. Farhi. 2019. "The Macroeconomic Impact of Microeconomic Shocks: Beyond Hulten's Theorem." *Econometrica* 87 (4): 1155–1203. https://doi.org/10.3982/ECTA15202.

———. 2020. "Productivity and Misallocation in General Equilibrium." *The Quarterly Journal of Economics* 135 (1): 105–163. https://doi.org/10.1093/qje/qjz030.

Bayer, C., P. K. Friz, M. Fukasawa, J. Gatheral, A. Jacquier, and M. Rosenbaum, eds. 2023. *Rough Volatility.* Philadelphia, PA: SIAM. https://doi.org/10.1137/1.9781611977783.

Bernanke, B. S., M. Gertler, and S. Gilchrist. 1996. "The Financial Accelerator and the Flight to Quality." *The Review of Economics and Statistics* 78 (1): 1–15. https://doi.org/10.2307/2109844.

Bialek, W., A. Cavagna, I. Giardina, T. Mora, O. Pohl, E. Silvestri, M. Viale, and A. M. Walczak. 2014. "Social Interactions Dominate Speed Control in Poising Natural Flocks Near Criticality." *Proceedings of the National Academy of Sciences* 111 (20): 7212–7217. https://doi.org/10.1073/pnas.1324045111.

Biroli, G., G. Bunin, and C. Cammarota. 2018. "Marginally Stable Equilibria in Critical Ecosystems." *New Journal of Physics* 20 (8): 083051. https://doi.org/10.1088/1367-2630/aada58.

Blanc, P., J. Donier, and J.-P. Bouchaud. 2017. "Quadratic Hawkes Processes for Financial Prices." *Quantitative Finance* 17 (2): 171–188. https://doi.org/10.1080/14697688.2016.1193215.

Bochud, T., and D. Challet. 2007. "Optimal Approximations of Power Laws with Exponentials: Application to Volatility Models with Long Memory." *Quantitative Finance* 7 (6): 585–589. https://doi.org/10.1080/14697680701278291.

Bouchaud, J.-P. 2001. "Power Laws in Economics and Finance: Some Ideas from Physics." *Quantitative Finance* 1 (1): 105. https://doi.org/10.1088/1469-7688/1/1/307.

———. 2011. "The Endogenous Dynamics of Markets: Price Impact, Feedback Loops and Instabilities." In *Lessons from the Credit Crisis,* 345–74. New York, NY: Risk Publications.

———. 2021. "Radical Complexity." *Entropy* 23 (12): 1676. https://doi.org/10.3390/e23121676.

Bouchaud, J.-P., J. Bonart, J. Donier, and M. Gould. 2018. "The Impact of Market Orders." In *Trades, Quotes and Prices: Financial Markets Under the Microscope,* edited by J.-P. Bouchaud, J. Bonart, J. Donier, and M. Gould, 208–228. Cambridge, UK: Cambridge University Press. https://doi.org/10.1017/9781316659335.017.

Bouchaud, J.-P., and D. Challet. 2017. "Why Have Asset Price Properties Changed so Little in 200 Years?" In *Econophysics and Sociophysics: Recent Progress and Future Directions,* edited by F. Abergel, H. Aoyama, B. K. Chakrabarti, A. Chakraborti, N. Deo, D. Raina, and I. Vodenska, 3–17. Cham, Switzerland: Springer. https://doi.org/10.1007/978-3-319-47705-3_1.

Bouchaud, J.-P., S. Ciliberti, Y. Lempérière, A. Majewski, P. Seager, and K. Sin Ronia. 2017. *Black was Right: Price is Within a Factor 2 of Value.* arXiv preprint: 1711.04717. https://doi.org/10.48550/arXiv.1711.04717.

Bouchaud, J.-P., J. D. Farmer, and F. Lillo. 2009. "How Markets Slowly Digest Changes in Supply and Demand." In *Handbook of Financial Markets: Dynamics and Evolution,* edited by T. Hens and K. R. Schenk-Hoppé, 57–160. Amsterdam, Netherlands: Elsevier. https://doi.org/10.1016/B978-012374258-2.50006-3.

Bouchaud, J.-P., Y. Gefen, M. Potters, and M. Wyart. 2003. "Fluctuations and Response in Financial Markets: The Subtle Nature of Random Price Changes." *Quantitative Finance* 4 (2): 176. https://doi.org/10.1080/14697680400000022.

Bouchaud, J.-P., and M. Mézard. 2000. "Wealth Condensation in a Simple Model of Economy." *Physica A: Statistical Mechanics and its Applications* 282 (3–4): 536–545. https://doi.org/10.1016/S0378-4371(00)00205-3.

Bouchaud, J.-P., and M. Potters. 2003. *Theory of Financial Risks.* Vol. 12. Cambridge, UK: Cambridge University Press.

Brémaud, P., and L. Massoulié. 2001. "Hawkes Branching Point Processes without Ancestors." *Journal of Applied Probability* 38 (1): 122–135. https://doi.org/10.1239/jap/996986648.

Brock, W. A., and C. H. Hommes. 1998. "Heterogeneous Beliefs and Routes to Chaos in a Simple Asset Pricing Model." *Journal of Economic Dynamics and Control* 22 (8-9): 1235–1274. https://doi.org/10.1016/S0165-1889(98)00011-6.

Brunnermeier, M. K., and L. H. Pedersen. 2009. "Market Liquidity and Funding Liquidity." *The Review of Financial Studies* 22 (6): 2201–2238. https://doi.org/10.1093/rfs/hhn098.

Bunin, G. 2017. "Ecological Communities with Lotka–Volterra Dynamics." *Physical Review E* 95 (4): 042414. https://doi.org/10.1103/PhysRevE.95.042414.

Cabrera, J. L., and J. G. Milton. 2004. "Human Stick Balancing: Tuning Lévy Flights to Improve Balance Control." *Chaos: An Interdisciplinary Journal of Nonlinear Science* 14 (3): 691–698. https://doi.org/10.1063/1.1785453.

———. 2012. "Stick Balancing, Falls and Dragon-Kings." *The European Physical Journal Special Topics* 205 (1): 231–241. https://doi.org/10.1140/epjst/e2012-01573-7.

Caccioli, F., P. Barucca, and T. Kobayashi. 2018. "Network Models of Financial Systemic Risk: A Review." *Journal of Computational Social Science* 1:81–114. https://doi.org/10.1007/s42001-017-0008-3.

Caccioli, F., J.-P. Bouchaud, and D. Farmer. 2012. "Impact-Adjusted Valuation and the Criticality of Leverage." *Risk* 25 (12): 74–77.

Caccioli, F., J. D. Farmer, N. Foti, and D. Rockmore. 2015. "Overlapping Portfolios, Contagion, and Financial Stability." *Journal of Economic Dynamics and Control* 51:50–63. https://doi.org/10.1016/j.jedc.2014.09.041.

Caccioli, F., M. Marsili, and P. Vivo. 2009. "Eroding Market Stability by Proliferation of Financial Instruments." *The European Physical Journal B* 71:467–479. https://doi.org/10.1140/epjb/e2009-00316-y.

Carlson, J. M., and J. Doyle. 1999. "Highly Optimized Tolerance: A Mechanism for Power Laws in Designed Systems." *Physical Review E* 60 (2): 1412. https://doi.org/10.1103/PhysRevE.60.1412.

Carvalho, V. M., and A. Tahbaz-Salehi. 2019. "Production Networks: A Primer." *Annual Review of Economics* 11:635–663. https://doi.org/10.1146/annurev-economics-080218-030212.

Castellano, C., and R. Pastor-Satorras. 2017. "Relating Topological Determinants of Complex Networks to Their Spectral Properties: Structural and Dynamical Effects." *Physical Review X* 7 (4): 041024. https://doi.org/10.1103/PhysRevX.7.041024.

Cavagna, A., I. Giardina, and G. Parisi. 1998. "Stationary Points of the Thouless–Anderson–Palmer Free Energy." *Physical Review B* 57 (18): 11251. https://doi.org/10.1103/PhysRevB.57.11251.

Challet, D., and M. Marsili. 2003. "Criticality and Market Efficiency in a Simple Realistic Model of the Stock Market." *Physical Review E* 68 (3): 036132. https://doi.org/10.1103/PhysRevE.68.036132.

Challet, D., M. Marsili, and Y.-C. Zhang. 2004. *Minority Games: Interacting Agents in Financial Markets.* Oxford, UK: Oxford University Press.

Cherkashin, D., J. D. Farmer, and S. Lloyd. 2009. "The Reality Game." *Journal of Economic Dynamics and Control* 33 (5): 1091–1105. https://doi.org/10.1016/j.jedc.2009.02.002.

Chialvo, D. R. 2010. "Emergent Complex Neural Dynamics." *Nature Physics* 6 (10): 744–750. https://doi.org/10.1038/nphys1803.

Chiarella, C. 1992. "The Dynamics of Speculative Behaviour." *Annals of Operations Research* 37 (1): 101–123. https://doi.org/10.1007/BF02071051.

Chicheportiche, R., and J.-P. Bouchaud. 2014. "The Fine-Structure of Volatility Feedback I: Multi-Scale Self-Reflexivity." *Physica A: Statistical Mechanics and its Applications* 410:174–195. https://doi.org/10.1016/j.physa.2014.05.007.

Cochrane, J. H. 1994. "Shocks: A Comment." *Carnegie–Rochester Conference Series on Public Policy* 41:295–364. https://doi.org/10.1016/0167-2231(94)00024-7.

Colon, C., and J.-P. Bouchaud. 2022. "The Radical Complexity of Rewiring Supplier–Buyer Networks." *SSRN Electronic Journal,* https://pure.iiasa.ac.at/19482.

Colon, C., and M. Ghil. 2017. "Economic Networks: Heterogeneity-Induced Vulnerability and Loss of Synchronization." *Chaos: An Interdisciplinary Journal of Nonlinear Science* 27 (12): 126703. https://doi.org/10.1063/1.5017851.

Cont, R. 2001. "Empirical Properties of Asset Returns: Stylized Facts and Statistical Issues." *Quantitative Finance* 1 (2): 223. https://doi.org/10.1080/713665670.

Cont, R., and L. Wagalath. 2013. "Running for the Exit: Distressed Selling and Endogenous Correlation in Financial Markets." *Mathematical Finance: An International Journal of Mathematics, Statistics and Financial Economics* 23 (4): 718–741. https://doi.org/10.1111/j.1467-9965.2011.00510.x.

Cutler, D. M., J. M. Poterba, and L. H. Summers. 1988. *What Moves Stock Prices?* Technical report 2538. Cambridge, MA: National Bureau of Economic Research. https://doi.org/10.3386/w2538.

Dall'Amico, L., A. Fosset, J.-P. Bouchaud, and M. Benzaquen. 2019. "How Does Latent Liquidity get Revealed in the Limit Order Book?" *Journal of Statistical Mechanics: Theory and Experiment* 2019 (1): 013404. https://doi.org/10.1088/1742-5468/aaf10e.

de Arcangelis, L., C. Perrone-Capano, and H. J. Herrmann. 2006. "Self-Organized Criticality Model for Brain Plasticity." *Physical Review Letters* 96 (2): 028107. https://doi.org/10.1103/PhysRevLett.96.028107.

Dekker, M. M., and D. Panja. 2021. "Cascading Dominates Large-Scale Disruptions in Transport Over Complex Networks." *PLoS One* 16 (1): e0246077. https://doi.org/10.1371/journal.pone.0246077.

Dessertaine, T., J. Moran, M. Benzaquen, and J.-P. Bouchaud. 2022. "Out-of-Equilibrium Dynamics and Excess Volatility in Firm Networks." *Journal of Economic Dynamics and Control* 138:104362. https://doi.org/10.1016/j.jedc.2022.104362.

di Santo, S., R. Burioni, A. Vezzani, and M. A. Muñoz. 2016. "Self-Organized Bistability Associated with First-Order Phase Transitions." *Physical Review Letters* 116 (24): 240601. https://doi.org/10.1103/PhysRevLett.116.240601.

Engle, R. F. 1982. "Autoregressive Conditional Heteroscedasticity with Estimates of the Variance of United Kingdom Inflation." *Econometrica: Journal of the Econometric Society,* 987–1007. https://www.jstor.org/stable/1912773.

————. 2001. "GARCH 101: The Use of ARCH/GARCH Models in Applied Econometrics." *Journal of Economic Perspectives* 15 (4): 157–168. https://doi.org/10.1257/jep.15.4.157.

Fair, R. C. 2002. "Events that Shook the Market." *The Journal of Business* 75 (4): 713–731. https://doi.org/10.1086/341640.

Farmer, J. D. 2002. "Market Force, Ecology and Evolution." *Industrial and Corporate Change* 11 (5): 895–953. https://doi.org/10.1093/icc/11.5.895.

————. 2024. *Making Sense of Chaos: A Better Economics for a Better World.* New Haven, CT: Yale University Press.

Farmer, J. D., and F. Lillo. 2004. "On the Origin of Power-Law Tails in Price Fluctuations." *Quantitative Finance* 4 (1): 7–11. https://doi.org/10.1088/1469-7688/4/1/C01.

Filimonov, V., D. Bicchetti, N. Maystre, and D. Sornette. 2014. "Quantification of the High Level of Endogeneity and of Structural Regime Shifts in Commodity Markets." *Journal of International Money and Finance* 42:174–192. https://doi.org/10.1016/j.jimonfin.2013.08.010.

Fisher, D. S., and D. A. Huse. 1991. "Directed Paths in a Random Potential." *Physical Review B* 43 (13): 10728–10742. https://doi.org/10.1103/PhysRevB.43.10728.

Fisher, F. M. 1989. *Disequilibrium Foundations of Equilibrium Economics.* Cambridge, UK: Cambridge University Press.

Flaschel, P. 2010. "The Classical Growth Cycle: Reformulation, Simulation and Some Facts." In *Topics in Classical Micro- and Macroeconomics,* 435–463. Amsterdam, Netherlands: Springer. https://doi.org/10.1007/978-3-642-00324-0_20.

Fosset, A., J.-P. Bouchaud, and M. Benzaquen. 2020. "Endogenous Liquidity Crises." *Journal of Statistical Mechanics: Theory and Experiment* 2020 (6): 063401. https://doi.org/10.1088/1742-5468/ab7c64.

————. 2022. "Non-Parametric Estimation of Quadratic Hawkes Processes for Order Book Events." *The European Journal of Finance* 28 (7): 663–678. https://doi.org/10.1080/1351847X.2021.1917441.

Franz, S., G. Parisi, M. Sevelev, P. Urbani, and F. Zamponi. 2017. "Universality of the SAT–UNSAT (Jamming) Threshold in Non-Convex Continuous Constraint Satisfaction Problems." *SciPost Physics* 2:019. https://doi.org/10.21468/SciPostPhys.2.3.019.

Frisch, U. 1995. *Turbulence: The Legacy of A. N. Kolmogorov.* Cambridge, UK: Cambridge University Press. https://doi.org/10.1017/CBO9781139170666.

Gabaix, X. 2009. "Power Laws in Economics and Finance." *Annual Review of Economics* 1 (1): 255–294. https://doi.org/10.1146/annurev.economics.050708.142940.

———. 2011. "The Granular Origins of Aggregate Fluctuations." *Econometrica* 79 (3): 733–772. https://doi.org/10.3982/ECTA8769.

Gabaix, X., P. Gopikrishnan, V. Plerou, and H. E. Stanley. 2003. "A Theory of Power-Law Distributions in Financial Market Fluctuations." *Nature* 423 (6937): 267–270. https://doi.org/10.1038/nature01624.

———. 2006. "Institutional Investors and Stock Market Volatility." *The Quarterly Journal of Economics* 121 (2): 461–504. https://www.jstor.org/stable/25098798.

Gai, P., and S. Kapadia. 2010. "Contagion in Financial Networks." *Proceedings of the Royal Society A: Mathematical, Physical and Engineering Sciences* 466 (2120): 2401–2423. https://doi.org/10.1098/rspa.2009.0410.

Garnier-Brun, J., M. Benzaquen, and J.-P. Bouchaud. 2024. "Unlearnable Games and 'Satisficing' Decisions: A Simple Model for a Complex World." *Physical Review X* 14 (2): 021039. https://doi.org/10.1103/PhysRevX.14.021039.

Garnier-Brun, J., M. Benzaquen, S. Ciliberti, and J.-P. Bouchaud. 2021. "A New Spin on Optimal Portfolios and Ecological Equilibria." *Journal of Statistical Mechanics: Theory and Experiment* 2021 (9): 093408. https://doi.org/10.1088/1742-5468/ac21d9.

Giardina, I., and J.-P. Bouchaud. 2003. "Bubbles, Crashes and Intermittency in Agent Based Market Models." *The European Physical Journal B—Condensed Matter and Complex Systems* 31:421–437. https://doi.org/10.1140/epjb/e2003-00050-6.

Gillemot, L., J. D. Farmer, and F. Lillo. 2006. "There's More to Volatility than Volume." *Quantitative Finance* 6 (5): 371–384. https://doi.org/10.1080/14697680600835688.

Glosten, L. R., and P. R. Milgrom. 1985. "Bid, Ask and Transaction Prices in a Specialist Market with Heterogeneously Informed Traders." *Journal of Financial Economics* 14 (1): 71–100. https://doi.org/10.1016/0304-405X(85) 90044-3.

Grossman, S. J., and J. E. Stiglitz. 1980. "On the Impossibility of Informationally Efficient Markets." *The American Economic Review* 70 (3): 393–408. https://www.jstor.org/stable/1805228.

Gualdi, S., J.-P. Bouchaud, G. Cencetti, M. Tarzia, and F. Zamponi. 2015a. "Endogenous Crisis Waves: Stochastic Model with Synchronized Collective Behavior." *Physical Review Letters* 114 (8): 088701. https://doi.org/10.1103/PhysRevLett.114.088701.

Gualdi, S., M. Tarzia, F. Zamponi, and J.-P. Bouchaud. 2015b. "Tipping Points in Macroeconomic Agent-Based Models." *Journal of Economic Dynamics and Control* 50:29–61. https://doi.org/10.1016/j.jedc.2014.08.003.

Haldane, A. G., and R. M. May. 2011. "Systemic Risk in Banking Ecosystems." *Nature* 469 (7330): 351–355. https://doi.org/10.1038/nature09659.

Hardiman, S. J., N. Bercot, and J.-P. Bouchaud. 2013. "Critical Reflexivity in Financial Markets: a Hawkes Process Analysis." *The European Physical Journal B* 86:1–9. https://doi.org/10.1140/epjb/e2013-40107-3.

Hardiman, S. J., and J.-P. Bouchaud. 2014. "Branching-Ratio Approximation for the Self-Exciting Hawkes Process." *Physical Review E* 90 (6): 062807. https://doi.org/10.1103/PhysRevE.90.062807.

Harras, G., C. J. Tessone, and D. Sornette. 2012. "Noise-Induced Volatility of Collective Dynamics." *Physical Review E—Statistical, Nonlinear, and Soft Matter Physics* 85 (1): 011150. https://doi.org/10.1103/PhysRevE.85.011150.

Harris, T. E. 1963. *The Theory of Branching Processes.* Vol. 6. Berlin, Germany: Springer.

Hawkins, D., and H. A. Simon. 1949. "Note: Some Conditions of Macroeconomic Stability." *Econometrica, Journal of the Econometric Society,* 245–248. https://www.jstor.org/stable/1905526.

Herskovic, B., B. Kelly, H. Lustig, and S. Van Nieuwerburgh. 2020. "Firm Volatility in Granular Networks." *Journal of Political Economy* 128 (11): 4097–4162. https://doi.org/10.1086/710345.

Hynes, W., B. D. Trump, A. Kirman, A. Haldane, and I. Linkov. 2022. "Systemic Resilience in Economics." *Nature Physics* 18 (4): 381–384. https://doi.org/ 10.1038/s41567-022-01581-4.

Jensen, H. J. 1998. *Self-Organized Criticality: Emergent Complex Behavior in Physical and Biological Systems.* Vol. 10. Cambridge, UK: Cambridge University Press. https://doi.org/10.1017/CBO9780511622717.

Joulin, A., A. Lefevre, D. Grunberg, and J.-P. Bouchaud. 2008. "Stock Price Jumps: News and Volume Play a Minor Role." *Wilmott Magazine* 46. https://doi. org/10.48550/arXiv.0803.1769.

Kaplan, G., B. Moll, and G. L. Violante. 2018. "Monetary Policy According to HANK." *American Economic Review* 108 (3): 697–743. https://doi.org/10. 1257/aer.20160042.

Keeler, J. D., and J. D. Farmer. 1986. "Robust Space–Time Intermittency and 1f Noise." *Physica D: Nonlinear Phenomena* 23 (1-3): 413–435. https://doi. org/10.1016/0167-2789(86)90148-X.

Khandani, A. E., and A. W. Lo. 2007. "What Happened to the Quants in August 2007? Evidence from Factors and Transactions Data." *Journal of Financial Markets* 14 (1): 1–46. https://doi.org/10.1016/j.finmar.2010.07.005.

Kinouchi, O., R. Pazzini, and M. Copelli. 2020. "Mechanisms of Self-Organized Quasicriticality in Neuronal Network Models." *Frontiers in Physics* 8:583213. https://doi.org/10.3389/fphy.2020.583213.

Krząkała, F., and J.-P. Bouchaud. 2005. "Disorder Chaos in Spin Glasses." *Europhysics Letters* 72 (3): 472. https://doi.org/10.1209/epl/i2005-10256-2.

Kyle, A. S. 1985. "Continuous Auctions and Insider Trading." *Econometrica: Journal of the Econometric Society,* 1315–1335. https://www.jstor.org/stable/1913210.

Laval, J. A. 2023. "Self-Organized Criticality of Traffic Flow: Implications for Congestion Management Technologies." *Transportation Research Part C: Emerging Technologies* 149:104056. https://doi.org/10.1016/j.trc.2023. 104056.

Leal, L., H. Mateen, M. Nirei, and J. A. Scheinkman. 2021. *Repricing Avalanches in the Billion-Prices Data.* Technical report. National Bureau of Economic Research. https://www.nber.org/papers/w29236.

LeRoy, S. F. 2008. "Excess Volatility Tests." In *The New Palgrave Dictionary of Economics,* 2nd, vol. 13. London, UK: Palgrave Macmillan. https://doi.org/10.1057/978-1-349-95121-5_2307-1.

LeRoy, S. F., and R. D. Porter. 1981. "The Present-Value Relation: Tests Based on Implied Variance Bounds." *Econometrica: Journal of the Econometric Society,* 555–574. https://doi.org/10.2307/1911512.

Lillo, F., S. Mike, and J. D. Farmer. 2005. "Theory for Long Memory in Supply and Demand." *Physical Review E—Statistical, Nonlinear, and Soft Matter Physics* 71 (6): 066122. https://doi.org/10.1103/PhysRevE.71.066122.

Lombardi, F., H. J. Herrmann, and L. de Arcangelis. 2017. "Balance of Excitation and Inhibition Determines 1/f Power Spectrum in Neuronal Networks." *Chaos: An Interdisciplinary Journal of Nonlinear Science* 27 (4). https://doi.org/10.1063/1.4979043.

Lombardi, F., O. Shriki, H. J. Herrmann, and L. de Arcangelis. 2021. "Long-Range Temporal Correlations in the Broadband Resting State Activity of the Human Brain Revealed by Neuronal Avalanches." *Neurocomputing* 461:657–666. https://doi.org/10.1016/j.neucom.2020.05.126.

Long, Jr., J. B., and C. I. Plosser. 1983. "Real Business Cycles." *Journal of Political Economy* 91 (1): 39–69. https://www.jstor.org/stable/1840430.

Lux, T. 2021. "Can Heterogeneous Agent Models Explain the Alleged Mispricing of the S&P 500?" *Quantitative Finance* 21 (9): 1413–1433. https://doi.org/10.1080/14697688.2021.1909744.

Lux, T., and M. Marchesi. 1999. "Scaling and Criticality in a Stochastic Multi-Agent Model of a Financial Market." *Nature* 397 (6719): 498–500. https://doi.org/10.1038/17290.

Majewski, A. A., S. Ciliberti, and J.-P. Bouchaud. 2020. "Co-existence of Trend and Value in Financial Markets: Estimating an Extended Chiarella Model." *Journal of Economic Dynamics and Control* 112:103791. https://doi.org/10.1016/j.jedc.2019.103791.

Marcaccioli, R., J.-P. Bouchaud, and M. Benzaquen. 2022. "Exogenous and Endogenous Price Jumps Belong to Different Dynamical Classes." *Journal of Statistical Mechanics: Theory and Experiment* 2022 (2): 023403. https://doi.org/10.1088/1742-5468/ac498c.

Marsili, M. 2014. "Complexity and Financial Stability in a Large Random Economy." *Quantitative Finance* 14 (9): 1663–1675. https://doi.org/10.1080/14697688. 2013.765061.

May, R. M. 1972. "Will a Large Complex System Be Stable?" *Nature* 238 (5364): 413–414. https://doi.org/10.1038/238413a0.

Mazzarisi, O., and M. Smerlak. 2024. *Beyond May: Complexity-Stability Relationships in Disordered Dynamical Systems.* ArXiv preprint:2403.11014. https://doi. org/10.48550/arXiv.2403.11014.

Minsky, H. P. 2015. *Can 'It' Happen Again?: Essays on Instability and Finance.* New York, NY: Routledge. https://doi.org/10.4324/9781315705972.

Minsky, H. P., and H. Kaufman. 2008. *Stabilizing an Unstable Economy.* Vol. 1. New York, NY: McGraw-Hill.

Moran, J., and J.-P. Bouchaud. 2019. "May's Instability in Large Economies." *Physical Review E* 100 (3): 032307. https://doi.org/10.1103/PhysRevE.100.032307.

Moran, J., F. P. Pijpers, U. Weitzel, J.-P. Bouchaud, and D. Panja. 2023. *Critical Fragility in Socio-Technical Systems.* ArXiv preprint:2307.03546. https://doi. org/10.48550/arXiv.2307.03546.

Moran, J., M. Romeijnders, P. Le Doussal, F. P. Pijpers, U. Weitzel, D. Panja, and J.-P. Bouchaud. 2024. "Timeliness Criticality in Complex Systems." *Nature Physics* 20:1352–1358. https://doi.org/10.1038/s41567-024-02525-w.

Moran, J., A. Secchi, and J.-P. Bouchaud. 2024. *Revisiting Granular Models of Firm Growth.* https://doi.org/10.48550/arXiv.2404.15226.

Müller, M., and M. Wyart. 2015. "Marginal Stability in Structural, Spin, and Electron Glasses." *Annual Review of Condensed Matter Physics* 6 (1): 177–200. https: //doi.org/10.1146/annurev-conmatphys-031214-014614.

Muzy, J., J.-F.and Delour, and E. Bacry. 2000. "Modelling Fluctuations of Financial Time Series: From Cascade Process to Stochastic Volatility Model." *The European Physical Journal B—Condensed Matter and Complex Systems* 17:537–548. https://doi.org/10.1007/s100510070131.

Muzy, J.-F., R. Baïle, and E. Bacry. 2013. "Random Cascade Model in the Limit of Infinite Integral Scale as the Exponential of a Nonstationary 1/f Noise: Application to Volatility Fluctuations in Stock Markets." *Physical Review E— Statistical, Nonlinear, and Soft Matter Physics* 87 (4): 042813. https://doi.org/ 10.1103/PhysRevE.87.042813.

Nirei, M., and J. A. Scheinkman. 2024. "Repricing Avalanches." *Journal of Political Economy* 132 (4): 1327–1388. https://doi.org/10.1086/727286.

Osorio, I., M. G. Frei, D. Sornette, J. Milton, and Y.-C. Lai. 2010. "Epileptic Seizures: Quakes of the Brain?" *Physical Review E* 82 (2): 021919. https://doi.org/10.1103/PhysRevE.82.021919.

Palmer, R. G., W. B. Arthur, J. H. Holland, B. LeBaron, and P. Tayler. 1994. "Artificial Economic Life: A Simple Model of a Stockmarket." *Physica D: Nonlinear Phenomena* 75 (1-3): 264–274. https://doi.org/10.1016/0167-2789(94)90287-9.

Parisi, G. 1999. "Complex Systems: A Physicist's Viewpoint." *Physica A: Statistical Mechanics and its Applications* 263 (1–4): 557–564. https://doi.org/10.1016/S0378-4371(98)00524-X.

———. 2007. "Physics Complexity and Biology." *Advances in Complex Systems* 10 (supp02): 223–232. https://doi.org/10.1142/S021952590700132X.

Patil, N., and J.-P. Bouchaud. 2024. *Emergent Inequalities in a Primitive Agent-Based Good-Exchange Model.* ArXiv preprint: 2405.18116. https://doi.org/10.48550/arXiv.2405.18116.

Patzelt, F., and K. Pawelzik. 2011. "Criticality of Adaptive Control Dynamics." *Physical Review Letters* 107 (23): 238103. https://doi.org/10.1103/PhysRevLett.107.238103.

———. 2013. "An Inherent Instability of Efficient Markets." *Scientific Reports* 3 (1): 2784. https://doi.org/10.1038/srep02784.

Perotti, J. I., O. V. Billoni, F. A. Tamarit, D. R. Chialvo, and S. A. Cannas. 2009. "Emergent Self-Organized Complex Network Topology out of Stability Constraints." *Physical Review Letters* 103 (10): 108701. https://doi.org/10.1103/PhysRevLett.103.108701.

Sachs, M. K., M. R. Yoder, D. L. Turcotte, J. B. Rundle, and B. D. Malamud. 2012. "Black Swans, Power Laws, and Dragon-Kings: Earthquakes, Volcanic Eruptions, Landslides, Wildfires, Floods, and SOC Models." *The European Physical Journal Special Topics* 205:167–182. https://doi.org/10.1140/epjst/e2012-01569-3.

Sato, Y., and K. Kanazawa. 2023. "Inferring Microscopic Financial Information from the Long Memory in Market-Order Flow: A Quantitative Test of the Lillo-Mike-Farmer Model." *Physical Review Letters* 131 (19): 197401. https://doi.org/10.1103/PhysRevLett.131.197401.

Scheinkman, J. A., and M. Woodford. 1994. "Self-Organized Criticality and Economic Fluctuations." *The American Economic Review* 84 (2): 417–421. https://www.jstor.org/stable/2117870.

Schmitt, N., and F. Westerhoff. 2021. "Trend Followers, Contrarians and Fundamentalists: Explaining the Dynamics of Financial Markets." *Journal of Economic Behavior & Organization* 192:117–136. https://doi.org/10.1016/j.jebo.2021.10.006.

Scholl, M. P., A. Calinescu, and J. D. Farmer. 2021. "How Market Ecology Explains Market Malfunction." *Proceedings of the National Academy of Sciences* 118 (26): e2015574118. https://doi.org/10.1073/pnas.2015574118.

Schwarzkopf, Y., and J. D. Farmer. 2010. "Empirical Study of the Tails of Mutual Fund Size." *Physical Review E—Statistical, Nonlinear, and Soft Matter Physics* 81 (6): 066113. https://doi.org/10.1103/PhysRevE.81.066113.

Sethna, J. P., K. A. Dahmen, and C. R. Myers. 2001. "Crackling Noise." *Nature* 410 (6825): 242–250. https://doi.org/10.1038/35065675.

Sharma, D., J.-P. Bouchaud, M. Tarzia, and F. Zamponi. 2021. "Good Speciation and Endogenous Business Cycles in a Constraint Satisfaction Macroeconomic Model." *Journal of Statistical Mechanics: Theory and Experiment* 2021 (6): 063403. https://doi.org/10.1088/1742-5468/ac014a.

Shiller, R. J. 1981. "Do Stock Prices Move Too Much to be Justified by Subsequent Changes in Dividends?" *American Economic Review* 71 (3): 421–436. https://www.jstor.org/stable/1802789.

———. 1987. "The Volatility of Stock Market Prices." *Science* 235 (4784): 33–37. https://doi.org/10.1126/science.235.4784.33.

Sornette, A., and D. Sornette. 1989. "Self-Organized Criticality and Earthquakes." *Europhysics Letters* 9 (3): 197. https://doi.org/10.1209/0295-5075/9/3/002.

Sornette, D. 1994. "Sweeping of an Instability: an Alternative to Self-Organized Criticality to get Powerlaws without Parameter Tuning." *Journal de Physique I* 4 (2): 209–221. https://doi.org/10.1051/jp1:1994133.

———. 2006. "Endogenous Versus Exogenous Origins of Crises." In *Extreme Events in Nature and Society*, edited by S. Albeverio, V. Jentsch, and H. Kantz, 95–119. Berlin, Germany: Springer. https://doi.org/10.1007/3-540-28611-X_5.

Sornette, D., and G. Ouillon. 2012. "Dragon-Kings: Mechanisms, Statistical Methods and Empirical Evidence." *The European Physical Journal Special Topics* 205 (1): 1–26. https://doi.org/10.1140/epjst/e2012-01559-5.

Squartini, T., I. van Lelyveld, and D. Garlaschelli. 2013. "Early-Warning Signals of Topological Collapse in Interbank Networks." *Scientific Reports* 3 (1): 1–9. https://doi.org/10.1038/srep03357.

Sterman, J. 1989. "Modeling Managerial Behavior: Misperceptions of Feedback in a Dynamic Decision Making Experiment." *Management Science* 35 (3): 321–339. https://doi.org/10.1287/mnsc.35.3.321.

Stone, L. 2018. "The Feasibility and Stability of Large Complex Biological Networks: A Random Matrix Approach." *Scientific Reports* 8 (1): 1–12. https://doi.org/10.1038/s41598-018-26486-2.

Taleb, N. N. 2014. *Antifragile: Things that Gain from Disorder.* Vol. 3. New York, NY: Random House.

Thurner, S., J. D. Farmer, and J. Geanakoplos. 2012. "Leverage Causes Fat Tails and Clustered Volatility." *Quantitative Finance* 12 (5): 695–707. https://doi.org/10.1080/14697688.2012.674301.

Watkins, N. W., G. Pruessner, S. C. Chapman, N. B. Crosby, and H. J. Jensen. 2016. "25 Years of Self-Organized Criticality: Concepts and Controversies." *Space Science Reviews* 198:3–44. https://doi.org/10.1007/s11214-015-0155-x.

Wehrli, A., and D. Sornette. 2022. "The Excess Volatility Puzzle Explained by Financial Noise Amplification from Endogenous Feedbacks." *Scientific Reports* 12:18895. https://doi.org/10.1038/s41598-022-20879-0.

Wheatley, S., A. Wehrli, and D. Sornette. 2019. "The Endo–Exo Problem in High Frequency Financial Price Fluctuations and Rejecting Criticality." *Quantitative Finance* 19 (7): 1165–1178. https://doi.org/10.1080/14697688.2018.1550266.

Wyart, M., J.-P. Bouchaud, J. Kockelkoren, M. Potters, and M. Vettorazzo. 2008. "Relation Between Bid–Ask Spread, Impact and Volatility in Order-Driven Markets." *Quantitative Finance* 8 (1): 41–57. https://doi.org/10.1080/14697680701344515.

Zumbach, G., and P. Lynch. 2001. "Heterogeneous Volatility Cascade in Financial Markets." *Physica A: Statistical Mechanics and its Applications* 298 (3-4): 521–529. https://doi.org/10.1016/S0378-4371(01)00249-7.

DATA-DRIVEN ECONOMIC AGENT-BASED MODELS

Marco Pangallo, CENTAI Institute; and
R. Maria del Rio-Chanona, University College London
and Complexity Science Hub

Abstract

Economic agent-based models (ABMs) are becoming more and more data-driven, establishing themselves as increasingly valuable tools for economic research and policymaking. We propose to classify the extent to which an ABM is data-driven based on whether agent-level quantities are initialized from real-world microdata and whether the ABMs dynamics track empirical time series. This chapter discusses how making ABMs data-driven helps overcome limitations of traditional ABMs and makes ABMs a stronger alternative to equilibrium models. We review state-of-the-art methods in parameter calibration, initialization, and data assimilation, and then present successful applications that have generated new scientific knowledge and informed policy decisions. This chapter serves as a manifesto for data-driven ABMs, introducing a definition and classification and outlining the state of the field, and as a guide for those new to the field.

Introduction

Economic modeling is essential to understand how the economy works and to address pressing global challenges. While econometric studies provide an initial understanding of causal effects, theoretical economic models enable policymakers to simulate scenarios like tax reforms, industrial policy, or economic crises and their impacts on employment, inflation,

GDP, and inequality. Theoretical models serve two additional purposes: They help us first anticipate how agents might respond in new scenarios like climate change or the rapid adoption of artificial intelligence, and second, they project the effects of policies that could reshape agent behaviors in ways not observed historically (Muth 1961).

Traditionally, economic models have relied on the concept of equilibrium, where the state of the economy is derived from solving a set of equations typically based on agents' maximizing behavior. This approach can model how agents respond to novel situations that may alter their behavior, as agents re-maximize in the new setting. Since equilibrium models are based on equations, they can often be solved analytically, making them clear and transparent. Furthermore, they provide a consistent and unifying framework that reduces ad hoc modeling assumptions. However, these benefits come at a cost: Equilibrium models rely on simplifications that can misrepresent real-world dynamics, heterogeneity, and human behavior. For example, the use of a representative agent limits the ability to capture diverse behaviors, while assumptions of perfect rationality can overlook heuristics common in decision-making. As a result, even though policy counterfactuals in equilibrium models are internally consistent, they may miss critical real-world phenomena, such as sudden crises, inequality, and bounded rationality, which limits their reliability for policy analysis.

In the last thirty years, agent-based models (ABMs) have emerged as an alternative to equilibrium models (Axtell and Farmer 2024). ABMs are nonequilibrium in the sense that all model variables are obtained as a map from previously computed values, rather than by solving equilibrium equations (Pangallo 2026). For instance, cash available to an agent may update following sales and purchases, but neither of these is determined by setting demand equal to supply. Nonequilibrium dynamics makes it possible to include all sort of realistic aspects. For example, an ABM can simulate decision-making through heuristics (Artinger, Gigerenzer, and Jacobs 2022)—such as households adjusting spending based on recent income changes rather than from market clearing conditions. As another example, ABMs can easily capture real-world dynamics and transient states, which is crucial for modeling sudden events like the COVID-19 pandemic. However, this flexibility comes with trade-offs: without equilibrium assumptions, ABMs may lose mathematical tractability and often lack closed-form solutions. The flexibility in modeling choices can also make ABM rules appear ad hoc or arbitrary, especially in earlier applications where ABMs served as thought experiments rather than calibrated models of actual economies, limiting their reliability for policy recommendations.

In this chapter, we argue that the recent trend toward data-driven ABMs is helping overcome their traditional limitations, making them a compelling alternative to equilibrium models. First, agent-level microdata enable us to observe behaviors directly, reducing the number of arbitrary assumptions. For example, in an economic-epidemic model simulating economic and mobility decisions (Pangallo *et al.* 2023), mobility data allow us to avoid detailed assumptions about daily activities like bringing kids to school and commuting, which are not crucial for an economic model. Second, ABMs can now faithfully represent a given economy by building synthetic populations that accurately reflect individual and household attributes, or by reconstructing the production networks linking firms. This detailed representation improves the quantitative reliability of policy counterfactuals, offering more grounded results than studies in abstract economies. Third, ABMs can now track empirical time-series data, allowing for in-sample calibration and out-of-sample testing. In some cases, data-driven ABMs have been able to forecast macroeconomic variables, even after major shocks like the COVID-19 pandemic (Poledna *et al.* 2023; Pichler *et al.* 2022). This focus on time-series fitting and forecasting increases the scientific validity of ABMs by allowing empirical testing of model assumptions. Assumptions that do not improve model performance can be streamlined or removed, simplifying models and making them more transparent.

The tendency to more closely link ABMs and data has been greatly accelerating in the last ten years. In the following, we first discuss in what precise sense economic ABMs are becoming more and more data-driven, and then review new methods and applications, including success stories of recent data-driven ABMs, and highlighting challenges and opportunities for future research.

Definitions

A TECHNICAL DEFINITION OF ABMs

Although there exist several conceptual definitions of ABM (Jennings 2000; Bonabeau 2002; de Marchi and Page 2014; Delli Gatti *et al.* 2018), to assess what makes an ABM data-driven we need a *technical* definition. Here is our proposal.

An ABM is composed of $N \gg 1$ model units, indexed by $i = 1, \ldots, N$, that for simplicity we call *agents*.[1] It is important that they are distinct discrete units, not a "continuum" of identical agents. Each agent is characterized by a vector of time-dependent *variables*, denoted by $\mathbf{x}_{i,t}$ at time t, and a vector of

---

[1] Model units could be agents, i.e., autonomous units with "agency," or features of the environment.

fixed *attributes* represented by $\mathbf{a}_i$. At the system level, we define a vector of *model-wide* variables $\mathbf{y}_t$, which evolve over time, and a vector of constant *parameters* θ that govern the behavior of the entire system. An ABM implicitly defines a probability distribution $\mathbb{P}$ that determines the current values of both the set of agent-specific vectors $\{\mathbf{x}_{i,t}\}_{i=1}^{N}$ and the model-wide vector $\mathbf{y}_t$, given the past values of these vectors, the agents' attributes, and the parameters, as follows[2]

$$\{\mathbf{x}_{i,t}\}_{i=1}^{N}, \mathbf{y}_t \sim \mathbb{P}\left(\{\mathbf{x}_{i,\tau<t}\}_{i=1}^{N}, \{\mathbf{y}_{\tau<t}\}, \{\mathbf{a}_i\}_{i=1}^{N}, \boldsymbol{\theta}\right). \quad (1)$$

This definition captures in what sense the model is nonequilibrium, as model variables can be obtained as a map of previously computed variables, rather than by solving equilibrium equations (Pangallo 2026). However, it characterizes many high-dimensional stochastic processes that are not necessarily ABMs, so one can add additional conditions such as that model units are heterogeneous[3] and interact.[4]

Examples.

Why is Schelling's model an ABM, but Brock and Hommes (1998) is not? The Schelling (1971) segregation model, one of the most famous ABMs in the social sciences, features N distinct agents, with N typically between tens and thousands. Each agent i has a position $\mathbf{x}_{i,t}$ and a race $\mathbf{a}_i$ that is fixed. At each time step, agents update their position depending on the fraction of neighboring agents with same race at the previous time step. The decision to move depends on a global

---

[2] While $N \gg 1$ requires that $\{\mathbf{x}_{i,t}\}_{i=1}^{N}$ is non-empty, other components in equation (1) can be empty sets. For example, the model may not include model-wide variables, making $\{\mathbf{y}_{\tau<t}\}$ an empty set. Similarly, some models may not have agent-specific fixed attributes or parameters.

[3] Typically, $\mathbf{x}_{i,t} \neq \mathbf{x}_{j,t}$ for some i, j and t.

[4] Formally, for some $i \neq j$, $\mathbf{x}_{i,t>0}$ is not independent of $\mathbf{x}_{j,\tau}$ for at least one $\tau \in \{0,\ldots,t\}$.

Component of the Model Obtained from Data

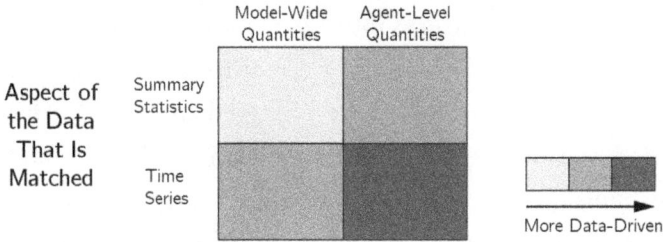

Figure 1. Data-driven ABMs classification diagram. We consider two dimensions to measure the extent to which an ABM is data-driven: On the vertical axis, whether the model reproduces real-world time series or only summary statistics, and on the horizontal axis, whether quantities obtained from data are model-wide or agent-specific. We consider all ABMs in the diagram except the ones in the top left quadrant to be data-driven. Darker shades indicate a more data-driven ABM.

tolerance threshold θ. This model meets all requirements of our definition. In contrast, while the Brock and Hommes (1998) asset pricing model[5] is often considered an ABM, it does not meet the definition since it models a continuum of agents and has only two to four trader *types*.

WHAT MAKES AN ABM DATA-DRIVEN?

Aren't all ABMs data-driven? If we consider that all ABMs have been built with an eye to explaining empirical patterns in real-world data, then, yes. For instance, Thomas Schelling (1971) built his famous ABM because he wanted to explain racial segregation patterns in US cities. But, does this make the Schelling model data-driven? To address this ambiguity, we need to distinguish between a qualitative and quantitative

---

[5]The Brock and Hommes (1998) paper is a seminal work that models a market where traders exchange a single asset using predefined strategies that adapt over time based on past performance. The model demonstrates that simple rules can lead to complex dynamics in asset pricing.

match with data. We can call ABMs that build theories to qualitatively match patterns in the data *theory-driven*.

For ABMs to be data-driven we require that some of the components of equation (1) are obtained by quantitatively matching the data. The extent to which models are data-driven depends both on what components of equation (1) are obtained from the data, and on what aspect of the data they aim to match (see fig. 1). Models that only calibrate parameters are less data-driven than models that also calibrate agent-level attributes and variables. Similarly, models that only match time-independent summary statistics (defined in the next section) are less data-driven than models that track empirical time series, either for in-sample fitting or for out-of-sample forecasting. Models that both obtain agent-specific quantities from data and track empirical time series are the most data-driven (see fig. 1).

Although all models in figure 1 are to some extent data-driven, to make sense of the developments in methods and applications that have occurred in the last ten years, we reserve the phrase "data-driven" only for some models. We say that an ABM is *data-driven* when at least one of the agent-specific variables $x_{i,t}$ or attributes a_i is obtained from data for all agents i, or when the model tracks at least one empirical time series. In other words, we consider all quadrants of figure 1 except the one on the top left.[6]

---

[6]In deciding whether a model is data-driven, the unit of analysis is the paper, not the ABM itself. It is in theory possible that an ABM is presented in an abstract way in one paper, and made data-driven in another paper. Although this is possible, we are not aware of any instance. A possible reason is that setting agent-level quantities from microdata or tracking empirical time series usually requires making specific assumptions while building the ABM that may not be following behavioral theories strictly. By contrast, theory-driven ABMs tend to prefer to follow economic and behavioral theories more closely.

Examples.

How did macroeconomic data-driven ABMs emerge? Most macroeconomic ABMs prior to 2015 (Dawid and Delli Gatti 2018) aimed at replicating stylized facts and summary statistics such as cross-correlations between macroeconomic variables and tail exponents of the firm size distribution. In these models, some parameters were directly taken from data, such as the 2% target inflation rate that has been used by central banks in the last thirty years. However, typically, agent-level quantities were initialized by sampling from theoretical distributions rather than from data. Moreover, the model time series were mostly compared to statistical features of real-world time series, such as matching an average or volatility, rather than attempting to track the series at each time step. While several of these models were fundamental for current data-driven ABMs, the relation to data places them in the top left quadrant of figure 1, so they do not fall within the data-driven ABMs classification.

Data-driven models started to emerge in the last ten years. For instance, the ABM in Georgios Papadopoulos (2019) tracks empirical time series even when agent-level quantities are not initialized from data, placing this model in the bottom left quadrant of figure 1. Conversely, Christian Otto *et al.* (2017) initialize the variables in their ABM from data, but do not track or forecast real-world time series, placing the model in the top right quadrant of figure 1. More recently, Sebastian Poledna *et al.* (2023) initialized their model with agent-level quantities from several real-world datasets, and attempted to track and forecast empirical time series. This model is in the bottom right quadrant of figure 1, making it most data-driven.

Why this definition? This definition of data-driven ABMs[7] comes from reviewing a large set of economic ABMs and aiming to capture the aspects that made ABMs more likely to resemble the real economy. While other definitions may be valid, several can be distilled into the two key dimensions we propose. For instance, one could measure how data-driven an ABM is by the number of data points used, irrespective of whether they are global parameters or agent-specific quantities. Yet, typically there are many more agent-specific quantities than parameters, so initializing at least one quantity for all agents is likely to use more data points than taking all parameters from data. Validation methods could be another criterion, with data-driven ABMs being those that can forecast rather than just reproduce stylized facts. However, a necessary condition for forecasting is to follow time series, and so this criterion reduces to a special case of our classification. This definition also highlights two important features of ABMs: heterogeneous agents and dynamics. Requiring agent-specific initialization emphasizes heterogeneity, as model-wide quantities alone would suffice if all agents were identical. Focusing on matching time series rather than summary statistics allows us to describe transient states in more detail—a feature distinguishing ABMs from equilibrium models.

---

[7] Is "data-driven ABM" the right terminology? In many cases, "data-driven" indicates statistical methods that do not make almost any assumption about the system being modeled. By contrast, ABMs are theoretical models (also known as mechanistic, or structural, or causal models depending on the discipline) that make lots of assumptions about agents' behavior and interactions. In fact, we like this terminology precisely because of the contrast it creates. Models are data-driven, because data availability dictates many important choices, but they are also agent-based, requiring the modeler to still develop a theory.

Methods

We use different methods depending on which component of equation (1) we want to obtain from the data. If we care about parameters $\boldsymbol{\theta}$, we do *calibration*. If we want to estimate agent-level attributes $\mathbf{a}_i$ and initial conditions $\mathbf{x}_{i,0}$, we do *initialization*. If we want to estimate the entire time series of agent-specific variables, $\mathbf{x}_{i,t}$, we do *data assimilation*.

PARAMETER CALIBRATION

The oldest and most developed method is parameter calibration.[8] Parameter calibration was developed for models in the top left quadrant of figure 1, which are not strictly data-driven, but calibration methods are also useful for parameters in more data-driven models.

When possible, parameters are directly derived from data. For instance, suppose that one of the parameters of the model is the percentage reduction in the asking price of a flat after it has not been sold for two months. We may look at housing market data or at the literature and pick a value, say, 5%. Alternatively, we can rely on previous estimations or expert knowledge. For example, Anton Pichler *et al.* (2022) used the parameters suggested by John Muellbauer (2020) out of educated guesses to set changes in consumption patterns during the COVID-19 pandemic. For parameters where there is no measurement or reliable expert knowledge, one can infer parameters from data in four steps: (i) sampling; (ii) choice of summary statistics; (iii) definition of the loss; and (iv) criterion to select the parameters. Most calibration methods are a combination of choices at these four steps.

---

[8]Sometimes we talk about parameter estimation. In principle, estimation is about finding the correct values of parameters, while calibration is about finding a value that makes the model match the data (whether it is the correct value or not), but the two terms are often used interchangeably. Here, to avoid confusion we just refer to calibration.

Sampling. A basic approach to sampling the parameter space is uniform random sampling within reasonable bounds. While this may suffice for smaller spaces, larger parameter spaces often require more sophisticated methods to ensure adequate coverage. Techniques like nearly orthogonal Latin hypercubes and low-discrepancy sequences, such as Halton sequences (Borgonovo *et al.* 2022), help by spreading the sampling points more evenly across the space, reducing gaps and clustering. However, when model evaluations are computationally expensive, these methods, which set sample before evaluating, may also become inefficient or impractical. In such cases, adaptive methods offer a more efficient alternative by iteratively selecting promising points based on previous evaluations. Examples include the Nelder–Mead simplex search algorithm (Franke 2009), as well as newer techniques that employ *meta-models* (also known as surrogate models or emulators) to predict performance at unsampled points (Lamperti, Roventini, and Sani 2018; Glielmo *et al.* 2023).

Summary statistics. A summary statistic is any scalar or vector that reduces the dimensionality of simulated and real data across time to make them easier to compare. A popular method that uses summary statistics is the simulated method of moments (Delli Gatti *et al.* 2018). Originally, this method consisted of calculating longitudinal statistical moments of economic variables, such as mean, variance, and skewness of unemployment. However, the terminology has been extended to any other scalar quantity that can be computed in both real and simulated data. For instance, a moment could also be the total number of firms that default in the considered period. However, one does need to restrict to statistical moments when studying steady states or pseudo-steady states, other choices include parameters of an auxiliary model (indirect inference) or the distribution of an important variable when the model reaches a pseudo-steady state (Kukacka and Barunik

2017), where variable distributions are stationary. One can be even more creative, and consider information-theoretic measures that capture up–down patterns in time series (Barde 2017) or similarities in causal structures (Guerini and Moneta 2017).

Definition of the loss. Once we have computed summary statistics in simulated and real data, we must combine them to obtain a total loss. This can be done in several ways: The most principled way is to use the method of simulated moments and weigh each moment by the inverse of its uncertainty (as captured by the covariance matrix), but equal weighting of all summary statistics is also common. When the summary statistic is the distribution at the pseudo steady state, the loss is the simulated likelihood (Kukacka and Barunik 2017).

Criterion to select the parameters. Once we have a total loss associated to each parameter combination, we may take the frequentist approach and pick the parameter combination that minimizes the loss, using any optimization algorithm, including gradient-based, gradient-free, simulated annealing, and genetic algorithms. Alternatively, we could take a Bayesian approach and select parameter combinations with a probability that is inversely proportional to the loss. Common Bayesian methods include approximate Bayesian computation (Pangallo *et al.* 2023), kernel density estimation (Grazzini, Richiardi, and Tsionas 2017), and neural posterior estimation (Dyer *et al.* 2024).

So, what calibration method should one choose? With several calibration methods available, it can be difficult to pick one. Unfortunately, there are very few systematic comparisons of which methods work best in which setting, and these comparisons are mostly inconclusive. For instance, Donovan Platt (2020) shows that Bayesian methods work better than frequentist methods in four simple models, but Ernesto Carrella (2021), considering 41 models, shows that no method clearly outperforms the others. To navigate the wilderness of calibration methods, our suggestion is twofold. First, we think that the choice of summary statistics, often overlooked, is crucial, as it strongly influences which parameters can be identified.[9]. Second, we suggest choosing the method that minimizes the combination of human and computer time that works best for the modeler, taking advantage of software packages such as Black-IT (Benedetti *et al.* 2022) or BlackBIRDS (Quera-Bofarull *et al.* 2023).

INITIALIZATION

Initialization refers to agent-level quantities, namely attributes a_i and initial conditions for variables, $x_{i,0}$.[10] Until few years ago, the typical approach in economic ABMs was to initialize variables and attributes at random. The last few years have seen more attempts to initialize them from data, with a mix of principled and unprincipled methods. Initialization requires making disparate sources of data compatible with themselves and with the model. This task requires lots of modeler attention, deep knowledge of economic statistics and national accounting, and many lines of code. In a data-driven ABM project, it is not unusual to develop more code on initialization than on the ABM itself. In the following, we give some examples of initialization procedures.

Generating synthetic populations. Data-driven ABMs often focus on a specific population of households, firms, banks, etc.

---

[9] For instance, in a labor market ABM, by using only aggregate unemployment rates and wage distributions, we might miss parameters that govern job search intensity or firm-specific wage setting. By including job search duration, vacancy fill rates, and wage negotiation outcomes in the summary statistics, we can better identify both worker search strategies and firm hiring behavior.

[10] We include attributes, which are fixed in time, in the scope of initialization, which generally refers to time-varying quantities, because the techniques used to determine a_i and $x_{i,0}$ are similar.

In most cases, the population belongs to a geographical area, but it could also correspond, for instance, to an industry (e.g., pharmaceutical firms). In either case, the synthetic population of agents must match the real population. The main difficulty is to respect the joint distribution of agents' attributes and initial conditions. Luckily, this is not a new problem, as it has already been extensively investigated in several social sciences including epidemiology, land use, urban science and transportation research (Arentze, Timmermans, and Hofman 2007), and in economics by the microsimulation community (Li and O'Donoghue 2013; Richiardi and van de Ven 2026). Principled methods, such as iterative proportional fitting and combinatorial optimization, are often used in conjunction with ad hoc, case-specific methods.

Network reconstruction. Attributes and initial conditions may also include the network connecting the agents. Here, most efforts in economics have focused on reconstructing production (Ialongo *et al.* 2022; Mungo *et al.* 2023) and financial (Anand *et al.* 2018) networks. There is also a large literature on regionalization of input–output models (Bonfiglio and Chelli 2008).

Compatibility with national accounts. National accounts are a great source of data to initialize data-driven ABMs, especially when "agents" are industries (Pichler *et al.* 2022). They also serve as a benchmark that agent-level data must be consistent with—for instance, the initial conditions for agent-level consumption must sum to total consumption. The problem is that agent-level data are often available as surveys, may not consider the same sample as national accounts, and may use a different classification system and aggregation level. Sticking to consumption as an example, in Pangallo *et al.* (2023) we used data from a US consumption survey that were available for detailed income and age subgroups, but were largely inconsistent with national accounts. For instance, imputed rents count as consumption, but the survey did not ask for this spending item. Pangallo *et al.* (2023) used bridge tables and ad-hoc assumptions to make the data compatible with national accounts, but more systematic attempts exist (Cazcarro *et al.* 2022).

So, what should I do regarding initialization? Roll up your sleeves, search for data, investigate in detail how the data are constructed, use principled methods when available, and your intuition for which approximations work best when better data and methods are not available. For instance, if data on consumption basket by household income are not available for a country, it may be a good strategy to use the data from another country that is similar. In our experience, approximations of this kind only change initialization of some agent-level variables by a few percentage points, barely affecting aggregate results.

DATA ASSIMILATION

Data assimilation is about estimating the entire time series of agent-specific or global variables, $x_{i,t}$ and y_t that are not observed, that is, *latent* variables.[11] This is particularly important when studying the timing of effects is of essence. For example, consider a macroeconomic ABM that focuses on business cycles. A well-calibrated ABM may be able to reproduce the frequency of recessions. However, to match the exact timing of recessions, we must know the value of relevant latent variables. For instance, to match the timing of the 2008 recession we must correctly estimate over time latent variables such as imbalances in the financial system. Conversely, data assimilation may not be important when

---

[11] While in deterministic models one only needs to estimate the initial condition $x_{i,0}$ and y_0, in ABMs, which are typically stochastic, this is not sufficient. Thus, data assimilation estimates the entire sequence of latent agent-specific and global variables.

tracking empirical time series is not crucial, for instance when focusing on scenarios in the future. In the following, we review the main data assimilation methods.

Kalman and particle filters. Rooted in Bayesian statistics, Kalman and particle filters are used to combine models and observations (Carrassi *et al.* 2018). These filters adjust the latent variables so that the model produces observations close to the real world. The ensemble Kalman filter, a specific type of Kalman filter that only requires simulating the model with no need to compute its Jacobian (as opposed to the standard Kalman filter), was introduced in ABMs by Jonathan Ward, Andrew Evans, and Nicholas Malleson (2016), and recently applied to inequality dynamics (Oswald and Malleson 2025). The particle filter, which is more flexible but also more computationally expensive than the ensemble Kalman filter, has been used in economics with heterogeneous agents in financial market ABMs (Lux 2018).

Probabilistic graphical models. Ensemble Kalman and particle filters treat the ABM as a black box. However, the dependency structure of model variables carries useful information. If it is possible to represent the ABM as a probabilistic graphical model, one can take advantage of conditional independence relations between variables to write a computationally tractable likelihood function of the latent variables. Corrado Monti, Gianmarco de Francisci Morales, and Francesco Bonchi (2020) applied this formalism first, to an opinion dynamics ABM, while Monti *et al.* (2023) apply it to an economic ABM of the housing market.

Heuristic methods. It is common to initialize the latent variables of the model (that cannot be initialized using the techniques in the previous section) by running the model for a transient period until it settles to a quasi-steady state, and then discard the transient period. In this way, the latent variables should become compatible with the observed variables (Geanakoplos *et al.* 2012). Moreover, when part of the aggregate variables $\mathbf{y}_t$ are observed, they can be taken directly from the data, so the model is fit with an exogenous trend (Papadopoulos 2019). This is particularly useful in case of abrupt disruptive events, such as natural disasters or pandemics. For instance, Pichler *et al.* (2022) impose exogenous trends on the COVID-related restrictions faced by different industries, matching the restrictions that were implemented by the UK government in the spring of 2020.

So, what data assimilation method is best? It is often a good idea to use exogenous trends, when they are available. But certain trends may be latent, for instance opinions or beliefs. In this case, the only option is to use one of the methods listed above. The literature on data assimilation in ABMs is not yet mature enough for general guidelines, but our conjecture is that filters work best when only aggregate data are observed, while the representation of an ABM as a probabilistic graphical model is particularly useful when some agent-level variables are observed.

Success Stories

We now discuss applications, highlighting the added value of making ABMs data-driven in obtaining domain-specific insights. Examples throughout are chosen to illustrate success stories of data-driven ABMs, rather than to provide an exhaustive review.

Housing markets. To understand what could have stopped the housing market bubble that led to the 2008 crisis, John Geanakoplos *et al.* (2012) built the first data-driven ABM of the housing market.[12] The key result is that raising interest rates would have done little to prevent the bubble, while imposing stricter limits on loan-to-value would have been much more

---

[12]See Axtell *et al.* (2014) for a more detailed description of the model.

successful. What makes this model's counterfactuals trustworthy is that the model was able to reproduce the dynamics of the Washington, DC housing market from 1997 to 2009 with actual central bank implemented policies. Several central banks have now used and extended this model, see for instance Bence Mérő *et al.* (2023) and András Borsos *et al.* (2026, ch. 18 in this volume).

More recently, spatially explicit data-driven models have been developed to assess how physical risk, often driven by climate change, affects the housing market (Filatova 2015; Pangallo *et al.* 2024). For instance, Toon Haer *et al.* (2017) find that under bounded rationality, households in the Netherlands will adopt mitigation measures that halve expected annual damage from floods compared to a no mitigation scenario. The quantitative reliability of this result hinges on using data from a specific area with detailed projections of how flood risk evolves under climate change.

Labor markets. Data-driven ABMs have modeled labor frictions using empirical data and network structures, and studied how net-zero policies and emerging technologies reshape labor markets. For instance, Robert L. Axtell, Omar A. Guerrero, and Eduardo López (2019) show that using empirical labor flow networks, which are heavy tailed, raises unemployment and shifts the Beveridge curve[13] toward higher unemployment for the same vacancy rate. At the occupational level, in del Rio-Chanona *et al.* (2021) we demonstrate that while some occupations face higher automation risk, others with fewer transition paths may actually experience greater unemployment. This model also shows that moving beyond equilibrium assumptions reproduces

---

[13] The Beveridge curve is an empirical relationship between job vacancies and unemployment, typically depicted as a negatively sloped curve. Shifts in the curve's position can signal changes in labor market efficiency, frictions, or other structural factors influencing the matching of workers to jobs.

the Beveridge curve's counterclockwise cyclical behavior during business cycles. Building on these approaches, Kathyrn Fair and Omar Guerrero (2025) incorporate industry and geographic mobility frictions into a data-driven ABM, and Anna Berryman *et al.* (2025) studied how geographic constraints shape growth and sustainable development scenarios in Brazil. See del Rio-Chanona *et al.* (2026) for complementary discussions on agent-based models and labor markets.

Economic impact of natural disasters and pandemics. Quantifying the impact of disasters is essential for policy making but challenging. Models require estimating both direct disruptions as well as indirect propagation of shocks across the production network. By leveraging industry input-output or firm-level supply-chain data, these models quantify the relative importance of indirect effects compared to direct effects. Stepháne Hallegatte (2008)'s pioneering work estimates the impact of hurricane Katrina on the economy of Louisiana. By initializing the input–output network connecting industries in the model with actual data, the author finds that the indirect impact resulting from the propagation of direct shocks amounts to 50% of the direct impact, a substantial propagation. This model was able to match the recovery profile of employment and industries for a year following the hurricane. Subsequent models have broadened this approach to incorporate international trade (Otto *et al.* 2017), firm-level heterogeneity (Inoue and Todo 2019), diverse households (Markhvida *et al.* 2020), and transportation networks (Colon, Hallegatte, and Rozenberg 2021).

A similar modeling framework has been used to understand the economic impact of the COVID-19 pandemic. Thanks to careful initialization on real-world data, Pichler *et al.* (2022) made a forecast on the UK economic recession in the second quarter of 2020 that turned out to be more accurate than those made by most public and private institutions. This showcased for the first time how data-driven ABMs could be used for forecasting ahead of time (that is, the forecast was made before data on the performance of the UK economy were released). Pangallo *et al.* (2023) extended this model and merged it with a data-driven epidemiological ABM. The resulting model simulates the epidemiological and economic decisions of a synthetic population of half a million people in the New York metropolitan area. The model accurately fits the dynamics of the first lockdown in spring 2020, reproducing a stylized fact typical of the COVID-19 recession that more traditional models had more difficulty replicating, namely that high-income households reduced their consumption more than low-income households, despite experiencing lower unemployment. In addition, the model uses counterfactuals to show that both epidemiological and economic impacts are similar whether policy makers impose lockdowns or the population spontaneously changes their behavior out of fear of infection, reducing contacts and consumption. Thus, under behavior change, it is the virus, not the lockdown, that takes a toll on the economy.

~327~

Macroeconomics. The first data-driven macroeconomic ABM that we are aware of is by Barbara R. Bergmann (1974).[14] This is a fully fledged macroeconomic ABM, with workers/consumers, firms, a bank, a government, and a central bank. The firms, representing industries, are connected through an input-output network initialized from real data. Bergmann shows that the model is capable of fitting US macroeconomic statistics from the beginning of 1967 to the end of 1970. This model was visionary and ahead of its time. Around thirty years later, agent-based macroeconomics started to develop, obtaining important

---

[14]The author uses the term "microsimulation," but the model could be described as an ABM *ante litteram* (Richiardi and van de Ven 2026).

theoretical results on growth and business cycles but without connecting much to real-world data (Dawid and Delli Gatti 2018). This trend has changed in recent years (Papadopoulos 2019; Kaszowska-Mojsa and Pipień 2020). The model with the greatest impact so far is from Poledna *et al.* (2023), who represented the Austrian economy at a one-to-one scale and showed that the ABM could compete with statistical vector autoregressive (VAR) models and traditional dynamic stochastic general equilibrium (DSGE) models at out of sample forecasting. This model has now been used for multiple applications. For instance, it showed that post-pandemic inflation in Canada was initially due to increases in input costs, then to expectations of future inflation, and finally by increases in demand by other firms as the economy reopened after lockdowns (Hommes *et al.* 2026, ch. 14 in this volume).

Challenges and Opportunities—Outlook and Ways Forward

How does the evolution towards data-driven ABMs affect the way we validate models, choose behavioral rules, and build counterfactuals? We discuss answers to these questions and challenges and opportunities lying ahead.

DATA, METHODS, AND VALIDATION

Developing new methods. We have covered several calibration, initialization and data assimilation methods that are now available for modelers. While calibration techniques have been developed for more than a decade, the initialization and data assimilation methods are fairly new for ABMs. A key challenge is enhancing these methods to adjust all microvariables in large-scale models, which is essential for macroeconomic ABMs to accurately track long-term time series. There is a growing literature exploring alternatives. For example, Daniele Grattarola, Lorenzo Livi, and Cesare Alippi (2021), Corneel Casert, Isaac Tamblyn, and Stephen Whitelam *2024*, and Francesco Cozzi *et al.* (2025) use neural network architectures as surrogate models when the ABMs are too complicated to estimate their latent variables. Developing these methods further presents significant opportunities to improve the performance and applicability of ABMs.

Dealing with massive datasets is time-consuming. A second, related, challenge is that initialization is time consuming, since it requires finding reliable data sources and then making them compatible with themselves and with the model. One way forward relies on governments and statistical offices facilitating access to data, as well as possible collaboration with the private sector for additional data (see Turrell 2026, for a more detailed discussion). Another suggestion, which has worked in our experience, is to work with large teams, where for instance two or three PhD students work on different aspects of the same project. This transformation towards a science-based lab model may already be happening in empirical economics (Athey 2018), and could be effective for data-driven ABMs as well.

Pushing validation to tracking time series. The main opportunity that using advanced methods and detailed data presents is to push forward the concept of validation. Traditionally, validation of ABMs has focused on replicating stylized facts and broad empirical regularities. Of course, this comes at the risk of overfitting, as complicated models with many parameters can easily fit a few stylized facts. Recently, more rigorous validation has been possible through out-of-sample forecasting. In this approach, modelers calibrate the model on some initial year(s) and then validate it on its ability to forecast economic dynamics in future years that have not been considered in the calibration process.

Although computationally and data demanding, several ABMs have succeeded in doing this. For example, Pichler *et al.* (2022) used their ABM to forecast the UK's industry-level output during the COVID-19 pandemic ahead of time. Poledna *et al.* (2023) demonstrated that their ABM competed with VAR and DSGE models in predicting Austria's GDP. Going forward, we consider that stylized facts should be the minimum bar for ABMs. When possible ABMs should try to seek validation through out-of-sample forecasting, or at least in sample-fitting time series. This validation benefits from fine-grained data and advanced statistical methods and hence developing these methods and acquiring data should be a priority for the field.

DATA, BEHAVIOR, AND THEORY

A general boundedly rational approach for modeling behavior. Modeling behavior is hard. The main advantage of the optimization framework traditionally used in economics is that it is a one-size-fits-all approach—one can use optimization to model decisions of consumers, firms, banks, politicians, even parents when making family choices. We currently lack an alternative general bounded rationality framework (Harstad and Selten 2013). This limitation led to criticism that ABMs rely on too many arbitrary "ad hoc" assumptions and free parameters.[15] Recent efforts have explored alternatives based on laboratory experiments (Hommes 2021), psychology (Roos 2018), economic theory (Gabaix 2019), and large language models (Horton 2023; del Rio-Chanona *et al.* 2024). These

---

[15] Arbitrary choices also exist in models using optimization, they are just hidden in the functions that are optimized. For example, DSGE models sometimes decide between preference structures, such as King–Plosser–Rebelo (KPR) or Greenwood–Hercowitz–Huffman (GHH) preferences, depending on practical modeling needs rather than realism. Similarly, when is it reasonable to use "MIT shocks"? How about "iceberg costs"?

approaches are promising and with further development of these methods could lead to one or a few standard frameworks in the future.

Data can replace unnecessary assumptions. Data-driven ABMs can address the "ad hoc" critique in two ways. First, they replace assumptions that are not critical to the model with real-world data. For example, in studying policies to reduce the economic impact of COVID-19, it is not essential to model detailed individual choices, such as how agents decide on commuting patterns or places to go out. Instead, one can use mobile phone data to infer people's mobility and contacts with others, and model the reduction in contacts due to the fear of infection with just one parameter (Pangallo *et al.* 2023). This results in fewer choices and free parameters.[16]

~331~

The contribution of assumptions to model validity can be quantitatively tested. A second opportunity for data-driven ABMs to address the "ad hoc-ness" critique is through measuring which assumptions really improve results, and drop unnecessarily complicated assumptions when they do not. Agent-based modelers are often proud of making "reasonable" assumptions. Doyne Farmer (2024) calls this the principle of *verisimilitude*: "assumptions should be plausible. Assumptions that seem wrong from the outset are more likely to lead to false conclusions than plausible assumptions." But who decides which assumptions are plausible and which assumptions are not? Inspired by machine learning, in the last few years psychologists have been

---

[16]This approach could be criticized by agent-based modelers who think that all results should be "grown" from first principles (Epstein and Axtell 1996). We agree that at least some assumptions should come from theory, but we also think that it is a fair compromise to draw less important assumptions from data.

making comparisons based on forecasting (Erev *et al.* 2017; Artinger, Gigerenzer, and Jacobs 2020). Data-driven agent-based modelers should follow in their footsteps, systematically testing how their assumptions may improve on the validity of the model, for instance its ability to do out-of-sample forecasting.

DATA, EXTERNAL VALIDITY, AND COUNTERFACTUALS

External validity should be addressed. One potential drawback of heavily relying on data is the criticism of external validity: How do you know that insights from your model also hold beyond the region/country/industry you have modeled? This is a fair criticism shared with all empirical work. To understand the extent to which a data-driven ABM can yield insights on other regions/countries/industries of interest, the modeler should study how data change between systems of interests. For example, input–output coefficients vary little between regions of the same country, but it may be useful to check that results do not critically depend on the specific values of input–output coefficients.

Counterfactuals are more reliable. The problem of external validity highlights a substantial opportunity for data-driven ABMs, compared to a purely abstract ABM that does not reproduce any specific system. By reproducing the dynamics of, say, a region, under the policy interventions that were in place, studying counterfactual policies yields much more reliable results (Geanakoplos *et al.* 2012; Pangallo *et al.* 2023). Data-driven ABMs should always aim at reproducing the actual dynamics as a baseline, and then test how alternative policies may have led to different outcomes in some specific historical episode.

Conclusion

The main takeaways of the chapter are that data-driven ABMs: (i) push validation standards from stylized facts to time-series tracking and forecasting; (ii) help model behavior in a general way by replacing unimportant assumptions with data and making it possible to systematically test the impact of behavioral assumptions on model outputs; and (iii) make counterfactuals more reliable by reproducing empirical dynamics with actual policies. These advantages address current limitations of agent-based models, and will promote their use in economic research and policy. Accurate models can inform policies that reduce unemployment, control inflation, and improve overall well-being.

Going forward, we must bear in mind that data-driven ABMs will not make economics value-free. While validation may protect models from arbitrary assumptions and increase objectivity, complete objectivity in economics is an illusion (Atkinson 2009; Coyle 2021). Economic decisions, models, and policies reflect underlying ethical and normative judgments, and economists should openly acknowledge these values and engage in transparent discussions about them. Incorporating insights from other disciplines, such as sociology, ethics, and political science, can further help address the ethical and societal implications of economic analysis. 🗲

Acknowledgments

We would like to thank the co-organizers and participants of the workshop "Data-Driven Economic Agent-Based Models" (https://sites.google.com/view/wddeabm), where we discussed many of the ideas presented in this chapter.

REFERENCES

Anand, K., I. van Lelyveld, Á. Banai, S. Friedrich, R. Garratt, G. Hałaj, J. Fique, *et al.* 2018. "The Missing Links: A Global Study on Uncovering Financial Network Structures from Partial Data." *Journal of Financial Stability* 35:107–119. https://doi.org/10.1016/j.jfs.2017.05.012.

Arentze, T., H. Timmermans, and F. Hofman. 2007. "Creating Synthetic Household Populations: Problems and Approach." *Transportation Research Record* 2014 (1): 85–91. https://doi.org/10.3141/2014-11.

Artinger, F. M., G. Gigerenzer, and P. Jacobs. 2020. "Labor Provision under Uncertainty." *SSRN Electronic Journal,* https://doi.org/10.2139/ssrn.3728515.

———. 2022. "Satisficing: Integrating Two Traditions." *Journal of Economic Literature* 60 (2): 598–635. https://doi.org/10.1257/jel.20201396.

Athey, S. 2018. "The Impact of Machine Learning on Economics." In *The Economics of Artificial Intelligence: An Agenda,* 507–547. Chicago, IL: University of Chicago Press.

Atkinson, A. B. 2009. "Economics as a Moral Science." *Economica* 76:791–804. https://doi.org/10.1111/j.1468-0335.2009.00788.x.

Axtell, R. L., and J. D. Farmer. 2024. "Agent-Based Modeling in Economics and Finance: Past, Present, and Future." *Journal of Economic Literature,* 1–101. https://doi.org/10.1257/jel.20221319.

Axtell, R. L., J. D. Farmer, J. Geanakoplos, P. Howitt, E. Carrella, B. Conlee, J. Goldstein, *et al.* 2014. "An Agent-Based Model of the Housing Market Bubble in Metropolitan Washington, DC." *SSRN Electronic Journal,* https://doi.org/10.2139/ssrn.4710928.

Axtell, R. L., O. A. Guerrero, and E. López. 2019. "Frictional Unemployment on Labor Flow Networks." *Journal of Economic Behavior & Organization* 160:184–201. https://doi.org/10.1016/j.jebo.2019.02.028.

Barde, S. 2017. "A Practical, Accurate, Information Criterion for Nth Order Markov Processes." *Computational Economics* 50 (2): 281–324. https://doi.org/10.1007/s10614-016-9617-9.

Benedetti, M., G. Catapano, F. De Sclavis, M. Favorito, A. Glielmo, D. Magnanimi, and A. Muci. 2022. "Black-It: A Ready-to-Use and Easy-to-Extend Calibration Kit for Agent-Based Models." *Journal of Open Source Software* 7 (79): 4622. https://doi.org/10.21105/joss.04622.

Bergmann, B. R. 1974. "A Microsimulation of the Macroeconomy with Explicitly Represented Money Flows." In *Annals of Economic and Social Measurement,* edited by S. V. Berg, 3:475–489. 3. Cambridge, MA: NBER. https://www.nber.org/system/files/chapters/c10173/c10173.pdf.

Berryman, A. K., J. Bücker, F. Senra de Moura, P. Barbrook-Johnson, M. Hanusch, P. Mealy, J. D. Farmer, and R. M. del Rio-Chanona. 2025. "Skill and Spatial Mismatches for Sustainable Development Pathways in Brazil." *arXiv: 2503.05310,* https://doi.org/10.48550/arXiv.2503.05310.

Bonabeau, E. 2002. "Agent-Based Modeling: Methods and Techniques for Simulating Human Systems." *Proceedings of the National Academy of Sciences* 99 (suppl 3): 7280–7287. https://doi.org/10.1073/pnas.082080899.

Bonfiglio, A., and F. Chelli. 2008. "Assessing the Behaviour of Non-Survey Methods for Constructing Regional Input–Output Tables through a Monte Carlo Simulation." *Economic Systems Research* 20 (3): 243–258. https://doi.org/10.1080/09535310802344315.

Borgonovo, E., M. Pangallo, J. Rivkin, L. Rizzo, and N. Siggelkow. 2022. "Sensitivity Analysis of Agent-Based Models: A New Protocol." *Computational and Mathematical Organization Theory* 28 (1): 52–94. https://doi.org/10.1007/s10588-021-09358-5.

Borsos, A., A. Carro, A. Glielmo, M. Hinterschweiger, J. Kaszowska-Mojsa, and A. Uluc. 2026. "Agent-Based Modeling at Central Banks: Recent Developments and New Challenges." In *The Economy as an Evolving Complex System IV,* edited by R. M. del Rio-Chanona, M. Pangallo, J. Bednar, E. D. Beinhocker, J. Kaszowska-Mojsa, F. Lafond, P. Mealy, A. Pichler, and J. D. Farmer. Santa Fe, NM: SFI Press.

Brock, W. A., and C. H. Hommes. 1998. "Heterogeneous Beliefs and Routes to Chaos in a Simple Asset Pricing Model." *Journal of Economic Dynamics and Control* 22 (8–9): 1235–1274. https://doi.org/10.1016/S0165-1889(98)00011-6.

Carrassi, A., M. Bocquet, L. Bertino, and G. Evensen. 2018. "Data Assimilation in the Geosciences: An Overview of Methods, Issues, and Perspectives." *Wiley Interdisciplinary Reviews: Climate Change* 9 (5): e535. https://doi.org/10.1002/wcc.535.

Carrella, E. 2021. "No Free Lunch when Estimating Simulation Parameters." *Journal of Artificial Societies and Social Simulation* 24 (2). https://doi.org/10.18564/jasss.4572.

Casert, C., I. Tamblyn, and S. Whitelam. 2024. "Learning Stochastic Dynamics and Predicting Emergent Behavior Using Transformers." *Nature Communications* 15 (1): 1875. https://doi.org/10.1038/s41467-024-45629-w.

Cazcarro, I., A. F. Amores, I. Arto, and K. Kratena. 2022. "Linking Multisectoral Economic Models and Consumption Surveys for the European Union." *Economic Systems Research* 34 (1): 22–40. https://doi.org/10.1080/09535314.2020.1856044.

Colon, C., S. Hallegatte, and J. Rozenberg. 2021. "Criticality Analysis of a Country's Transport Network via an Agent-Based Supply Chain Model." *Nature Sustainability* 4 (3): 209–215. https://doi.org/10.1038/s41893-020-00649-4.

Coyle, D. 2021. *Cogs and Monsters: What Economics Is, and What It Should Be.* Princeton, NJ: Princeton University Press.

Cozzi, F., M. Pangallo, A. Perotti, A. Panisson, and C. Monti. 2025. *Learning Individual Behavior in Agent-Based Models with Graph Diffusion Networks.* NeurIPS 2025. arXiv preprint: 2505.21426 [cs.AI]. https://doi.org/10.48550/arXiv.2505.21426.

Dawid, H., and D. Delli Gatti. 2018. "Agent-Based Macroeconomics." In *Handbook of Computational Economics,* edited by C. Hommes and B. LeBaron, 4:63–156. Amsterdam, Netherlands: North Holland.

de Marchi, S., and S. E. Page. 2014. "Agent-Based Models." *Annual Review of Political Science* 17 (1): 1–20. https://doi.org/10.1146/annurev-polisci-080812-191558.

del Rio-Chanona, R. M., M. R. Frank, P. Mealy, E. Moro, and L. Nedelkoska. 2026. "Beyond Efficiency: Labor-Market Resilience in an Age of AI and Net Zero." In *The Economy as an Evolving Complex System IV,* edited by R. M. del Rio-Chanona, M. Pangallo, J. Bednar, E. D. Beinhocker, J. Kaszowska-Mojsa, F. Lafond, P. Mealy, A. Pichler, and J. D. Farmer. Santa Fe, NM: SFI Press.

del Rio-Chanona, R. M., P. Mealy, M. Beguerisse-Díaz, F. Lafond, and J. D. Farmer. 2021. "Occupational Mobility and Automation: A Data-Driven Network Model." *Journal of the Royal Society Interface* 18 (174): 20200898. https://doi.org/10.1098/rsif.2020.0898.

del Rio-Chanona, R. M., M. Pangallo, P. Mishkin, and C. Hommes. 2024. "Market Dynamics of Price Expectations with Generative AI Agents." In preparation.

Delli Gatti, D., G. Fagiolo, M. Gallegati, M. Richiardi, and A. Russo. 2018. *Agent-Based Models in Economics: A Toolkit.* Cambridge, UK: Cambridge University Press.

Dyer, J., P. Cannon, J. D. Farmer, and S. M. Schmon. 2024. "Black-Box Bayesian Inference for Agent-Based Models." *Journal of Economic Dynamics and Control* 161:104827. https://doi.org/10.1016/j.jedc.2024.104827.

Epstein, J. M., and R. Axtell. 1996. *Growing Artificial Societies: Social Science from the Bottom Up.* Cambridge, MA: MIT Press.

Erev, I., E. Ert, O. Plonsky, D. Cohen, and O. Cohen. 2017. "From Anomalies to Forecasts: Toward a Descriptive Model of Decisions under Risk, under Ambiguity, and from Experience." *Psychological Review* 124 (4): 369. https://doi.org/10.1037/rev0000062.

Fair, K. R., and O. A Guerrero. 2025. "Emerging Labour Flow Networks." *EPJ Data Science* 14 (1). https://doi.org/10.1140/epjds/s13688-025-00539-9.

Farmer, J. D. 2024. *Making Sense of Chaos: A Better Economics for a Better World.* London, UK: Penguin.

Filatova, T. 2015. "Empirical Agent-Based Land Market: Integrating Adaptive Economic Behavior in Urban Land-Use Models." *Computers, Environment and Urban Systems* 54:397–413. https://doi.org/10.1016/j.compenvurbsys.2014.06.007.

Franke, R. 2009. "Applying the Method of Simulated Moments to Estimate a Small Agent-Based Asset Pricing Model." *Journal of Empirical Finance* 16 (5): 804–815. https://doi.org/10.1016/j.jempfin.2009.06.006.

Gabaix, X. 2019. "Behavioral Inattention." In *Handbook of Behavioral Economics: Applications and Foundations 1,* 2:261–343. https://doi.org/10.1016/bs.hesbe.2018.11.001.

Geanakoplos, J., R. L. Axtell, J. D. Farmer, P. Howitt, B. Conlee, J. Goldstein, M. Hendrey, N. M. Palmer, and C.-Y. Yang. 2012. "Getting at Systemic Risk via an Agent-Based Model of the Housing Market." *American Economic Review* 102 (3): 53–58. https://doi.org/10.1257/aer.102.3.53.

Glielmo, A., M. Favorito, D. Chanda, and D. Delli Gatti. 2023. "Reinforcement Learning for Combining Search Methods in the Calibration of Economic ABMs." In *Proceedings of the Fourth ACM International Conference on AI in Finance,* 305–313. New York, NY: Association for Computing Machinery.

Grattarola, D., L. Livi, and C. Alippi. 2021. "Learning Graph Cellular Automata." In *NIPS '21 Proceedings of the 35th International Conference on Neural Information Processing Systems,* edited by M. Ranzato, A. Beygelzimer, Y. Dauphin, P. S. Liang, and J. Wortman Vaughan, 34:20983–20994.

Grazzini, J., M. G. Richiardi, and M. Tsionas. 2017. "Bayesian Estimation of Agent-Based Models." *Journal of Economic Dynamics and Control* 77:26–47. https://doi.org/10.1016/j.jedc.2017.01.014.

Guerini, M., and A. Moneta. 2017. "A Method for Agent-Based Models Validation." *Journal of Economic Dynamics and Control* 82:125–141. https://doi.org/10.1016/j.jedc.2017.06.001.

Haer, T., W. J. Wouter Botzen, H. de Moel, and J. C. J. H. Aerts. 2017. "Integrating Household Risk Mitigation Behavior in Flood Risk Analysis: An Agent-Based Model Approach." *Risk Analysis* 37 (10): 1977–1992. https://doi.org/10.1111/risa.12740.

Hallegatte, S. 2008. "An Adaptive Regional Input–Output Model and Its Application to the Assessment of the Economic Cost of Katrina." *Risk Analysis* 28 (3): 779–799. https://doi.org/10.1111/j.1539-6924.2008.01046.x.

Harstad, R. M., and R. Selten. 2013. "Bounded-Rationality Models: Tasks to Become Intellectually Competitive." *Journal of Economic Literature* 51 (2): 496–511. https://doi.org/10.1257/jel.51.2.496.

Hommes, C. 2021. "Behavioral and Experimental Macroeconomics and Policy Analysis: A Complex Systems Approach." *Journal of Economic Literature* 59 (1): 149–219. https://doi.org/10.1257/jel.20191434.

Hommes, C., S. Kozicki, S. Poledna, and Y. Zhang. 2026. "How an Agent-Based Model Can Support Monetary Policy in a Complex Evolving Economy." In *The Economy as an Evolving Complex System IV,* edited by R. M. del Rio-Chanona, M. Pangallo, J. Bednar, E. D. Beinhocker, J. Kaszowska-Mojsa, F. Lafond, P. Mealy, A. Pichler, and J. D. Farmer. Santa Fe, NM: SFI Press.

Horton, J. J. 2023. *Large Language Models as Simulated Economic Agents: What Can We Learn from Homo Silicus?* Technical report. National Bureau of Economic Research. https://doi.org/10.3386/w31122.

Ialongo, L. N., C. de Valk, E. Marchese, F. Jansen, H. Zmarrou, T. Squartini, and D. Garlaschelli. 2022. "Reconstructing Firm-Level Interactions in the Dutch Input–Output Network from Production Constraints." *Scientific Reports* 12 (1): 11847. https://doi.org/10.1038/s41598-022-13996-3.

Inoue, H., and Y. Todo. 2019. "Firm-Level Propagation of Shocks through Supply-Chain Networks." *Nature Sustainability* 2 (9): 841–847. https://doi.org/10.1038/s41893-019-0351-x.

Jennings, N. R. 2000. "On Agent-Based Software Engineering." *Artificial Intelligence* 117 (2): 277–296. https://doi.org/10.1016/S0004-3702(99)00107-1.

Kaszowska-Mojsa, J., and M. Pipień. 2020. "Macroprudential Policy in a Heterogeneous Environment—An Application of Agent-Based Approach in Systemic Risk Modelling." *Entropy* 22 (2): 129. https://doi.org/10.3390/e22020129.

Kukacka, J., and J. Barunik. 2017. "Estimation of Financial Agent-Based Models with Simulated Maximum Likelihood." *Journal of Economic Dynamics and Control* 85:21–45. https://doi.org/10.1016/j.jedc.2017.09.006.

Lamperti, F., A. Roventini, and A. Sani. 2018. "Agent-Based Model Calibration Using Machine Learning Surrogates." *Journal of Economic Dynamics and Control* 90:366–389. https://doi.org/10.1016/j.jedc.2018.03.011.

Li, J., and C. O'Donoghue. 2013. "A Survey of Dynamic Microsimulation Models: Uses, Model Structure and Methodology." *International Journal of Microsimulation* 6 (2): 3–55. https://doi.org/10.34196/ijm.00082.

Lux, T. 2018. "Estimation of Agent-Based Models Using Sequential Monte Carlo Methods." *Journal of Economic Dynamics and Control* 91:391–408. https://doi.org/10.1016/j.jedc.2018.01.021.

Markhvida, M., B. Walsh, S. Hallegatte, and J. Baker. 2020. "Quantification of Disaster Impacts through Household Well-Being Losses." *Nature Sustainability* 3 (7): 538–547. https://doi.org/10.1038/s41893-020-0508-7.

Mérő, B., A. Borsos, Z. Hosszú, Z. Oláh, and N. Vágó. 2023. "A High-Resolution, Data-Driven Agent-Based Model of the Housing Market." *Journal of Economic Dynamics and Control* 155:104738. https://doi.org/10.1016/j.jedc.2023.104738.

Monti, C., G. De Francisci Morales, and F. Bonchi. 2020. "Learning Opinion Dynamics from Social Traces." In *Proceedings of the 26th ACM SIGKDD International Conference on Knowledge Discovery & Data Mining,* 764–773. New York, NY: Association for Computing Machinery. https://doi.org/10.1145/3394486.3403119.

Monti, C., M. Pangallo, G. De Francisci Morales, and F. Bonchi. 2023. "On Learning Agent-Based Models from Data." *Scientific Reports* 13 (1): 9268. https://doi.org/10.1038/s41598-023-35536-3.

Muellbauer, J. 2020. "The Coronavirus Pandemic and US Consumption." *VoxEU.org, 11 April*, https://voxeu.org/article/coronavirus-pandemic-and-us-consumption.

Mungo, L., F. Lafond, P. Astudillo-Estévez, and J. D. Farmer. 2023. "Reconstructing Production Networks Using Machine Learning." *Journal of Economic Dynamics and Control* 148:104607. https://doi.org/10.1016/j.jedc.2023.104607.

Muth, J. F. 1961. "Rational Expectations and the Theory of Price Movements." *Econometrica* 29:315–335. https://doi.org/10.2307/1909635.

Oswald, K., Y. Suchak, and N. Malleson. 2025. "Agent-Based Models of the United States Wealth Distribution with Ensemble Kalman Filter." *Journal of Economic Behavior & Organization* 229:106820. https://doi.org/10.1016/j.jebo.2024.106820.

Otto, C., S. N. Willner, L. Wenz, K. Frieler, and A. Levermann. 2017. "Modeling Loss-Propagation in the Global Supply Network: The Dynamic Agent-Based Model Acclimate." *Journal of Economic Dynamics and Control* 83:232–269. https://doi.org/10.1016/j.jedc.2017.08.001.

Pangallo, M. 2026. "Equations vs. Maps: Complexity, Equilibrium, Disequilibrium." In *The Economy as an Evolving Complex System IV,* edited by R. M. del Rio-Chanona, M. Pangallo, J. Bednar, E. D. Beinhocker, J. Kaszowska-Mojsa, F. Lafond, P. Mealy, A. Pichler, and J. D. Farmer. Santa Fe, NM: SFI Press.

Pangallo, M., A. Aleta, R. M. del Rio-Chanona, A. Pichler, D. Martín-Corral, M. Chinazzi, F. Lafond, *et al.* 2023. "The Unequal Effects of the Health–Economy Trade-Off During the COVID-19 Pandemic." *Nature Human Behaviour,* 1–12. https://doi.org/10.1038/s41562-023-01747-x.

Pangallo, M., M. Coronese, F. Lamperti, G. Cervone, and F. Chiaromonte. 2024. "Climate-Change Attitudes in a Data-Driven Agent-Based Model of the Housing Market." In preparation.

Papadopoulos, G. 2019. "Income Inequality, Consumption, Credit and Credit Risk in a Data-Driven Agent-Based Model." *Journal of Economic Dynamics and Control* 104:39–73. https://doi.org/10.1016/j.jedc.2019.05.002.

Pichler, A., M. Pangallo, R. M. del Rio-Chanona, F. Lafond, and J. D. Farmer. 2022. "Forecasting the Propagation of Pandemic Shocks with a Dynamic Input–Output Model." *Journal of Economic Dynamics and Control,* 104527. https://doi.org/10.1016/j.jedc.2022.104527.

Platt, D. 2020. "A Comparison of Economic Agent-Based Model Calibration Methods." *Journal of Economic Dynamics and Control* 113:103859. https://doi.org/10.1016/j.jedc.2020.103859.

Poledna, S., M. G. Miess, C. Hommes, and K. Rabitsch. 2023. "Economic Forecasting with an Agent-Based Model." *European Economic Review* 151:104306. https://doi.org/10.1016/j.euroecorev.2022.104306.

Quera-Bofarull, A., J. Dyer, A. Calinescu, J. D. Farmer, and M. Wooldridge. 2023. "BlackBIRDS: Black-Box Inference for Differentiable Simulators." *Journal of Open Source Software* 8 (89). https://doi.org/10.21105/joss.05776.

Richiardi, M., and J. van de Ven. 2026. "Back to the Future: Agent-Based Modeling and Dynamic Microsimulation." In *The Economy as an Evolving Complex System IV,* edited by R. M. del Rio-Chanona, M. Pangallo, J. Bednar, E. D. Beinhocker, J. Kaszowska-Mojsa, F. Lafond, P. Mealy, A. Pichler, and J. D. Farmer. Santa Fe, NM: SFI Press.

Roos, M. W. M. 2018. *Values, Attitudes and Economic Behavior.* Ruhr Economic Papers, No. 777. Leibniz-Institut für Wirtschaftsforschung, Essen. https://doi.org/10.4419/86788905.

Schelling, T. C. 1971. "Dynamic Models of Segregation." *Journal of Mathematical Sociology* 1 (2): 143–186. https://doi.org/10.1080/0022250X.1971.9989794.

Turrell, A. 2026. "Cutting Through Complexity: How Data Science Can Help Policymakers Understand the World." In *The Economy as an Evolving Complex System IV,* edited by R. M. del Rio-Chanona, M. Pangallo, J. Bednar, E. D. Beinhocker, J. Kaszowska-Mojsa, F. Lafond, P. Mealy, A. Pichler, and J. D. Farmer. Santa Fe, NM: SFI Press.

Ward, J. A., A. J. Evans, and N. S. Malleson. 2016. "Dynamic Calibration of Agent-Based Models Using Data Assimilation." *Royal Society Open Science* 3 (4): 150703. https://doi.org/10.1098/rsos.150703.

CUTTING THROUGH COMPLEXITY: HOW DATA SCIENCE CAN HELP POLICYMAKERS UNDERSTAND THE WORLD

Arthur Turrell, Bank of England

Abstract

Economies are fundamentally complex and becoming more so, but the new discipline of data science—which combines programming, statistics, and domain knowledge—can help cut through that complexity, potentially with productivity benefits to boot. This chapter looks at examples of where innovations from data science and artificial intelligence are cutting through the complexities faced by policymakers in measurement, allocating resources, monitoring the natural world, making predictions, and more. These examples show the promise and potential of data science to aid policymakers, and point to where actions may be taken that would support further progress in this space.

Introduction

Economies are fundamentally complex and becoming more so, for a host of reasons. Declining response rates from surveys (once the bedrock of official statistics), the switch from tangible to intangible capital, mismeasurement of economic activity, the switch from production to services, the need to track different types of assets (for example, natural assets), increasingly specialized supply lines, and a demand for ever-more-granular insights are just some of the headwinds facing policymakers as they try to navigate this complexity.

The challenge is made even greater as policymakers are now directly responsible for a greater share of GDP than ever before—for the thirteen countries in Europe for which the IMF has long-run data, government expenditure as a percent of GDP grew by a mean of 26.6 percentage points between 1960 and 2022. And many of the noted challenges, especially with respect to measurement, are made harder in the context of the public sector because of its diverse aims, and because the inputs, production, outputs, and outcomes are hard to "count."

Policymakers are particularly exposed to these complexities because they must try to look at the whole system, whether it be for the purposes of statistical production, the management of natural resources, in order to determine who has what and why, or even to decide on who *should* have what and why.

Data science provides powerful tools to chip away at this complexity. Some see this emerging field as a mishmash of other topics neglectfully crumpled together. The worry is that its adherents naively wade into other domains with impractical solutions. And that is a risk. But the examples in this chapter show that, at its best, data science can offer new ways to manage the explosion of information that policymakers are exposed to in the course of their decision making. And they show that, in addition to the precedents that already exist, there are numerous other high-impact applications just around the corner.

Others have called out the potential for data science to aid policymaking. Zeynep Engin and Philip Treleaven (2019) look at a number of case studies in what they call GovTech, examining applications that might improve the workings of government. Mauricius M. de Medeiros, Norberto Hoppen, and Antonio C. G. Maçada (2020) look more broadly, at businesses, and find that the increased agility with which insights may be obtained and the management of organizational performance are leading benefits of using data science. Studies have suggested that data science and the use of big data might improve productivity as much as 7% (Brynjolfsson, Hitt, and Kim 2011; Müller, Fay, and vom Brocke 2018). These productivity benefits are likely to come from automating processes, improving the quality of decision-making through better or more timely insights, and, via large language models, through providing aids to people—whether that's for writing, for coding, or even tutoring for subjects like mathematics. Jonathan Bright *et al.* (2019) examine the potential for UK local government to benefit from data science but find that severe budget constraints and a lack of appetite for innovative projects, especially those that could fail, has held back progress. A report (Misuraca and van Noordt 2020) that brings together a number of examples of the use of artificial intelligence in public services in Europe is clear that there is limited evidence of large-scale benefits to date, though it does highlight some useful early applications, including recognizing whether or not agricultural land has been mowed in Estonia and predicting which child daycare facilities it would be most efficient to inspect next in Belgium.

Data Science and What it Can Do

Data science is a relatively new discipline that unites software engineering, analysis, statistics, domain knowledge, and the scientific method to leverage the maximum value from data of all types and sizes—from billions of rows of numerical data to text to images: any sort of recorded information that can be imagined. Its practitioners typically use code to do this and are proficient in one or more programming languages. The combined knowledge of statistics, mathematics, algorithms, and coding is what allows them to extract value from data, for example in the form of insights, predictions, or even just smoother and more efficient operations. Predictions may be arrived at through the use of machine-learning algorithms, and include the outputs from large language models and other types of generative artificial intelligence (AI). While data is the focus, the coding skills also mean that those trained in data science are able to automate digitized processes very effectively: for example, producing a report based on the latest data on a schedule without the need for human intervention. An additional skill of those trained in the discipline is communication, particularly of the insights from data and especially achieving this through visualizations—graphical representations of the data along with key messages derived from it. The best practitioners are comfortable communicating insights from data to technical and nontechnical audiences alike. As an example, a firm might employ a data scientist to make sense of the success of marketing campaigns, to create dashboards of key performance indicators, or to run a machine-learning algorithm that makes recommendations to its customers of what show to watch next. Public-sector institutions might employ data scientists to provide rapid policy analysis, for example, monitoring the positions and contents of ships during a supply-chain crisis.

As a discipline, data science is often complementary to other topic areas—not only does it draw on many other disciplines, it only reaches its full potential when being integrated with a particular domain. As such, when it has been applied without considering the domain context, it has gone awry. There have been naive applications of machine learning to problems where the relationships between variables have shifted considerably—Google Flu Trends, which I will cover in more detail later on in the chapter, is a classic example. Ignoring domain-specific context, not thinking about causality, and, relatedly, falling foul of the Lucas, Jr. (1976) critique, which says that we should not expect relationships in data to be invariant to policy or incentive changes, are all ways that data science applications can go wrong. In particular, data science is much more effective when its practitioners have a solid understanding of how data are being generated and what the causal links behind that generating process are. Bringing in deep domain expertise, either through collaborative working with existing experts or by training data scientists in the domain, mitigates these risks and can unlock the full and enormous potential for problem solving that this new discipline offers.

As the field matures, data science is becoming smarter and data scientists are generating breakthroughs they can very much call their own. Causality plays an ever-larger role in the discipline, both in the development of algorithms that combine causal inference with the power of data-science algorithms (for example, double machine learning; see Chernozhukov *et al.* 2018), and in recognizing that modeling the data-generating process is essential to reducing the amount of "model shift" (Kaur, Kiciman, and Sharma 2022). A reduction in model shift means higher-quality predictions for longer before a model has to be retrained on more recent data. Some breakthroughs are recognizable as being primarily achievements in data science. Data visualization is one such example: some of the authors of the widely used graphical plotting package **matplotlib** (Hunter 2007) developed entirely new color schemes that are perceptually uniform to humans, and therefore more accurate in representing numerical information in charts (Nuñez, Anderton, and Renslow 2018). Generative AI in general, and large language models in particular, are perhaps the most well-known major breakthroughs, and ones that have reached the general public with record speed. Protein-folding predictions are another good example that, although based in a specific domain, could only have been possible using artificial intelligence. Proteins are molecules that are responsible for many of the processes fundamental to life at the cellular level, including converting food into energy and carrying oxygen around in blood. Predicting the three-dimensional structures of proteins based on the linear sequence of amino acids that they are made from is key to both understanding their role in life and in developing medical innovations, such as new drug targets, and had been a goal of biologists for decades. However, determining the three-dimensional structure of even a single protein previously took months to years of effort. A groundbreaking machine-learning model called AlphaFold (Jumper *et al.* 2021) has been developed that has a prediction accuracy competitive with experimentally determined protein structures, enormously cutting down the time and resources required to solve protein-folding problems.

It's clear that data science can cut through complexity in multiple ways. For example, for understanding modern economies, the following attributes make data science useful:

- It can turn unstructured information (text, images, audio) into tabular data, and those recorded data may ultimately be useful in measuring a phenomenon or aspect of activity. The YOLO (you only look once) machine-learning algorithm for object detection is an extremely successful example of this; out of the box, it can detect and classify eighty objects from videos or still images (Wang, Yeh, and Liao 2024).

- It can navigate complexity between inputs and outputs with predictions. For example, to predict the weather over the next ten days from current information historically required complex physics models, which were expensive because they needed to be run on supercomputers; newer "deep" learning (a type of machine learning) models, however, are able to learn the mapping between conditions now and in ten days' time, even though the weather is the canonical example of a chaotic system in which small changes in inputs can have vastly different outcomes (Lam *et al.* 2023).

- It can programmatically deal with vast datasets and complex networks. For example, in a single month, June 2023, UK citizens made 2.5 billion purchase transactions, but modern data-science tools can crunch through this and produce a range of analytical insights (Hoolohan and Colliass 2023). And, as another example, using network data science to capture the labor market as a series of related skills, rather than as mutually exclusive skills, can arguably give a better indication of which jobs and tasks are truly at risk of automation (Mealy and Coyle 2022).

- It can reduce high-dimensional information to a smaller amount of salient information. For example, in recommender systems—algorithms that might be used to suggest what media to consume next—the matrix of existing consumption of products by individual consumers might be very sparse. Dealing with large, sparse matrices is computationally costly, and recommender systems that work directly with them perform poorly. To avoid the curse of dimensionality, reduction algorithms that capture a smaller number of latent factors can be used, and the recommendation problem solved on them. While singular value decomposition is often used in recommender systems, other algorithms that have found wide use for dimensional reduction include autoencoders (Tschannen, Bachem, and Lucic 2018) and uniform manifold approximation and projection (McInnes, Healy, and Melville 2020).

Cutting Through Complexity

CUTTING THROUGH MEASUREMENT COMPLEXITIES

What we measure tells the story, to some extent, of what we as a society care about. We have long measured economic activity because, imperfect as it is, it is consistently linked with better quality of life for most people. This interest hasn't changed, but the nature of economic activity has and is, and so the way we measure the economy must change also. But it's not just about what we produce or make available through services anymore, though that is a challenge. What we measure also reflects what we value—and, increasingly, we wish to make measurements of much that is not produced directly through economic activity: the natural world, happiness, and even how we spend our time. And we care far more about making our measurements reflect distributions, as well as aggregates, now—albeit partly because modern computing makes this more possible than before. In all of these cases, our values are reflected and, if we are not careful, our biases, too. For example, and as we shall see in more detail in this section, the inflation faced by those buying a budget basket of goods could be quite different to those buying an average basket of goods, and reflecting the experience of the former requires a different—and perhaps more difficult—measurement.

In this section, we shall see how data science can help with these measurement complexities. However, data science is no panacea—it can still be misled by the biases in, and values of, society. Biases can easily creep into machine-learning models that are trained on the text that humanity has produced—if the original text tends to be misogynistic then unfortunately so will large language models trained on it, at least, they will without further intervention and correction. Another cautionary tale is that of the app where residents can report finding a pothole in order to get their local government to fix it—which resulted in potholes mainly getting fixed in affluent areas where people could afford smartphones that would support the app (Crawford 2013). Biases like these can, again, be corrected but they are a very real, if conquerable, risk for these applications of data science to measurement.

Measuring Complex Economies

Large, advanced, service-based economies tend to be more complex to measure, understand, and improve than those based on agriculture and manufacturing because it is harder to count services, service improvements, and intangibles, than it is to count the rate of production of physical goods. Statistical agencies face numerous challenges (Luiten, Hox, and de Leeuw 2020) in trying to get a handle on what's happening with economic activity—especially given response rates to surveys declining by tens of percentage points since the 1970s (Stedman *et al.* 2019).

Data science enables new types of measurement, with greater granularity, to be made. As an example, at the UK's Office for National Statistics (ONS), prices have traditionally been collected by field agents visiting supermarkets with clipboards. This is an extremely reliable method of data collection but the number of prices that can be obtained is necessarily limited because information is manually recorded by humans. In January 2022, the UK was experiencing rapid rises in the cost of living. Food campaigner Jack Monroe used social media to draw attention to the fact that price statistics as published (via the ONS' Consumer Price Index) did not describe the change in the *lowest cost*, or budget, items that consumers might purchase in their grocery shop. This is true by design: this snippet from the ONS' website describes how to think about the aggregate price statistics: "A convenient way to understand the nature of these statistics is to envisage a very large shopping basket comprising all the different kinds of goods and services bought by a typical household." But, Jack's point went, a barely-managing household is not a typical household.

Given the rapidly increasing public interest in the question of price rises for the poorest households, there was pressure for ONS to act. However, the traditional price data it collects was for the items purchased by the typical household only, and couldn't tell a comprehensive story about changes in the lowest prices.

Data scientists, led by myself and a colleague, were able to step in and web scrape price data from supermarkets' websites—thereby collecting information on the price of *every* good available, including the lowest-cost ones. Web scraping is rarely trivial, and was not in this case due to the need to extract product and price information from very differently structured websites. As well as legal agreements from the owners of the websites, the operation required advanced programming skills—and so could not have been done by statisticians alone. Furthermore, to construct a price index based on the lowest-cost items, the data scientists needed to be able to automatically create categories of products based on the names and descriptions of products, which can be achieved with data-science techniques such as clustering, and to be able to track products in those categories over time too. Additionally, these relatively large data needed to be organized and stored in a database. By bringing these skills together, data scientists at ONS were able to inform the public debate—and show that, indeed, the lowest available prices of some goods had increased by 20 or 50%, far higher than an equivalent "typical" grocery index of 6% (Casey, Banks, and King 2022). This is an example of quality deepening—it provided a more granular economic measurement—but it's also an example of data science unlocking new insights: the data also showed that the *next-cheapest* product was often substantially more (20%+) than the cheapest, making clear the importance of product line-up to those with constrained budgets.

Developments in data science can also deliver improvements in the accuracy and timeliness of statistics. The US Census Bureau's work on their construction indicator is a good example of this, as it's currently based on a survey, which is expensive and takes time to compile. So they are also using machine learning on satellite images to classify whether construction on an approved site has started or not using known permit locations. To do this, they use a convolutional neural network trained to classify satellite images as having no construction yet, having started construction, or having completed construction. In the pilot, the accuracy of this approach was 92% across all categories and there is a plan to extend it to other types of information that can be classified from satellite images. Given declining survey-response rates, the primary other method of collecting these data, and the almost instantly available nature of satellite imagery, this approach delivers significant benefits. The Census Bureau worked with Statistics Canada on this project, and the latter has used the same approach to determine if construction is commercial or residential, and what crop types are under cultivation (Erman *et al.* 2022).

On numerous occasions, data science has helped to *measure* the modern economy. In the United States, where driving to shops is common, satellite data has been used to estimate the extent to which individual retailers' stores have customers visiting them based on counts of parked cars near those stores (Bonelli and Foucault 2023). Data scientists have been quick to adopt so-called *naturally occurring data* too, that is, data that are generated naturally through the course of economic activity. These data are less directly suitable for plugging into statistical processes, like the compilation of national accounts, because they were never developed for that purpose. However, they cannot suffer from declining response rates as they reflect genuine economic activity. Vacancies as posted on online jobs boards are an example: the availability of these online job adverts, and the fact that they constitute the vast majority of all job adverts in the economy, means they can be transformed from a signal between a firm and prospective employees to a measure of labor demand (Turrell *et al.* 2019), and potentially one that can tell us about that demand in a much more granular way than surveys—for example, employers' attempts to attract women to jobs (Duchini *et al.* 2022) and the extent of flexible working arrangements (Draca *et al.* 2022; Adams-Prassl, Balgova, and Qian 2020). Many national statistical offices are considering how to incorporate other kinds of naturally occurring data into official statistics: two examples that require data-science skills are anonymized card payments, which constitute extremely big data (tens of millions of transactions a day in the UK) for national accounts and anonymized and partially aggregated mobile-phone location data for internal and external migration.

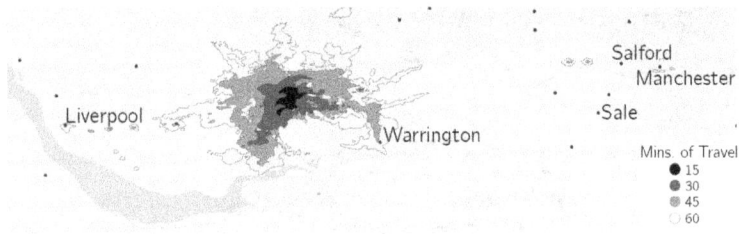

Figure 1. The feasible commuting area starting from St Helens, UK, based on starting between 7:15am and 9:15am, traveling by a combination of foot, bus, and rail.

Services remain challenging for statistical agencies to measure. In the United Kingdom, the ability to access public and private services (for example, hospitals, schools, good-quality local shops, and good-quality local restaurants) has been recognized by recent "leveling-up" policies, and especially by a commitment to increase people's satisfaction with their local high street (Gove and Haldane 2022). Similarly, good public transport opens up employment opportunities—a transport service that reliably delivers its passengers before the start of the working day is going to enable people to take jobs that an unreliable service cannot. Measuring the ability of the populace to access these kinds of in-person services is made difficult by the extremely heterogeneous means by which those living in different households could travel to them. However, modern data science techniques are able to handle both the complexity of the routing problem and the scale of the data required to analyze, hypothetically, how long it would take a person starting at arbitrary location A to get to an in-person service at location B. To help inform the debate about access to services, staff at the ONS were able to compute the area within the United Kingdom that people can feasibly travel to using public transport starting from hundreds of thousands of points across the United Kingdom—enabling anyone to identify groups of forty to 150 households that might not be able to access services in towns and cities that are nearby (Banks 2023). An example of such a feasible commuting area is shown in figure 1. This work is a foundation for more granular access-to-services and access-to-jobs measures.

The complexity of the modern economy is present in goods too, as the vast range of products on offer today, and their subtle differences, can make it challenging to accurately record changes in prices once quality is taken into account. While hedonic quality adjustments have long existed (Goodman 1998), the ability of data science and machine learning to capture unstructured data is enabling those subtle differences to be likewise captured, and also reduced from a vast amount of information into a single number (for example, a quality factor). This is a very direct example of both dimensional reduction and dealing with unstructured data. As an example, to help with Amazon's pricing strategies for clothing, Patrick Bajari *et al.* (2023) developed hedonic models that can combine text, images, prices, and quantities, and output hedonic price estimates. To achieve this, they use machine-learning models to generate arbitrary numerical factors that capture both text and images. The price predictions they are able to make with this model give a strong accuracy of over 80%. The strength of this approach is that some features that distinguish the quality of goods may only be apparent to viewers of the item, and, previously, this has been entirely outside of the ability of economists or statisticians to capture or explain. Data-science methods allow these features to be tracked and recorded. There is a lot of potential for this approach to be used to make quality adjustments in other contexts.

Complexity can arise in different ways. In some countries, it's less about a shift to services, and more about the difficulties of collecting data. Where countries have fewer resources to dedicate to the activities of their statistical offices, or conditions are dangerous because of a lack of infrastructure or the threat of conflict, data science can provide complementary statistics. As an example, since 2012, researchers have been using the nighttime light emissions of developing countries as a proxy for GDP (Henderson, Storeygard, and Weil 2012), which is especially useful when official statistics are more difficult to collect or have a long lag. There is now a rich literature on using satellite data as a complement to traditional, survey-based statistics in developing countries where agriculture, which is discernible from space, tends to represent a larger fraction of economic activity.

Measuring the Natural World

In an increasingly post-industrial world, what society cares about has come to encompass much more than providing the immediate needs for survival and, happily, has expanded to include preserving the natural environment as well as benefiting from it. Rapid environmental destruction and climate change have furthered the interest, and there is a strong need for statisticians and economists to turn their measuring tools to the natural world and its assets. This presents significant measurement challenges—it may be easy to count widgets coming out of a factory, but far harder to track the reduction in habitat of a moss. Data science provides peerless tools to measure, and even react to, the natural world.

There are a number of efforts that use remote sensors and computer vision—machine-learning algorithms that can turn an image or video into counts of what is seen—to monitor the distribution of animals. One such example is Mélisande Teng *et al.* (2023), who use citizen science–recorded bird location data to help train a model that uses satellite images to predict whether a range of bird species are present in an area. The model has some success at this prediction problem, with accuracies of more than 70% for predicting which of the 684 species have been observed in a given region. Recent high-profile machine-learning competitions have addressed the conservation of the Great Barrier Reef, of an endangered species of beluga whale, and of turtles by using facial recognition.

One complexity with Earth's changing climate is that the rapid change in the likelihood of dramatic and damaging events can cause disruption to well-functioning insurance markets, because uncertainty makes pricing challenging. Wildfires, which are unfortunately becoming more frequent and more intense in US states like California, make this more pressing as they combine large population centers with conditions favorable to the ignition and rapid spread of fires. In 2018 alone, the damage from wildfires was equivalent to 1.5% of the state's GDP. Efficient and timely prediction of where fires might break out is an important part of assessing risk. Yaling Liu *et al.* (2023) build a machine-learning model that uses anthropogenic factors (such as power lines), topography, climate, and vegetation, and the US Forest Service's database of historical ignitions, to give annual estimates of the likelihood of ignition. Given the unfortunate fact of the increasing number of climate change–related events, models that are able to assess these kinds of risks will be increasingly useful.

Of course, climate change is also putting pressure on natural resources that humanity uses—water being a prime example. In 2022, the US Bureau of Reclamation sponsored a $500,000 USD prize for estimating, in real-time, the amount of water held in snow on mountains (which will eventually enter the water supply). The competition was particularly exciting because submissions were tested against data as it arrived in real-time. The winning model explained 70% of the variation in the water volume.

BETTER INFORMED DECISIONS

Although the examples of improved or new measurement are compelling in themselves, it's no good if they do not actually improve the quality of decision-making. In this section, I present examples of where data science has cut through complexity to clearly and directly improve the ability of leaders to make informed policy decisions.

Forecasting

Forecasting is hardly new, but data science is bringing powerful new methods to bear on the problem, and these are, in relative terms, especially powerful in complex contexts. I already introduced one successful example of forecasting using data science: very much the canonical example, the weather, by Remi Lam *et al.* (2023). But higher-resolution, state-of-the-art, two-hour-ahead rain fore- and now-casts have also been achieved using machine learning (Ravuri *et al.* 2021).

There has been an explosion of new open-source software tools for time-series analysis and prediction coming out of data science in recent years, many of them developed by big technology firms, including two from Meta's Prophet (Taylor and Letham 2017) and Kats, LinkedIn's GreyKite (Hosseini *et al.* 2021), Uber's Orbit, Amazon's GluonTS, and Salesforce's Merlion.

The ability of new methods to incorporate non-numeric information and reduce the dimensionality of data to retain just the salient information has meant the development of new types of informative inputs for forecasts. For example, in Eleni Kalamara *et al.* (2022), the extent to which the economy is being spoken about positively or negatively in newspapers is transformed into indicators using various methods. These signals, which can be identified with animal spirits, are extremely real-time, and provide forecasting power for a range of macroeconomic variables above and beyond what can be gleaned from models that only incorporate traditional indicators. This kind of work provides a new and apparently informative measurement of economic sentiment. Another example is Google Trends. These data track the most popular searches of hundreds of millions of people and have been called a "database of intentions." There is little reason for an individual to perform a purposefully misleading search, so the combined searches of many individuals over time and in particular regions can be usefully informative about macroeconomic trends (Choi and Varian 2012). For example, one could use a search term related to claiming out of work benefits to anticipate rises in benefit claimants, and the OECD (Organisation for Economic Cooperation and Development) has incorporated the popularity of some search terms—such as "luggage" and "bankruptcies"—into their real-time economic activity tracker (Woloszko 2020). .

It should also be noted that the use of these data for forecasting is not without pitfalls. Simply counting the occurrences of the words "financial crisis" from newspapers and putting that in a machine learning model would give a very misleading impression of the ability of text to forecast economic outcomes if applied today to the period 2007–2009. Back in 2007, you wouldn't have known to include that term in your analysis. Google Trends was famously used to track seasonal flu but, as a paper looking into why the model had predicted *twice* the incidence of the disease put it, what was developed was part flu detector, part winter detector (Lazer *et al.* 2014).

Regardless of the pitfalls, it seems clear that there are gains to be had from relatively easy-to-use and accurate forecasting tools. The potential in, say, local government to predict seasonal demands on services based on a range of factors could play a big role in helping allocate resources more effectively.

Getting Information to Policymakers

There are a range of topics where getting timely or detailed information to policymakers is likely to significantly improve the quality of decision-making.

During the COVID-19 pandemic, there was a great need for high frequency measures of human movement and mobility because mobility is a key determinant of how easily a virus is able to spread, and because existing official statistics of mobility were released with a substantial lag. Data scientists at the Office for National Statistics were challenged to develop faster indicators of within-country movement. In the United Kingdom, there are a large number of publicly owned, publicly accessible closed-circuit television (CCTV) feeds. So staff set up a cloud-based automatic data-processing pipeline that woke up thousands of virtual computers every ten minutes to draw down stills from CCTV cameras all over the country (Chen *et al.* 2021). Without direct oversight from any human, these images were then run through an anonymization algorithm that blurred any faces or vehicle number plates that could be used to identify individuals. Next, the images were sent through a region-based recurrent neural network (Ren *et al.* 2017), a type of machine-learning algorithm. This was trained to recognize pedestrians, cars, vans, and cyclists and create counts of each of them. With so many different cameras in play, issues such as broken feeds, artifacts on screenshots, and dropped out connections were common, so there was some postcount imputation of missing values. Finally, the mobility counts were aggregated into type and region.

This new monitoring system, making use of unstructured data and a set of existing public assets, was able to deliver real-time estimates of the extent to which people were moving around the country—and inform policymakers whether or not further mobility-related measures were necessary or not. Figure 2 shows

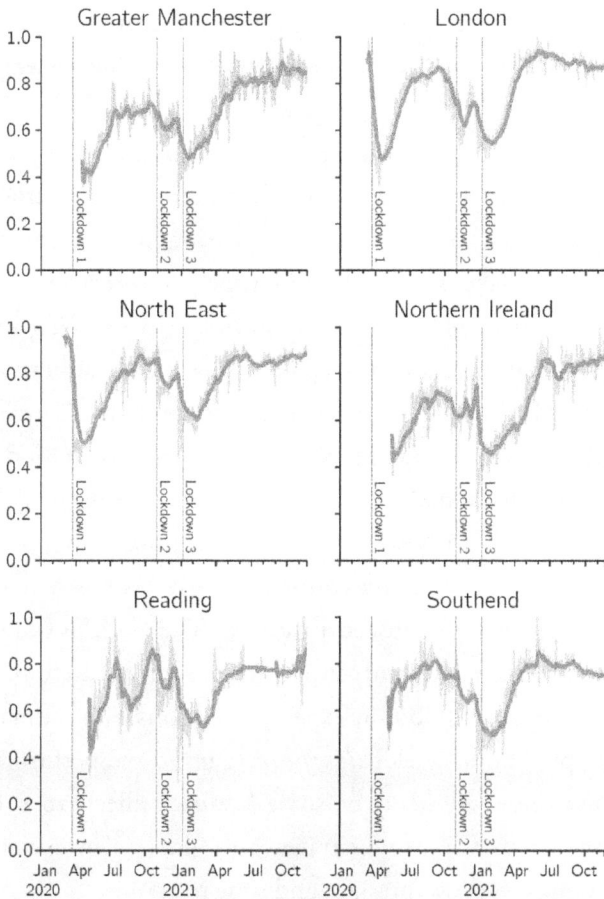

Figure 2. Activity of cars based on piping still frames from publicly accessible CCTV cameras to a series of machine-learning algorithms; see Chen *et al.* (2021) for more information. The anticipatory effects of lockdowns are apparent.

some results from this algorithm, and in particular that not only did lockdowns severely reduce mobility by car, but also that there were anticipatory effects that saw the reduction begin before policy interventions were fully in place. This pipeline is estimated to have cost ONS around GBP20 a day to run.

Supply chain issues were also a prominent feature of the COVID-19 pandemic, and it became clear that monitoring systems that could let policymakers know what goods might be lacking in the near future would be extremely useful for informing actions. Few goods are as critical as food, and many countries rely on imports to meet a substantial portion of their needs. In emergencies, having accurate and timely information on the supply and distribution of food is critical to reducing human suffering. Ananth Balashankar, Lakshminarayanan Subramanian, and Samuel P. Fraiberger (2023) took a new approach to providing information that was more timely in predicting when food crises would occur in countries with a history of food insecurity. They developed a food crisis early warning system that uses newspaper articles published in at-risk countries to give as much as twelve months' notice of food insecurity events. The model is trained on 11 million news articles and outperforms expert predictions alone (and, as is typical, combining expert predictions with the machine learning predictions yields *significantly* improved performance). Although there will likely be some feedback effects to contend with, the approach demonstrates how data science can lend decision-makers extra foresight and time to act.

Perhaps the most urgent occasions when policymakers must make decisions quickly are when natural disasters hit. This can be especially challenging in countries where there are fewer resources available to go around when a disaster occurs. Earthquakes are a good example of where data science, rapidly deployed, can help assess where, and how badly, building damage has occurred in order to distribute rescue parties and other first responders effectively. Approaches that have been introduced include using geographically precise data on earthquake intensity alongside detailed building characteristics (such as number of stories and age) to predict building damage (Mangalathu *et al.* 2020) in Nepal, and using pre- and post-earthquake satellite images, along with vision (in this case YOLO) and segmentation models, to quickly label damaged buildings on maps that were sent to first responders in Turkey.

Summary and Conclusion

The examples presented in this chapter show the great potential of data science to help policymakers. But it's likely these are only the start of the story—a 2019 report found that few UK local government organizations were using any machine learning in production (Bright *et al.* 2019). It seems clear that this relatively new discipline could provide a whole host of further improvements, many yet to be imagined. Because of its ability to make processes more efficient or deliver new insights, data science may provide a path to improving the productivity of the public sector too: and this is significant given the relative size of the public sector in many economies.

For data science to continue to improve policymakers' understanding of the world, the conditions need to be right. Much of data science is built on free and open-source software, which, in aggregate, is estimated to be worth 8.8 million million USD, which is what it could cost to replace it (Hoffmann, Negle, and Zhou 2024). But for free and open-source software to thrive, workers at firms, in the public sector, and in academia need to be given the right incentives to produce it.

Data too are important: as a former colleague put it, "you can't data science without data," and there are challenges with regards to this input to the data science production function. While, historically, statistical offices had access to the data (typically surveys) that went into creating national accounts, the measurement challenges of today, and the need for greater granularity, mean that some key data sources are in private hands.

For example, for a real-time view of consumption in nations with large economies, there are few data sources better than credit and debit card data—but the payment systems that track these data are usually privately held, and require national statistical offices to either buy the data or enter into complicated legal agreements to access them. Such data need careful treatment to protect privacy too, though there's nothing new in that. There are great success stories where private data has been acquired, but it involves much more coordination, planning, and goodwill than running a survey. The difficulty in setting up these agreements affects the rate of progress—while national statistical offices may have the purchasing power to buy these datasets, or can rely on legislation to acquire them, academics who would ideally be working on improving how these data, say, map into official statistics typically have neither, and no access. We know that when datasets have been made public (in the case that they do not contain sensitive data), they deliver huge value—John Loomis *et al.* (2015) show that the US government releasing Landsat images without cost resulted in a net efficiency gain worth tens of millions of USD per year. In order to keep providing improvements through data science, it's likely that the public sector will need to find ways to allow the wider community to connect to, and develop on, its sensitive datasets securely, either through cloud platforms or the provision of synthetic data.

And acquiring *some* data isn't always enough. Machine-learning models scale aggressively with compute and data ingested—but most institutions do not have access to the resources required to feed in enough data to create the most powerful models. It's a problem that is likely to get worse as the early providers of, say, large language models, have a first mover advantage: they can train on their customers' queries and so bootstrap the performance of their models in a way that is hard for others to replicate. Potential solutions include nationally supported foundation models and government-supported coordination on improving a small number of the most promising open-source models.

Data is not the only input. Computing capital matters enormously too, and institutions need to provide the right tooling for data scientists to do their work. Today, many firms and public sector organizations have legacy IT that greatly limits what is possible, or they do not sufficiently empower workers to use computing services effectively. It is typical for a person with a laptop and a cloud computing account acting privately to have more flexibility, power, choice, and, ultimately, productivity in their data science tooling than an individual using a large institution's locked-down and centralized IT platform. But, of course, the most useful data for public policy analysis tend to only be available on the latter and, somehow, we need to bring the best inputs (data) and the best capital (tooling) together if we are to get the best outputs. Fortunately, due to Amazon Web Services, Google Cloud, Azure, and other providers, there is a clear path for organizations to get the best of their institutional data and provide best-in-class tooling, but it does require them to empower workers with comprehensive access to these cloud services.

Despite these challenges, the examples in this chapter show the potential of data science to help policymakers. The discipline offers a compelling set of tools for cutting through the complexities of measurement and decision-making, and holds out the prospect of productivity improvements to boot.

REFERENCES

Adams-Prassl, A., M. Balgova, and M. Qian. 2020. *Flexible Work Arrangements in Low Wage Jobs: Evidence from Job Vacancy Data*. IZA Discussion Paper no. 13691. IZA Institute of Labor Economics. https://doi.org/10.2139/ssrn.3695392.

Bajari, P., Z. Cen, V. Chernozhukov, M. Manukonda, S. Vijaykumar, J. Wang, R. Huerta, *et al.* 2023. *Hedonic Prices and Quality Adjusted Price Indices Powered by AI*. arXiv preprint: 2305.00044. https://doi.org/10.48550/arXiv.2305.00044.

Balashankar, A., L. Subramanian, and S. P. Fraiberger. 2023. "Predicting Food Crises Using News Streams." *Science Advances* 9 (9): eabm3449. https://doi.org/10.1126/sciadv.abm3449.

Banks, A. 2023. *Using Open Data to Understand Hyperlocal Differences in UK Public Transport Availability | Data Science Campus*. https://datasciencecampus.ons.gov.uk/using-open-data-to-understand-hyperlocal-differences-in-uk-public-transport-availability/.

Bonelli, M., and T. Foucault. 2023. *Displaced by Big Data: Evidence from Active Fund Managers*. HEC Paris, HEC Research Papers Series No. 1491, 4527672. https://doi.org/10.2139/ssrn.4527672.

Bright, J., B. Ganesh, C. Seidelin, and T. Vogl. 2019. *Data Science for Local Government*, 33701217. https://doi.org/10.2139/ssrn.3370217. eprint: 33701217.

Brynjolfsson, E., L. M. Hitt, and H. H. Kim. 2011. *Strength in Numbers: How Does Data-Driven Decision-Making Affect Firm Performance?*, 1819486. https://doi.org/10.2139/ssrn.1819486.

Casey, A., A. Banks, and A. King. 2022. *Tracking the Price of the Lowest-Cost Grocery Items, UK, Experimental Analysis — Office for National Statistics*. https://www.ons.gov.uk/economy/inflationandpriceindices/articles/trackingthelowestcostgroceryitemsukexperimentalanalysis/april2021toapril2022.

Chen, L., I. Grimstead, D. Bell, J. Karanka, L. Dimond, P. James, L. Smith, and A. Edwardes. 2021. "Estimating Vehicle and Pedestrian Activity from Town and City Traffic Cameras." *Sensors* 21 (13): 4564. https://doi.org/10.3390/s21134564.

Chernozhukov, V., D. Chetverikov, M. Demirer, E. Duflo, C. Hansen, W. Newey, and J. Robins. 2018. "Double/Debiased Machine Learning for Treatment and Structural Parameters." *The Econometrics Journal* 21 (1): C1–C68. https://doi.org/10.1111/ectj.12097.

Choi, H., and H. Varian. 2012. "Predicting the Present with Google Trends." *Economic Record* 88 (s1): 2–9. https://doi.org/10.1111/j.1475-4932.2012.00809.x.

Crawford, K. 2013. "The Hidden Biases in Big Data." *Harvard Business Review*, https://www.microsoft.com/en-us/research/publication/the-hidden-biases-in-big-data/.

de Medeiros, M. M., N. Hoppen, and A. C. G. Maçada. 2020. "Data Science for Business: Benefits, Challenges and Opportunities." *The Bottom Line* 33 (2): 149–163. https://doi.org/10.1108/BL-12-2019-0132.

Draca, M., E. Duchini, R. Rathelot, A. Turrell, and G. Vattuone. 2022. *Revolution in Progress? The Rise of Remote Work in the UK.* CAGE working paper, 616. https://warwick.ac.uk/fac/soc/economics/research/centres/cage/publications/workingpapers/2022/revolution_in_progress_the_rise_of_remote_work_in_the_uk/.

Duchini, E., S. Simion, A. Turrell, and J. Blundell. 2022. *Pay Transparency and Gender Equality,* arXiv:2006.16099. https://doi.org/10.48550/arXiv.2006.16099. eprint: 2006.16099.

Engin, Z., and P. Treleaven. 2019. "Algorithmic Government: Automating Public Services and Supporting Civil Servants in Using Data Science Technologies." *The Computer Journal* 62 (3): 448–460. https://doi.org/10.1093/comjnl/bxy082.

Erman, S., E. Rancourt, Y. Beaucage, and A. Loranger. 2022. "The Use of Data Science in a National Statistical Office." *Harvard Data Science Review* 4 (4). https://doi.org/10.1162/99608f92.13e1d60e.

Goodman, A. C. 1998. "Andrew Court and the Invention of Hedonic Price Analysis." *Journal of Urban Economics* 44 (2): 291–298. https://doi.org/10.1006/juec.1997.2071.

Gove, M., and A. Haldane. 2022. *Levelling Up the United Kingdom.* White Paper 604. Department for Levelling Up, Housing and Communities. https://www.gov.uk/government/publications/levelling-up-the-united-kingdom.

Henderson, J. V., A. Storeygard, and D. N. Weil. 2012. "Measuring Economic Growth from Outer Space." *American Economic Review* 102 (2): 994–1028. https://doi.org/10.1257/aer.102.2.994.

Hoffmann, M., F. Nagle, and Y. Zhou. 2024. *The Value of Open Source Software,* 4693148. https://doi.org/10.2139/ssrn.4693148.

Hoolohan, V., and J. Colliass. 2023. *Regional Consumer Card Spending, UK.* https://www.ons.gov.uk/economy/economicoutputandproductivity/output/articles/regionalconsumercardspendinguk/2019to2023.

Hosseini, R., K. Yang, A. Chen, and S. Patra. 2021. *A Flexible Forecasting Model for Production Systems,* arXiv preprint: 2105.01098. https://doi.org/10.48550/arXiv.2105.01098.

Hunter, J. D. 2007. "Matplotlib: A 2D Graphics Environment." *Computing in Science & Engineering* 9 (03): 90–95. https://doi.org/10.1109/MCSE.2007.55.

Jumper, J., R. Evans, A. Pritzel, T. Green, M. Figurnov, O. Ronneberger, K. Tunyasuvunakool, *et al.* 2021. "Highly Accurate Protein Structure Prediction with AlphaFold." *Nature* 596 (7873): 583–589. https://doi.org/10.1038/s41586-021-03819-2.

Kalamara, E., A. Turrell, C. Redl, G. Kapetanios, and S. Kapadia. 2022. "Making Text Count: Economic Forecasting Using Newspaper Text." *Journal of Applied Econometrics* 37 (5): 896–919. https://doi.org/10.1002/jae.2907.

Kaur, J. N., E. Kiciman, and A. Sharma. 2022. *Modeling the Data-Generating Process is Necessary for Out-of-Distribution Generalization.* ArXiv preprint: 2206.07837. https://doi.org/10.48550/arXiv.2206.07837.

Lam, R., A. Sanchez-Gonzalez, M. Willson, P. Wirnsberger, M. Fortunato, F. Alet, S. Ravuri, *et al.* 2023. "Learning Skillful Medium-Range Global Weather Forecasting." *Science,* https://doi.org/10.1126/science.adi2336.

Lazer, D., R. Kennedy, G. King, and A. Vespignani. 2014. "The Parable of Google Flu: Traps in Big Data Analysis." *Science* 343 (6176): 1203–1205. https://doi.org/10.1126/science.1248506.

Liu, Y., S. Le, Y. Zou, M. Sadeghi, Y. Chen, N. Andela, and P. Gentine. 2023. "A Simplified Machine Learning Based Wildfire Ignition Model from an Insurance Perspective." In *ICLR 2023 Workshop on Tackling Climate Change with Machine Learning.*

Loomis, J., S. Koontz, H. Miller, and L. Richardson. 2015. "Valuing Geospatial Information: Using the Contingent Valuation Method to Estimate the Economic Benefits of Landsat Satellite Imagery." *Photogrammetric Engineering & Remote Sensing* 81 (8): 647–656. https://doi.org/10.14358/PERS.81.8.647.

Lucas, Jr., R. E. 1976. "Econometric Policy Evaluation: A Critique." In *The Phillips Curve and Labor Markets*, edited by K. Brunner and A. H. Meltzer, 1:19–46. Amsterdam, Netherlands: North-Holland.

Luiten, A., J. Hox, and E. de Leeuw. 2020. "Survey Nonresponse Trends and Fieldwork Effort in the 21st Century: Results of an International Study across Countries and Surveys." *Journal of Official Statistics* 36 (3): 469–487. https://doi.org/10.2478/jos-2020-0025.

Mangalathu, S., H. Sun, C. C. Nweke, Z. Yi, and H. V. Burton. 2020. "Classifying Earthquake Damage to Buildings Using Machine Learning." *Earthquake Spectra* 36 (1): 183–208. https://doi.org/10.1177/8755293019878137.

McInnes, L., J. Healy, and J. Melville. 2020. *UMAP: Uniform Manifold Approximation and Projection for Dimension Reduction.* arXiv preprint: 1802.03426. https://doi.org/10.48550/arXiv.1802.03426.

Mealy, P., and D. Coyle. 2022. "To Them That Hath: Economic Complexity and Local Industrial Strategy in the UK." *International Tax and Public Finance* 29 (2): 358–377. https://doi.org/10.1007/s10797-021-09667-0.

Misuraca, G., and C. van Noordt. 2020. *AI Watch — Artificial Intelligence in Public Services.* EUR 30255 EN. JRC120399. Luxembourg, Belgium: Publications Office of the European Union. https://doi.org/10.2760/039619.

Müller, O., M. Fay, and J. vom Brocke. 2018. "The Effect of Big Data and Analytics on Firm Performance: An Econometric Analysis Considering Industry Characteristics." *Journal of Management Information Systems* 35 (2): 488–509. https://doi.org/10.1080/07421222.2018.1451955.

Nuñez, J. R., C. R. Anderton, and R. S. Renslow. 2018. "Optimizing Colormaps with Consideration for Color Vision Deficiency to Enable Accurate Interpretation of Scientific Data." *PLOS One* 13 (7): e0199239. https://doi.org/10.1371/journal.pone.0199239.

Ravuri, S., K. Lenc, M. Willson, D. Kangin, R. Lam, P. Mirowski, M. Fitzsimons, *et al.* 2021. "Skilful Precipitation Nowcasting Using Deep Generative Models of Radar." *Nature* 597 (7878): 672–677. https://doi.org/10.1038/s41586-021-03854-z.

Ren, S., K. He, R. Girshick, and J. Sun. 2017. "Faster R-CNN: Towards Real-Time Object Detection with Region Proposal Networks." *IEEE Transactions on Pattern Analysis and Machine Intelligence* 39 (6): 1137–1149. https://doi.org/10.1109/TPAMI.2016.2577031.

Stedman, R. C., N. A. Connelly, T. A. Heberlein, D. J. Decker, and S. B. Allred. 2019. "The End of the (Research) World As We Know It? Understanding and Coping With Declining Response Rates to Mail Surveys." *Society & Natural Resources* 32 (10): 1139–1154. https://doi.org/10.1080/08941920.2019.1587127.

Taylor, S. J., and B. Letham. 2017. "Forecasting at Scale." *PeerJ Preprints,* no. e3190v2, https://doi.org/10.7287/peerj.preprints.3190v2.

Teng, M., A. Elmustafa, B. Akera, H. Larochelle, and D. Rolnick. 2023. *Bird Distribution Modelling Using Remote Sensing and Citizen Science Data.* arXiv preprint:2305.01079. https://doi.org/10.48550/arXiv.2305.01079.

Tschannen, M., O. Bachem, and M. Lucic. 2018. *Recent Advances in Autoencoder-Based Representation Learning.* arXiv preprint: 1812.05069. https://doi.org/10.48550/arXiv.1812.05069.

Turrell, A., B. J. Speigner, J. Djumalieva, D. Copple, and J. Thurgood. 2019. *Transforming Naturally Occurring Text Data Into Economic Statistics: The Case of Online Job Vacancy Postings.* Working Paper, 25837. https://doi.org/10.3386/w25837. eprint: 25837.

Wang, C.-Y., I.-H. Yeh, and H.-Y. M. Liao. 2024. *YOLOv9: Learning What You Want to Learn Using Programmable Gradient Information.* arXiv preprint: 2402.13616. https://doi.org/10.48550/arXiv.2402.13616.

Woloszko, N. 2020. *Tracking Activity in Real Time with Google Trends.* Technical report. Paris: OECD. https://doi.org/10.1787/6b9c7518-en.

ON THE EMERGENCE OF ZIPFIAN SIZE DISTRIBUTIONS: COUPLED STOCHASTIC GROWTH BIASED TOWARD LARGER SIZES

Robert Axtell, George Mason University and Santa Fe Institute; and
Omar Guerrero, University of Helsinki

Abstract

Stochastic models for the growth of firms and cities are typically written in terms of individual firms or cities subject to exogenous shocks. In their simplest form such models yield asymptotic size distributions that deviate from empirical data (e.g., producing nonstationary size distributions). More elaborate random-growth models can get closer to the empirical results but introduce new problems, such as the wrong value for the tail exponent or excessive fluctuations. We review such models, pointing out their main weaknesses. An alternative model of coupled growth is then elaborated and illustrated for firm sizes through the movement of workers between firms, that is, when one firm grows by a single employee the firm that worker left behind shrinks by one. This model specification remedies many of the problems associated with unconstrained stochastic growth. Furthermore, it is shown, both analytically and computationally, that biasing worker flows toward larger firms is sufficient to generate a Pareto distribution of firm sizes having exponent of exactly unity, asymptotically, that is, a Zipf distribution, closely approximating empirical data. The model is calibrated using data from the US and Finland. The general specification is also applicable to city sizes.

Coupling the Growth of Firms through Job-to-Job Flows of Workers

The sciences of complexity are fundamentally concerned with the study of large numbers of diverse entities and their interactions, focusing on what emerges from those interactions at higher levels of aggregation or organization (Simon 1962: 468; Anderson 1972). In this spirit, complexity economics involves the study of heterogeneous populations of purposive agents who interact directly with one another, the goal of such analyses being to understand the kinds of phenomena that can emerge at mesoscopic levels (e.g., within multi-agent groups or organizations) and at the macroscopic (i.e., macroeconomic or societal) level.[1]

This approach to economic organizations, viewing them as being composed of many individuals, is different from how businesses are treated in the textbooks of economics, where firms are usually considered to be unitary actors—e.g., they are modeled as maximizing profits or minimizing costs, with few, if any, "frictions" when it comes to coordinating the activities of their employees. While the lack of methodological individualism in modeling the behavior of firms has been lamented (e.g., Winter 1993), and the emerging field of "organizational economics" (Gibbons and Roberts 2012) has grown up, in part, to address such lacunae, a further problem is that the actions of distinct business entities are often viewed as being independent, or coupled primarily through product markets, as when firms have buyer–supplier relationships with one another, or through intra-industry competition when firms are direct competitors.

---

[1]For more on what constitutes "complexity economics," see Axtell *et al.* (2016) and the introduction to this volume.

An important way in which firms are coupled, both directly and indirectly, is through labor markets. Some workers may possess industry-specific skills that make them valuable to competing firms—think Silicon Valley and its computer scientists and engineers. Some workers may change industries when they move to a new job in order to remain in some specific geographical region. In any case, the flow of workers between firms is large in the US (Fallick and Fleischman 2001, 2004; Davis, Faberman, and Haltiwanger 2006), involving millions of people monthly. Focusing either on labor markets while neglecting the dynamics of firms, or on firm dynamics and ignoring labor flows, leaves out half the story. Complexity economics sees firms and workers as coupled.

A further motivation for studying firm and labor dynamics together is that much richer empirical data on job-to-job flows are available these days than were accessible to previous generations of researchers. Comprehensive firm size (e.g., Axtell 2001) and growth rate (e.g., Teitelbaum and Axtell 2005; Kim *et al.* 2024) data—all workers, all firms—can be found in administrative records and provide clear "targets" that analytical models must strive to "hit" if they are to claim empirical relevance. From Pareto-distributed firm sizes with exponent near unity—the so-called Zipf distribution—to log growth rates that are Subbotin-distributed (Perline, Axtell, and Teitelbaum 2006; Bottazzi *et al.* 2011), these are heavy-tailed distributions. Now, while such distributions have had a persistent presence in economics since the end of the nineteenth century, when they were introduced to describe cross-sections of wealth and income (Pareto 1896; Mandelbrot 1961a), and were later used to describe city sizes (Auerbach 1913; Zipf 1949; Mandelbrot 1965), firm sizes (Gibrat 1931; Simon and Bonini 1958), and certain statistics in financial markets (Mandelbrot

1963, 1964), it has been argued that they were long neglected in economics, especially in finance (Mackenzie 2006).

When it comes to firm sizes, multiplicative random-growth models are a class of stochastic processes capable of generating heavy-tailed distributions that were seemingly similar to the patterns found in various kinds of empirical data. Pioneered by the Dutch astronomer Jacobus Kapteyn (1903) in the context of natural phenomena, these were adopted to the study of inequality by Robert Gibrat (1931) in his doctoral thesis and applied across a wide variety of social science phenomena (cf. Kleiber and Kotz 2003). The simplest multiplicative random-growth models are problematical in ways described below, but over time were modified in order to be relevant empirically. Stochastic growth models similar to but different from multiplicative growth models later appeared (Simon 1955), solving some problems while introducing others, as surveyed below.[2] Here we recapitulate some of these mechanisms in order to point out their deficiencies, on our way toward describing a different class of stochastic growth models that seems to be much more relevant to the empirics of firm and labor dynamics that we want to explain.

Background on Multiplicative Random Growth Models

The study of firm dynamics from the perspective of random growth was pioneered by Gibrat (1931), whose "law of proportional effect" implies that firms *grow* in proportion to their size and that their growth *rate* is independent of size. This model and comparable multiplicative growth specifications are widely used in a number of fields, from physics (Redner 1990; Sornette 2006) and network science (Bornholdt and Ebel 2001; Albert and Barabási 2002) to ecology (Lewontin and Cohen

---

[2] An overview of such models can be found in Michael Mitzenmacher (2004).

1969) and organizations (Harrison 2004). In the context of firms, the original Gibrat model has been subject to empirical tests for at least seventy-five years, with mixed support (Kalecki 1945; Mansfield 1962; Sutton 1997; Fujiwara *et al.* 2004; Reichstein *et al.* 2010; Axioglou and Christodoulakis 2020; Yakubu 2020). The basic idea of proportional growth has been extended in various ways to address limitations of the original model, including firm entry and exit (Ijiri and Simon 1977), growth rates that depend on size (de Wit 2005), and serial correlation of growth rates. We briefly review Gibrat's multiplicative random-growth model and several variants and point out certain problems common to them, ultimately arguing that a different approach is needed. We then develop a stochastic theory of firm growth that is not multiplicative but rather involves the movement of labor between firms as workers seek better employment opportunities.

THE GIBRAT PROCESS: NONSTATIONARY LOGNORMAL FIRM SIZES

Consider a population of firms and note the size of the ith at time t by $S_i(t)$, a random variable.[3] Each period the firm's size changes by a factor that varies randomly, $G_i(t)$, the growth *rate* of firm i at time t, as:

$$S_i(t+1) = G_i(t) S_i(t). \tag{1}$$

Sizes can be measured in terms of employees, output, sales, receipts, and so on. Note that in this model the growth *rate*, $S_i(t+1)/S_i(t) = G_i(t)$, does not depend on size, whereas growth, $S_i(t+1) - S_i(t) = (G_i(t) - 1) S_i(t)$, is *proportional*

---

[3]For alternative derivations of the basic Gibrat model see Steindl (1965) or John Sutton (1997). Note that the exact nature of S is not specified, i.e., it could be either an input to a firm or an output of firm operations, e.g., revenue or profit.

to size, thus the subtitle of Gibrat's (1931) book: the law of proportionate growth. Over time,

$$S_i\left(t+1\right) = G_i\left(t\right) G_i\left(t-1\right) G_i\left(t-2\right) \cdots G_i\left(0\right) S_i\left(0\right)$$

$$= S_i\left(0\right) \prod_{\tau=0}^{t} G_i\left(\tau\right).$$

Such multiplicative growth processes can be readily related to the central limit theorem. Divide both sides by $S_i(0)$ and take logs, noting $ln\ G_i(t)$ by $g_i(t)$, to obtain

$$ln\left[\frac{S_i\left(t+1\right)}{S_i\left(0\right)}\right] = \sum_{\tau=0}^{t} g_i\left(\tau\right).$$

In reality, growth rates for any particular firm are tied up with the firm's lifecycle (Klepper and Simons 1997) and that of its industry (Klepper and Graddy 1990). Gibrat's formalism abstracts from lifecycle considerations and stipulates that log growth rates are independent realizations from some distribution, γ_i. In such a case, the left-hand side of this equation is a random variable, and the central limit theorem applies (Krapivsky, Redner, and Ben-Naim 2010; Saichev, Malevergne, and Sornette 2010). For all firms having the same growth rate distribution the subscript i can be dropped and the first moment of γ by μ_γ. Then, if the variance of γ, σ_γ^2, is finite, the left-hand side is normally distributed, implying the size ratio $S_i(t+1)/S_i(0)$ is lognormally distributed with *probability density function*

$$\frac{1}{s\sigma_\gamma\sqrt{2\pi t}} exp\left[-\frac{\left(ln\left(s\right)-\mu_\gamma t\right)^2}{2\sigma_\gamma^2 t}\right].$$

Note that the limiting distribution does *not* depend on the specific shape of γ, only its first two moments. Gibrat did not have access to anything like comprehensive data on French firms,[4] so the fact that this stochastic process yields results that are independent of the shape of γ was an attractive feature, indeed, perhaps its main strength.

However, this is a nonstationary distribution with both the mean and variance increasing over time,[5] as shown in figure 1 for $\sigma_\gamma = 0.1$. In effect, as some firms get large and others shrink, this distribution gets *stretched*. This is a highly unrealistic result *vis-à-vis* the empirics of firm growth, despite claims to the contrary (e.g., Mata 2008). A second problem with this distribution is that, in reality, firms have different ages and initial sizes, making the overall size distribution a mixture of lognormal distributions.[6] Another important problem with the basic Gibrat model is that, despite a long history of firm size data being described by lognormal distributions (e.g., Stanley *et al.* 1995), these results are almost surely artifacts of sampling processes that underrepresent small firms. When samples are large or unbiased, such as comprehensive data on all firms in a country, it is common to find that a Pareto distribution with exponent near unity well describes firm sizes empirically (Ramsden and Kiss-Haypál 2000; Axtell 2001; Gaffeo, Gallegati, and Palestrini 2003; Fujiwara *et al.* 2004; Kang *et al.* 2011; Bee, Riccaboni, and Schiavo 2017; da Silva *et al.* 2018).

These three problems with Gibrat's model—(a) the wrong limiting distribution empirically, (b) nonstationarity, and (c)

---

[4] Gibrat's data have been reanalyzed using modern methods by Sherzod Akhundjanov and Alexis Toda (2020), who find somewhat different results from those originally reported.

[5] Kalecki (1945) seems to have been the first to point out this problem and proposed ways to recover stationarity.

[6] Elliott Montroll and Michael Shlesinger (1982) demonstrate that exponential mixtures of lognormal sizes can lead to distributions that *appear* to be lognormal in their body with a power-law tail; see also Bernardo Huberman and Lada Adamic (1999).

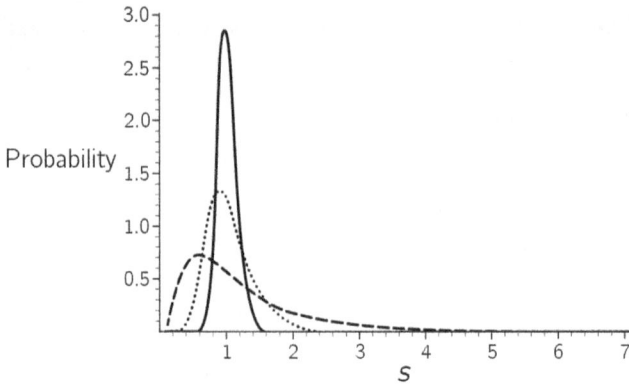

Figure 1. The nonstationary lognormal distribution of firm sizes implied by Gibrat's law of proportional growth, for $S_i(0) = 1$ and $\sigma_\gamma = 0.1$, at $time = 2$ (solid), 10 (dotted), and 50 (dashed).

inability to aggregate across initial sizes and ages—can be partially remedied.

FIXING GIBRAT? MINIMUM FIRM SIZE, RANDOM WALKS, AND SINGLE-SERVER QUEUES

One way to repair many of the problems with Gibrat's specification recognizes that all real firms have some minimum viable size, s_{min}, and thus the growth process of equation 1 1 needs to be modified accordingly (Champernowne 1953; Gabaix 1999a). Rewriting equation 1 with a size floor gives

$$S_i(t+1) = max\,[s_{min}, G_i\,(t)\,S_i\,(t)]\,. \qquad (2)$$

This intuitively reasonable specification, which is roughly equivalent to adding a small amount of noise to the Gibrat growth process, is sufficient to produce stationary Pareto-distributed sizes (Levy and Solomon 1996). Further, because Pareto distributions with the same exponent can be aggregated, this ostensibly fixes most of the problems with the Gibrat specification. It only remains to determine the exponent, α. It has been claimed (Sornette and Cont 1997; Gabaix 2009, 2016) that $\alpha \sim 1$ due to first moment restrictions. The argument is that since the mean of a Pareto distribution is

$$\bar{S} = \frac{\alpha}{\alpha - 1} s_{min}, \qquad (3)$$

which is clearly only meaningful for α greater than unity, solving for α gives

$$\alpha = \frac{1}{1 - s_{min}/\bar{S}}. \qquad \text{~379~}$$

For $s_{min} << \bar{S}$, α is slightly larger than 1, in agreement with the data. While this argument seems reasonable, there is a problem with it. For any finite population, the support of the distribution is finite, so the mean firm size exists for *all* α. Specifically, for a Pareto distribution having bounded support on $[s_{min}, s_{max}]$, the mean firm size is (Malcai, Biham, and Solomon 1999)

$$\bar{S} = \frac{\alpha}{\alpha - 1} \frac{s_{min}s_{max}^{\alpha} - s_{max}s_{min}^{\alpha}}{s_{max}^{\alpha} - s_{min}^{\alpha}},$$

which for $s_{max} >> s_{min}$ becomes

$$\bar{S} \approx \frac{\alpha s_{min}}{\alpha - 1} \left[1 - \left(\frac{s_{max}}{s_{min}} \right)^{1-\alpha} \right]. \qquad (4)$$

This expression is positive for $\alpha > 0$ with a hole at $\alpha = 1$, so the mean exists generically, not just for $\alpha > 1$. While it is not possible to analytically solve for α in closed form as a function of the other variables, this is shown numerically in figure 2. Clearly, α can be either greater or less than 1 for finite s_{max}, and there is no privileged status for $\alpha \approx 1$.[7] Indeed, for US firms with

---

[7] The empirical relevance of Pareto exponents *less than* unity has been debated since the 1950s (Simon 1955; Mandelbrot 1959; Simon 1960; Mandelbrot 1961a, 1961b; Simon 1961a, 1961b), and we do not wish to resuscitate this old debate, which has produced more heat than light despite the back and forth between two giants of twentieth-century science.

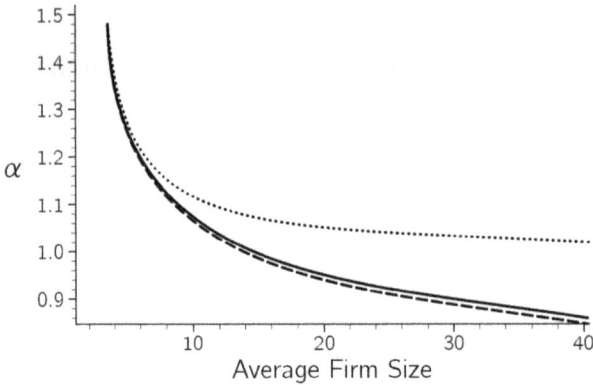

Figure 2. Pareto exponent from equation 4 as a function of average firm size: dependence of alpha on $\bar{S}$ for $s_{min} = 1$ and $s_{max} = 10^6$ (dashed), $s_{max} = 2x10^6$ (solid), and $s_{max} = \infty$ (dotted).

employees, $s_{min} = 1$ and $\bar{S} \approx 25$ are reasonable parameters, yielding $\alpha \approx 1.04$ from equation 3, while for $s_{max} \approx 2 \times 10^6$ (Walmart has some two million employees at present), $\alpha \approx 0.92$ via equation 4, in opposition to the data. Thus, the argument of equation 3 is an artifact of the *assumption* that firm sizes are unbounded.[8]

Stipulating a minimum firm size fixes most of the problems with the basic Gibrat process, but does the fact that the exponent is *not* right mean there is something wrong with this explanation? It turns out that the random growth process of equation 2 may also be interpreted as a discrete-time, continuous-position random walk on $[ln\ s_{min}, \infty)$, that is, with a lower boundary. The behavior of such walks has been well understood since at least the middle of the last century (Spitzer 1964; Feller, 1968 [1950], 1971). Remarkably, these random walks are intimately related to queueing problems, studied even

---

[8]Xavier Gabaix (2009: 268) asserts that finite size effects may matter and reproduces a version of equation 4, but attributes discrepancies between equations 4 and 3 to noncommuting limits instead of assumptions about the first moment being finite and the support being infinite.

earlier. Specifically, the distribution of waiting times in single-server queues with independent arrival and arbitrary service times (Lindley 1952; Kingman 1961, 1962), that is, the GI/G/1 queue, coincides with the distribution produced by the random walk of equation 2 (Grimmett and Stirzaker 1992). In these literatures, standard results are (i) a stationary distribution for *ln* $S_i(t)$ exists asymptotically if and only if γ has negative drift, that is, $\mu_\gamma < 0$; (ii) the distribution of *ln* $S_i(t)$ is exponential; and (iii) with mean of $-\sigma_\gamma^2/2\mu_\gamma$. Note that the mean makes no sense unless $\mu_\gamma < 0$ and that it can be large or small, depending on γ (Gross *et al.* 2008). In terms of firm growth, (i) keeps firms from growing arbitrarily large, while (ii) can be interpreted using a standard characterization from statistics (Johnson, Kotz, and Balakrishnan 1994): a *random variable* is Pareto-distributed when its logarithm is exponentially distributed. Specifically, for $S \sim$ *Pareto* with exponent α and support on $[S_{min}, \infty)$, ln $S \sim$ exponential with mean $1/\alpha$ and support on [ln $S_m in, \infty$). Combining this with (iii) gives

~ 381 ~

$$\alpha = -\frac{2\mu_\gamma}{\sigma_\gamma^2}.$$

Clearly α can take on *any positive value* ($\mu_\gamma < 0$), depending on the underlying growth rate distribution γ. For US establishments, γ is known with some precision, as shown in figure 3.

These data reveal that growth rate distributions depend on size to a modest extent (Perline, Axtell, and Teitelbaum 2006), in opposition to Gibrat's original specification. The moments of the distributions are shown in figure 4. Specifically, μ_γ is slightly *positive* for small firms and *negative* for firms having more than five employees, as can be seen on the left panel, with μ_γ in the range $[-0.10, -0.05]$ for firms larger than five. From the right panel, aggregating over all firms, $\sigma_\gamma^2 \approx 0.20$. Overall, if

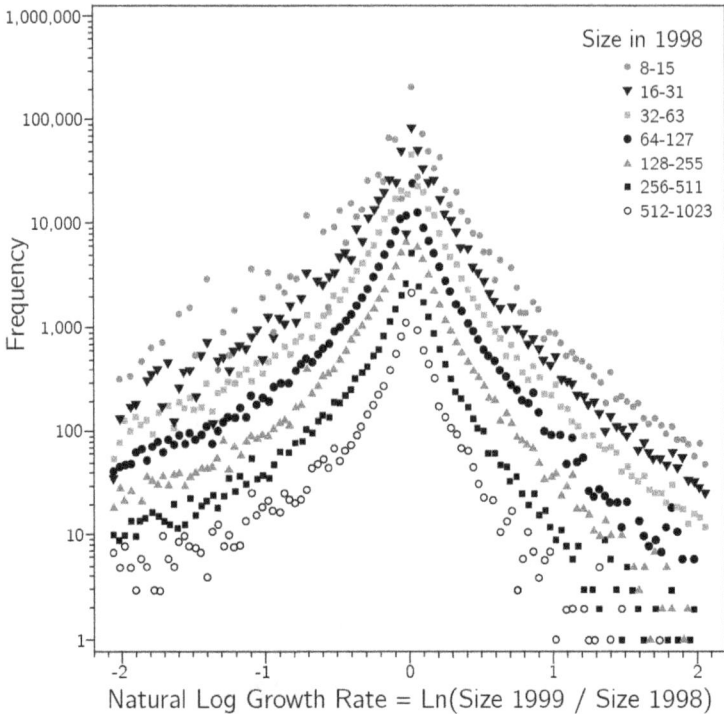

Figure 3. Histogram of annual US growth rates of all firms, from 1998 to 1999. *Source: Census (Perline, Axtell, and Teitelbaum 2006).*

these moments did not depend on size it would be fair to suggest that the exponent, α, is around 1.

Summarizing, the addition of a firm size floor to Gibrat's law leads away from a nonstationary lognormal distribution and to a stationary Pareto distribution whose exponent depends on the growth-rate distribution, γ. The argument that the exponent is bounded from below by 1 results from *assuming* that firm sizes can be arbitrarily large; correcting this misspecification yields smaller exponents, outside of the empirical range of values. Aggregate data on γ gives very nearly

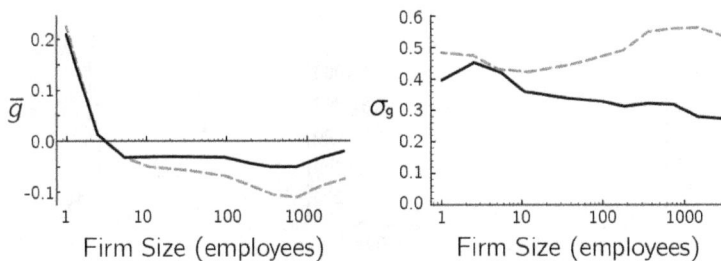

Figure 4. Dependence of average annual g on size (dashed) and excluding the most extreme log growth rates (black), with size by employees. *Source: Census (Perline, Axtell and Teitelbaum 2006).*

the correct exponent despite the fact that its moments are not independent of size.

OTHER ISSUES WITH MULTIPLICATIVE GROWTH MODELS

So far, we have only considered random growth with a fixed number of firms. Models with firm entry and exit and proportional growth have been investigated analytically (Simon 1955; Hart and Prais 1956; Steindl 1965; Richiardi 2004; de Wit 2005; Saichev, Malevergne, and Sornette 2010). It turns out that these also remedy some of the basic problems with the Gibrat process, producing (a) stationary distributions that (b) can be aggregated (c) with power-law tails having nearly the correct exponent. One problem with many of these models is that they get the tail behavior about right but tend to overestimate the number of very small firms *vis-à-vis* the empirical data. Computational models with entry and exit have also been put forth by a variety of authors (McCloughan 1995; Harrison 2004; Richiardi 2004).

From the time of Gibrat (1931) and continuing in the work of Michał Kalecki (1945), Herbert Simon (1955), Josef Steindl (1965), and others, little was known about *actual* firm

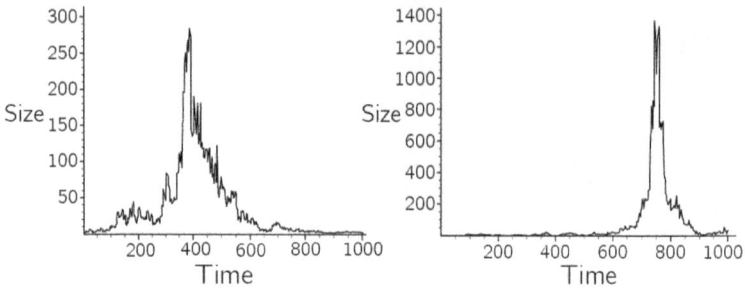

Figure 5. Typical size evolution of individual firms in multiplicative random growth models, 2 examples; such rapid growth and collapse (i.e., large volatility), while empirically possible, is not a good description of firm dynamics in general and is thus not realistic.

growth rate distributions. Therefore, the fact that equations 1 and 2 work for *any* nondegenerate growth distribution was considered to be a strength of these theories. However, today we can use firm-level growth-rate data, as shown in figure 2. So far, only a few multiplicative growth models have attempted to use these data or others like it (Bottazzi *et al.* 2001; Bottazzi, Cefis, and Dosi 2002; Bottazzi *et al.* 2007; Bottazzi *et al.* 2011). Such data can be used to explicitly compute firm sizes corresponding to such log growth rates. When we do this, we get individual firm-size time series that can be *very* volatile. Typical plots of this type are shown in figure 5 for two separate realizations. In both of these figures, we see firms are small for some time, followed by a period of explosive if irregular growth, and then rather rapid collapse. This highly volatile growth, while possible for particular firms, is unrealistic overall.

Next, looking at a *population* of firms growing in this highly volatile way, we discover that fluctuations in the total size of all firms—the private sector—is even *more* volatile. Figure 6 plots the aggregate sizes of a million firms growing in this multiplicative way. In essence, the outliers in individual firm growth dominate aggregate growth, producing excessive

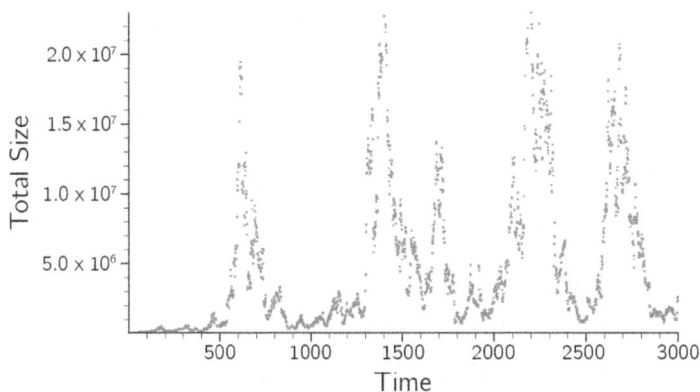

Figure 6. Total size of all firms in an economy composed of 1 million firms under multiplicative random growth; this degree of volatility at the level of an entire economy is highly unrealistic.

volatility that does not resemble actual economic growth processes in any meaningful way. The origin of this problem is the lack of constraints on growth. For firm size measured by employees, for instance, the number of workers varies by orders of magnitudes over these growth "flare-ups." In analytic derivations, such fluctuations are assumed away by normalization of the total size of the firm population. Such highly erratic dynamics are not realistic since the number of workers, the amount of capital, and other inputs and outputs are constrained on all but the longest time scales, effectively eliminating such large fluctuations. This problem is *catastrophic* for multiplicative random growth (i.e., Gibrat-like) processes, as there is no easy way to remedy it.

Another stream of literature attempts to add economic motivation to models of firm growth, including some that focus on general equilibrium (Luttmer 2007, 2011, 2018) and others that look at technological change at the firm level (Klette and Kortum 2004; Chatterjee and Rossi-Hansberg 2012). What these approaches have in common is some coupling between firms, at least *indirectly*. Such coupling remedies an obvious flaw evident from equations 1 and 2 that individual firms have no relation to any other firms. It is as if each firm lives on its own island, not influenced by either peers or competitors, acting autonomously and oblivious to the ups and downs elsewhere in the economy—a kind of *Swiss Family Robinson* if not *Robinson Crusoe* (Samuelson 1952). What is missing from Gibrat-type growth specifications analyzed so far is *direct* coupling between firms, so that the growth of one comes at the expense of another, for example. In reality, as the fortunes of specific firms ebb and flow, new opportunities and challenges arise for other firms, whether due to buyer–supplier relationships (e.g., Atalay *et al.* 2011), job-to-job worker flows (e.g., Guerrero and Axtell 2013), or macroeconomic linkages (Acemoglu *et al.* 2012; Anthonisen 2015).

A further problem with stochastic growth models, as well as some of the models mentioned previously, is they depart from the norm in economics of explaining the actions of economic entities in terms of the individual agents who make up the economic process under investigation (Winter 1993). That is, most models of firm dynamics neglect labor dynamics. We next move on to develop a theory of firm growth and size that takes a step toward representing individuals while avoiding the manifold problems encountered above.

Beyond Multiplicative Growth: Coupling Firms via the Flow of Workers

As an alternative stochastic-growth model, we move away from the idea that firm growth happens in isolation, without reference to other firms. In its place we develop a model in which workers move between firms, seeking better jobs, so the growth of one firm is *negatively* correlated with that of another. This makes giant fluctuations, of the kind on display in figures 5 and 6, impossible.

A SIMPLE MODEL OF JOB-TO-JOB FLOWS BIASED TOWARD LARGER FIRMS

Consider an economy with a fixed, finite number of workers, W, employed at some number of firms, $F(t)$, which can vary with time t. That is, firm entry and exit can occur but this is not essential to the subsequent derivations. Call $f_s(t)$ the number of firms employing exactly s workers at time t and $\lambda(t)$ the largest firm size at t. By definition,

$$F(t) \equiv \sum_{s=1}^{\lambda(t)} f_s(t)$$

$$W = \sum_{s=1}^{\lambda(t)} s f_s(t)$$

for all t.[9] (Firms without employees can be considered by changing the lower bound to 0, but we will not pursue this here.) Note that $\{f_s(t)\}$ is the histogram of firm sizes in terms of employees, and $W/F(t)$ is the average firm size at t.[10] At $t = 0$ there is some specific distribution of firms by size, $f_s(0)$, for all sizes from 1 to $\lambda(0)$, and the size of the largest firm, $\lambda(t)$, will generally change over time, although this plays no role in what follows. As time advances, workers migrate between firms in search of better job opportunities in a specific

~387~

---

[9] A constant number of firms and workers is a reasonable assumption over relatively short time periods. Indeed, in the past twenty years, $W \sim 115$–120 million for the US private sector, with those employees arranged in 5–6 million firms.

[10] While the number of firms can be time varying, for purposes of the derivations to come we consider only the case of a fixed number of firms. Firm entry and exit can be added to the theory without difficulty, but the formalism is then somewhat less simple and intuitive, so we have elected not to present that more advanced formulation here.

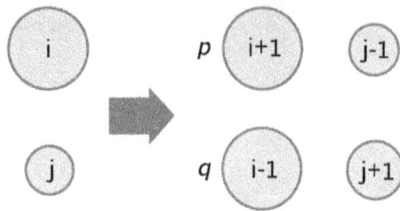

Figure 7. Schematic of the stochastic worker rearrangement process for two firms of sizes i and j, with $i > j$; with probability p the larger firm gains a worker while the smaller firm loses a worker, with probability q the reverse happens, and no change occurs with probability $1-p-q$.

manner to be described. In the US economy, such flows are quite large, involving millions of employees each month (Fallick and Fleischman 2004; Bjelland *et al.* 2011). Consider these movements as occurring one at a time, although this is merely an analytical convenience and not necessary in general.

The particular mechanism to be analyzed goes as follows: two firms, having sizes i and j, are selected at random to exchange a worker. If $i \geq j$ then the new sizes become:

a. $i + 1$ and $j-1$ with probability p;

b. $i-1$ and $j + 1$ with probability $q < p$, $p + q \leq 1$;

c. i and j, that is, their sizes are unchanged, with probability $1-p-q$.[11]

This situation is represented graphically in figure 7.

This means that workers coming from smaller firms are more likely to take new jobs at larger firms, that larger firms tend to hire more from smaller firms, but also that large firms

---

[11] The case of $p = q$, that is, equal probability of moving, yields exponential distributions of sizes and is discussed in many books on probability, stochastic processes, network dynamics, and related fields.

can lose workers to smaller firms. An equivalent representation of this process is:

$$(i, j) \to (i + 1, j - 1) \text{ with probabilities } \begin{cases} p, & i \geq j \\ q, & i < j \end{cases} .$$

This asymmetry in transition probabilities, that is, $p > q$, could arise for many reasons, such as the fact that large firms pay more (Brown and Medoff 1989), because larger firms have more opportunities for professional advancement, and so on. This is a model of *one firm growing at the expense of another*, effectively eliminating the possibility of large fluctuations such as those shown in figures 5 and 6. Note that this is *not* a "preferential attachment" model, since movements of workers between firms are not *proportional* to firm size.[12] Proportionality is a kind of "bias." The $p > q$ "bias" differentiates it from other "exchange" models (e.g., Yakovenko and Rosser 2009), and is different from other stochastic mechanisms capable of producing power laws in systems of constant size (e.g., Metzig and Colijn 2019). Superficially, it might seem that the biased flow of workers to larger forms would lead to all agents in one giant firm. But the flow of workers to larger and smaller firms depends not only on p and q but also on how many firms of various sizes are present in the economy, since firms are selected at random to acquire and give up workers, as will be shown below. For a sufficiently small number of big firms the biased movement of workers to larger firms is balanced by flows to smaller ones. Indeed, once the flows of workers up and down the firm size distribution is balanced in this way the firm size distribution reaches a steady-state, that is, it is stationary.

---

[12] Arguably, it is a kind of generalization of the preferential (really proportional) attachment mechanism, which is also a "biased" mechanism, insofar as all we require is some amount of bias toward larger size, not literal proportionality.

In what follows we derive expressions for the steady-state distribution of firm sizes when workers follow the process described heuristically above.[13] To accomplish such derivations it will be convenient to work not with the raw firm counts, but rather with the corresponding probability mass and cumulative distribution functions. These are defined in standard ways as, respectively,

$$\phi_s(t) \equiv \frac{f_s(t)}{F(t)},$$

$$\Phi_s(t) = \sum_{i=1}^{s} \phi_i(t).$$

We start by developing expressions for the probability that firms lose or gain workers, depending on their size. Specifically, when a job opening occurs in some firm in the economy, the chance it will be filled by an agent currently working for a firm of size $s > 1$ depends on the relative number of firms larger and smaller than s. If there are lots of firms as large as or larger than s, then there is a good chance the firm in question will lose a worker, depending on the exact value of p. However, if most firms are smaller than s, then there is a lesser chance, based on q. Summing across all firm sizes, from 1 to $\lambda(t)$, we get the following expression for the probability a size $s > 1$ firm *loses* an employee:

$$\Pr[s \to s - 1, t] = p(1 - \Phi_{s-1}(t)) + q\Phi_{s-1}(t)$$
$$= p - (p - q)\Phi_{s-1}(t).$$

Analogously, the probability a firm of size s *attracts* a worker from another firm is, with the stipulation that the size 1 firms

---

[13] The derivations are similar to certain models of binary random interactions and redistributions (Ree 2006) and network evolution in which links are perpetually rearranged based on vertex degree (Ree 2007).

cannot lose their sole employees,

$$\Pr[s \to s+1, t] = p(\Phi_s(t) - \phi_1(t)) + q(1 - \Phi_s(t))$$
$$= q - p\phi_1(t) + (p - q)\Phi_s(t).$$

Note these probabilities more compactly as

$$\Pr[s \to s-1, t] \equiv \Pr[-|s, t]$$

$$\Pr[s \to s+1, t] \equiv \Pr[+|s, t]$$

and

$$\Pr[s \to s, t] \equiv \Pr[0|s, t] = 1 - \Pr[-|s, t] - \Pr[+|s, t]$$

For each size class, a differential equation can be written for how the fraction of firms of that size changes with time, with the entire set of equations called the "master equation." Because the number of agents is finite, the largest firm is finite (although fluctuating in size), and the overall system of ordinary differential equations is finite:

$$\frac{d\phi_1(t)}{dt} = \phi_2(t)\Pr[-|2, t] - \phi_1(t)\Pr[+|1, t]$$

$$\frac{d\phi_2(t)}{dt} = \phi_1(t)\Pr[+|1, t] + \phi_3(t)\Pr[-|3, t] - \phi_2(t)\{1 - \Pr[0|2, t]\}$$

$$\vdots$$

$$\frac{d\phi_s(t)}{dt} = \phi_{s-1}(t)\Pr[+|s-1, t] + \phi_{s+1}(t)\Pr[-|s+1, t]$$
$$- \phi_s(t)\{1 - \Pr[o|s, t]\}$$

$$(5)$$

$$\vdots$$

$$\frac{d\phi_\lambda(t)}{dt} = \phi_{\lambda-1}(t)\Pr[+|\lambda-1, t] - \phi_\lambda(t)\Pr[-|\lambda, t]$$

Consider the expression for $d\phi_s(t)/dt$. The first two terms on the right-hand side represent how the number of size s

firms increases, while the last represents losses of such firms. Rearranging the right-hand side so that the first and second terms are the exchange of firms between size categories s and $s - 1$, while the third and fourth represent the change between sizes s and $s + 1$, there results

$$\frac{d\phi_s(t)}{dt} = \phi_{s-1}(t)\mathbb{P}\mathrm{r}[+|s-1,t] - \phi_s(t)\mathbb{P}\mathrm{r}[-|s,t] -$$

$$\{\phi_s(t)\mathbb{P}\mathrm{r}[+|s,t] - \phi_{s+1}(t)\mathbb{P}\mathrm{r}[-|s+1,t]\}.$$

It is standard in stochastic systems to call the terms on the right-hand side "fluxes" (van Kampen 2007) and to note the flux from size s to $s+1$, $J_{s,s+1}$, at time t as

$$J_{s,s+1}(t) \equiv \phi_s(t)\mathbb{P}\mathrm{r}[+|s] - \phi_{s+1}(t)\mathbb{P}\mathrm{r}[-|s+1,t].$$

In this notation the equations (5) for the evolution of the firm size distribution evolve as

$$\frac{d\phi_1(t)}{dt} = -J_{1,2}(t)$$

$$\frac{d\phi_2(t)}{dt} = J_{1,2}(t) - J_{2,3}(t)$$

$$\vdots$$

$$\frac{d\phi_s(t)}{dt} = J_{s-1,s}(t) - J_{s,s+1}(t) \tag{6}$$

$$\vdots$$

$$\frac{d\phi_\lambda(t)}{dt} = J_{\lambda-1,\lambda}(t).$$

From this representation of the dynamics, note that a negative flux term in one equation appears in the subsequent equation as a positive flux. Thus, for the system as a whole, it must be true that

$$\sum_{s=1}^{\lambda(t)} \frac{d\phi_s(t)}{dt} = 0$$

for all t. This simply means that the total probability is invariant, a direct result of the total number of workers being constant.

STATIONARY LABOR FLOWS, GROWTH RATES, AND FIRM SIZES

At all times, some firms are growing while others are shrinking. There is never an equilibrium at the level of individual workers or firms, that is, no workers stay in one firm forever, so firms do not have a fixed set of workers. Rather, we investigate aggregate steady states in which the number of firms moving up the size distribution is exactly balanced by the number moving down, and the number of firms and workers in each size category is constant. Here we derive this steady-state distribution as a function of the underlying parameters, p and q.

It is tempting to try to solve this system by simply equating the two distinct fluxes that sit on the right-hand side of the expression for $d\phi_s(t)/dt$, that is, $J_{s-1,s}(t) = J_{s,s+1}(t)$. Nonzero fluxes are not economically meaningful, however, since they involve either all firms shrinking to size 1 or all workers ending up as employees of a single large firm. Economically meaningful solutions are obtained by setting the flux terms to 0, meaning $J_{s-1,s}(t) = J_{s,s+1}(t) = 0$. Clearly this amounts to

$$\phi_s \mathrm{Pr}[+|s] = \phi_{s+1} \mathrm{Pr}[-|s+1]$$

or

$$\phi_{s+1} = \frac{\mathrm{Pr}[+|s]}{\mathrm{Pr}[-|s+1]} \phi_s$$

and all time arguments have been suppressed since this is a steady state. This expression can be rewritten in terms of p and q as

$$\phi_{s+1} = \frac{q - p\phi_1 + (p-q)\Phi_s}{p - (p-q)\Phi_s} \phi_s.$$

One parameter can be eliminated by dividing both the numerator and denominator on the right-hand side by p. Noting q/p by $\theta < 1$, there results

$$\phi_{s+1} = \frac{\theta - \phi_1 + (1-\theta)\Phi_s}{1 - (1-\theta)\Phi_s} \phi_s. \tag{7}$$

This is a nonlinear recurrence relation and appears difficult to solve in closed form. Note that the right-hand side depends only on terms having size index strictly below that on the left-hand side. If the solution is to be a distribution that is monotone decreasing in s, it must be the case that the fraction on the right-hand side is less than unity for all s. That is,

$$\frac{\theta - \phi_1 + (1 - \theta)\Phi_s}{1 - (1 - \theta)\Phi_s} \leq 1.$$

Solving for Φ_s there results

$$\Phi_s \leq \frac{1}{2}(1 + \frac{\phi_1}{1 - \theta}).$$

Since this expression must be true for all s and because Φ_s is bounded from above by 1, it must be the case that $\phi_1 \geq 1 - \theta$. Plugging $\phi_1 = 1 - \theta$ into equation 7 and writing out the first several terms, one immediately sees that a general expression for ϕ_s is

$$\phi_s = \frac{\theta(1 - \theta)}{[s - (s - 1)\theta][(s - 1) - (s - 2)\theta]}$$
$$= \frac{\theta(1 - \theta)}{[(1 - \theta)s + \theta][(1 - \theta)(s - 1) + \theta]}$$

Tedious algebra reveals this to be a valid probability density function. For large s, it is clear that

$$\phi_s \approx \frac{\theta}{(1 - \theta)s^2}.$$

This is asymptotically a Pareto distribution (Johnson, Kotz, and Balakrishnan 1994) with exponent *exactly* unity, that is, a Zipf distribution. The cumulative distribution function can be written in closed form as

$$\Phi_s = \frac{(1 - \theta)s}{s - (s - 1)\theta}$$
$$= \frac{(1 - \theta)s}{(1 - \theta)s + \theta}$$
$$= 1 - \frac{\theta}{(1 - \theta)s + \theta},$$

making the complementary cumulative distribution function

$$1 - \Phi_s = \frac{\theta}{(1-\theta)s + \theta}$$
$$\approx \frac{\theta}{(1-\theta)s},$$

for large s, confirming exact Zipfian behavior asymptotically. This is also true for finite firm sizes, that is, for finite support. Note that this result does not depend in any way on the initial distribution of firm sizes, that is, Zipfian firm sizes result from any initial distribution.

~395~

We have confirmed these analytical results computationally. The speed of convergence can be slow for either p only slightly larger than q or for both p and q small so that most of time no job changes happen. But otherwise the numerical results broadly support the mathematical ones, always yielding power-law firm sizes with exponent year 1.

This model, coupling firms through the labor flows between them, fixes all the problems with the basic Gibrat process: It yields a stationary Pareto distribution with the correct exponent and without large-scale fluctuations. It remains only to empirically specify the model with data, to see if the basic mechanism—biased movement toward larger firms—is credible. In what follows we will do this in two different ways, one a calibration at the aggregate level, using US data, and the other an estimation using comprehensive data on individual firms and workers in Finland, since such data are not available for the US.

SIMPLE CALIBRATION OF θ USING DATA ON US FIRMS

The firm size distribution in the US is very nearly stationary, as can be seen from a decade's worth of data, superimposed,

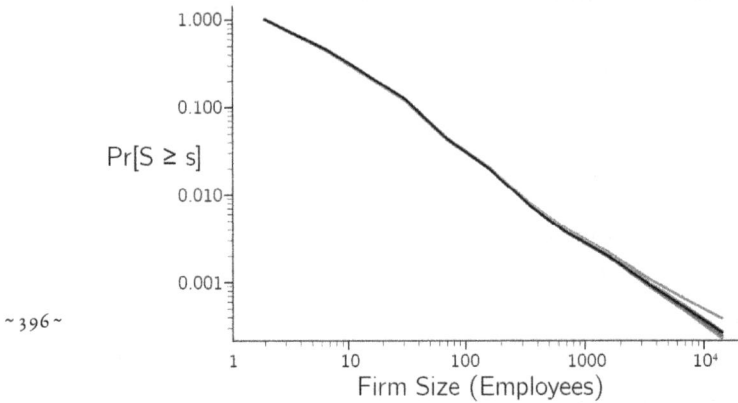

Figure 8. US firm size distribution 2003–2012, superimposed.

in figure 8. Individual firms move up and down the size distribution over time, but the overall structure is invariant, even though such a significant event as the financial crisis of 2007–2009 and subsequent Great Recession. Since the Zipfian power law results for any value of θ, it is not possible to infer θ from this plot alone. However, moment relations can be used to pin down θ. While the pure Zipf distribution (Pareto law with $\alpha = 1$) does not have finite moments, real data always have finite support and thus moments that can be explicitly computed. Specifically, for an average firm size of approximately twenty employees in the US private sector and a maximum firm size of one million, figure 9 is a plot of average firm size as a function of θ. From this plot we can infer that $\theta \approx 0.61$, meaning $q \approx 0.61p$, implying $p \approx 1.64q$, or US workers are some 64% more likely to move to a larger firm than a smaller one.

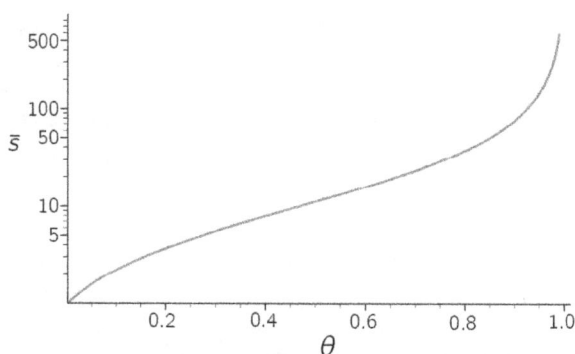

Figure 9. Dependence of average firm size of the model on θ.

EMPIRICAL ESTIMATION OF θ BASED ON JOB-TO-JOB FLOWS IN FINLAND

The FLEED panel from Statistics Finland contains employer–employee matched records covering the universe of firms and workers for twenty-five years. The data consist of annual registries in which each worker is linked to its employer at that time. If, across two years, a worker is linked to different firms, i and j, we consider that they moved from a job at i to one at j. Flows that occur within a single year and unemployment spells are not observable. We hope to show that $\theta < 1$ and that it is stable.

Consider the movement of a worker between firms as a random variable produced by sampling two firms, i and j. Firm i has an employee who wants to leave, and firm j has a job opening that the worker is considering. If $s_i \leq s_j$, then the transition takes place with probability p, while if $s_i > s_j$, it happens with probability q. Let J_{sb} denote the observed flux from smaller to equal or bigger firms and J_{bs} the observed flux from bigger to smaller firms. J_{sb} is just the probability, ϕ_{s_i}, of sampling an origin firm of size s_i times the probability of sampling a destination firm of equal or larger size, $1 - \Phi_{s_i}$, times p, and times the number of attempts, A, to change jobs, summed over all origin sizes:

$$J_{sb} = A \sum_{s_i=1}^{\lambda} \phi_{s_i}(1 - \Phi_{s_i})p = Ap(1 - \mathbb{E}[\Phi]).$$

Similarly, the flux to smaller firms is

$$J_{bs} = A \sum_{s_i=1}^{\lambda} \phi_{s_i} \Phi_{s_i} q = Aq\mathbb{E}[\Phi].$$

The number of attempts to change jobs is not directly observable. However, we can take the ratio of the fluxes in order to obtain

$$\frac{J_{bs}}{J_{sb}} = \frac{q\mathbb{E}[\Phi]}{p(1 - \mathbb{E}[\Phi])} = \theta \frac{\mathbb{E}[\theta]}{(1 - \mathbb{E}[\Phi])}.$$

This allows us to directly compute θ from the data, giving $\theta \approx$ 0.41, averaged over the past twenty-five years, with a standard deviation of 0.05. Thus, $q \approx 0.41p$, implying $p \approx 2.44q$, or Finnish workers are more than twice as likely to move to a larger firm than a smaller one.

Summary and Conclusions

The conventional view of firm growth is problematic. In its pure Gibrat variant, it yields the wrong answer—a nonstationary lognormal distribution that resists aggregation. The variation involving minimum firm sizes yields stationary power laws but derives approximately correct exponents by assumptions— that is, that firm size distributions have infinite support—that cannot be justified for finite economies. These specifications have in common a focus on the exogenous random growth of a single firm—there are no interactions between firms in these models. Other approaches that get the tail exponent right (e.g., permitting entry and exit) have their own problems. Indeed, it is possible to add entry and exit to our model, but we have resisted doing this on the grounds that the extant model gets the right answer without needing this other mechanism. While the main results have been illustrated for firms, they are also applicable to cities.

BEYOND EXOGENOUS MULTIPLICATIVE RANDOM GROWTH

~399~

We have argued that this view of firm growth is inadequate on a variety of grounds and suggested that an important class of alternative models *couples* the growth of firms through employee migration between firms. Specifically, we have articulated a theory in which firms gain or lose workers *vis-à-vis* other firms, thus "building in" negative correlation between the growth events of two distinct firms that employ the same person in different periods. These interfirm migrations of workers— job-to-job flows—are large and persistent in the US economy, accounting for some three million events/month, averaged over the past twenty years (Fallick and Fleischman 2004; Bjelland *et al.* 2011). Such flows are a kind of zero-sum game in which one firm's gain is another's loss, a process very different from Gibrat- style proportional growth processes. Among the many merits of this new model of firm growth is its ability to produce the Zipf law exactly.

While we have interpreted random growth processes exclusively in the context of firms, it also applies to city growth, among other areas. In fact, because the theoretical derivations above were for a fixed number of firms, it goes over directly to the city-size case, since the formation of new cities is a rare event, or has been for the past hundred years. A wide variety of authors have used multiplicative random-growth processes to model city sizes (Simon 1955; Berry and Garrison 1958; Makse *et al.* 1998; Gabaix 1999a, 1999b; Gabaix and Ioannides 2004; Batty 2005; Berry and Okulicz-Kozaryn 2011).

TOWARD REALISM IN FIRM GROWTH MODELS

Purely stochastic explanations of social phenomena that are innocent of human behavioral and process details may reasonably be viewed as crude, approximate theories, suitable at the beginning of scientific exploration. But as science progresses, the need develops for models that more closely reflect the actual social and economic processes responsible for observed aggregate phenomena. Nearly ninety years after Robert Gibrat's initial forays into quantitative modeling of firm growth, it is time to move beyond random exogenous shocks and toward models of interactive firm dynamics. ✤

REFERENCES

Acemoglu, D., V. M. Carvalho, A. Ozdaglar, and A. Tahbaz-Salehi. 2012. "The Network Origins of Aggregate Fluctuations." *Econometrica* 80:1977–2016. https://doi.org/10.3982/ECTA9623.

Akhundjanov, S. B., and A. A. Toda. 2020. "Is Gibrat's 'Economic Inequality' Lognormal?" *Empirical Economics* 59:2071–2091. https://doi.org/10.1007/s00181-019-01719-z.

Albert, R., and A.-L. Barabási. 2002. "Statistical Mechanics of Complex Networks." *Reviews of Modern Physics* 74:47–97. https://doi.org/10.1103/RevModPhys.74.47.

Anderson, P. W. 1972. "More is Different." *Science* 177 (4047): 393–396. https://doi.org/10.1126/science.177.4047.393.

Anthonisen, N. 2015. "Microeconomic Shocks and Macroeconomic Fluctuations in a Dynamic Network Economy." *Journal of Macroeconomics* 47 (B): 233–254. https://doi.org/10.1016/j.jmacro.2015.11.001.

Atalay, E., A. Hortaçsu, J. Roberts, and C. Syverson. 2011. "Network Structure of Production." *Proceedings of the National Academy of Sciences* 108 (13): 5199–5202. https://doi.org/10.1073/pnas.1015564108.

Auerbach, F. 1913. "Das Gesetz der Bevölkerungskonzentration." *Petermanns Geographische Mitteilungen* 59:74–76.

Axioglou, C., and N. Christodoulakis. 2020. "Which Firms Survive in a Crisis? Investigating Gibrat's Law in Greece 2001–2014." *Journal of Industrial and Business Economics* 48:159–217. https://doi.org/10.1007/s40812-020-00176-5.

Axtell, R. L. 2001. "Zipf Distribution of U.S. Firm Sizes." *Science* 293 (5536): 1818–1820. https://doi.org/10.1126/science.1062081.

Axtell, R. L., A. Kirman, I. D. Couzin, D. Fricke, T. Hens, M. E. Hochberg, J. E. Mayfield, P. Schuster, and R. Sethi. 2016. "Challenges of Integrating Complexity and Evolution into Economics." In *Complexity and Evolution: Toward a New Synthesis for Economics,* edited by D. S. Wilson and A. Kirman. Cambridge, MA: MIT Press.

Batty, M. 2005. *Cities and Complexity: Understanding Cities with Cellular Automata, Agent-Based Models, and Fractals.* Cambridge, MA: MIT Press.

Bee, M., M. Riccaboni, and S. Schiavo. 2017. "Where Gibrat Meets Zipf: Scale and Scope of French Firms." *Physica A* 481:265–275. https://doi.org/10.1016/j.physa.2017.04.012.

Berry, B. J. L., and W. L. Garrison. 1958. "Alternative Explanations of Urban Rank-Size Relationships." *Annals of the Association of American Geographers* 48 (1): 83–91. https://doi.org/10.1111/j.1467-8306.1958.tb01559.x.

Berry, B. J. L., and A. Okulicz-Kozaryn. 2011. "The City Size Distribution Debate: Resolution for US Urban Regions and Megalopolitan Areas." *Cities* 29 (Supplement 1): S17–S23. https://doi.org/10.1016/j.cities.2011.11.007.

Bjelland, M., B. C. Fallick, J. C. Haltiwanger, and E. McEntarfer. 2011. "Employer-to-Employer Flows in the United States: Estimates Using Linked Employer–Employee Data." *Journal of Business and Economic Statistics* 29 (4): 493–505. https://doi.org/10.1198/jbes.2011.08053.

Bornholdt, S., and H. Ebel. 2001. "World Wide Web Scaling Exponent from Simon's 1955 Model." *Physical Review E* 64 (3): 035104. https://doi.org/10.1103/PhysRevE.64.035104.

Bottazzi, G., E. Cefis, and G. Dosi. 2002. "Corporate Growth and Industrial Structures: Some Evidence from the Italian Manufacturing Industry." *Industrial and Corporate Change* 11 (4): 705–723. https://doi.org/10.1093/icc/11.4.705.

Bottazzi, G., E. Cefis, G. Dosi, and A. Secchi. 2007. "Invariances and Diversities in the Patterns of Industrial Evolution: Some Evidence from Italian Manufacturing Industries." *Small Business Economics* 29:137–159. https://doi.org/10.1007/s11187-006-0014-y.

Bottazzi, G., A. Coad, N. Jacoby, and A. Secchi. 2011. "Corporate Growth and Industrial Dynamics: Evidence from French Manufacturing." *Applied Economics* 43 (1): 103–116. https://doi.org/10.1080/00036840802400454.

Bottazzi, G., G. Dosi, M. Lippi, F. Pammolli, and M. Riccaboni. 2001. "Innovation and Corporate Growth in the Evolution of the Drug Industry." *International Journal of Industrial Organization* 19 (7): 1161–1187. https://doi.org/10.1016/S0167-7187(01)00068-6.

Brown, C., and J. Medoff. 1989. "The Employer Size–Wage Effect." *Journal of Political Economy* 97 (5): 1027–1059. https://doi.org/10.1086/261642.

Champernowne, D. G. 1953. "A Model of Income Distribution." *The Economic Journal* 63 (250): 318–351. https://doi.org/10.2307/2227127.

Chatterjee, S., and E. Rossi-Hansberg. 2012. "Spinoffs and the Market for Ideas." *International Economic Review* 53 (1): 53–93. https://doi.org/10.1111/j.1468-2354.2011.00671.x.

da Silva, S., R. Matsushita, R. Giglio, and G. Massena. 2018. "Granularity of the Top 1,000 Brazilian Companies." *Physica A* 512:68–73. https://doi.org/10.1016/j.physa.2018.08.027.

Davis, S. J., R. J. Faberman, and J. C. Haltiwanger. 2006. "The Flow Approach to Labor Markets: New Data Sources and Micro–Macro Links." *Journal of Economic Perspectives* 20 (3): 3–26. https://doi.org/10.1257/jep.20.3.3.

de Wit, G. 2005. "Firm Size Distributions: An Overview of Steady-State Distributions Resulting from Firm Dynamics Models." *International Journal of Industrial Organization* 23 (5–6): 423–450. https://doi.org/10.1016/j.ijindorg.2005.01.012.

Fallick, B. C., and C. A. Fleischman. 2001. "The Importance of Employer-to-Employer Flows in the U.S. Labor Market." *Finance and Economics Discussion Series* 2001 (April): 1–44. https://doi.org/10.17016/FEDS.2001.18.

———. 2004. *Employer-to-Employer Flows in the U.S. Labor Market: The Complete Picture of Gross Worker Flows.* Finance and Economics Discussion Series 2004-34. Board of Governors of the Federal Reserve System (U.S.)

Feller, W. 1971. *An Introduction to Probability Theory and Its Applications.* Vol. II. New York, NY: John Wiley & Sons.

———. 1968 [1950]. *An Introduction to Probability Theory and Its Applications.* Vol. I. New York, NY: John Wiley & Sons.

Fujiwara, Y., C. Di Guilmi, H. Aoyama, M. Gallegati, and W. Souma. 2004. "Do Pareto–Zipf and Gibrat Laws Hold True? An Analysis with European Firms." *Physica A: Statistical Mechanics and its Applications* 335 (1–2): 197–216. https://doi.org/10.1016/j.physa.2003.12.015.

Gabaix, X. 1999a. "Zipf's Law and the Growth of Cities." *American Economic Review* 89 (2): 129–132. https://doi.org/10.1257/aer.89.2.129.

———. 1999b. "Zipf's Law for Cities: An Explanation." *Quarterly Journal of Economics* 114 (3): 739–767. https://doi.org/10.1162/003355399556133.

———. 2009. "Power Laws in Economics and Finance." *Annual Review of Economics* 1:255–293. https://doi.org/10.1146/annurev.economics.050708.142940.

———. 2016. "Power Laws in Economics: An Introduction." *Journal of Economic Perspectives* 30 (1): 185–206. https://doi.org/10.1257/jep.30.1.185.

Gabaix, X., and Y. M. Ioannides. 2004. "The Evolution of City Size Distributions." In *Handbook of Urban and Regional Economics,* edited by V. J. Henderson and J.-F. Thisse, 4:2341–2378. Amsterdam, Netherlands: North-Holland.

Gaffeo, E., M. Gallegati, and A. Palestrini. 2003. "On the Size Distribution of Firms: Additional Evidence from the G7 Countries." *Physica A: Statistical Mechanics and its Applications* 324 (1–2): 117–123. https://doi.org/10.1016/S0378-4371(02)01890-3.

Gibbons, R., and J. Roberts, eds. 2012. *The Handbook of Organizational Economics.* Princeton, NJ: Princeton University Press.

Gibrat, R. 1931. *Les Inégalités Économiques.* Paris: Librairie du Recueil Sirey.

Grimmett, G. R., and D. R. Stirzaker. 1992. *Probability and Random Processes.* Oxford, UK: Oxford University Press.

Gross, D., J. F. Shortle, J. M. Thompson, and C. M. Harris. 2008. *Fundamentals of Queueing Theory.* New York, NY: Wiley-Interscience.

Guerrero, O. A., and R. L. Axtell. 2013. "Employment Growth through Labor Flow Networks." *PLoS ONE* 8 (5): e60808. https://doi.org/10.1371/journal.pone. 0060808.

Harrison, J. R. 2004. "Models of Growth in Organizational Ecology: A Simulation Assessment." *Industrial and Corporate Change* 13 (1): 243–261. https://doi. org/10.1093/icc/13.1.243.

Hart, P. E., and S. J. Prais. 1956. "The Analysis of Business Concentration: A Statistical Approach." *Journal of the Royal Statistical Society, Series A* 119:150–191. https://doi.org/10.2307/2342882.

Huberman, B. A., and L. A. Adamic. 1999. "Growth Dynamics of the World Wide Web." *Nature* 401:131. https://doi.org/10.1038/43604.

Ijiri, Y., and H. A. Simon. 1977. *Skew Distributions and the Sizes of Business Firms.* New York, NY: North-Holland.

Johnson, N. L., S. Kotz, and N. Balakrishnan. 1994. *Continuous Univariate Distributions.* New York, NY: John Wiley & Sons.

Kalecki, M. 1945. "On the Gibrat Distribution." *Econometrica* 13:161–170. https: //doi.org/10.2307/1907013.

Kang, S. H., Z. Jiang, C. Cheong, and S.-M. Yoon. 2011. "Changes of Firm Size Distribution: The Case of Korea." *Physica A* 390:319–327. https://doi.org/ 10.1016/j.physa.2010.10.007.

Kapteyn, J. C. 1903. *Skew Frequency Curves in Biology and Statistics.* Groningen, NY: Noordhoff.

Kim, J. D., J. Choi, N. Goldschlag, and J. Haltiwanger. 2024. *High-Growth Firms in the United States: Key Trends and New Data Opportunities.* FEDS Working Paper 2024-74. Federal Reserve Board. https://doi.org/10.17016/FEDS.2024. 074.

Kingman, J. F. C. 1961. "The Single Server Queue in Heavy Traffic." *Proceedings of the Cambridge Philosophical Society* 57:902–904. https://doi.org/10.1017/ S0305004100036094.

————. 1962. "On Queues in Heavy Traffic." *Journal of the Royal Statistical Society, Series B (Methodological)* 24:383–392. https : / / www . jstor . org / stable / 2984229.

Kleiber, C., and S. Kotz. 2003. *Statistical Size Distributions in Economics and Actuarial Sciences.* New York, NY: Wiley.

Klepper, S., and E. Graddy. 1990. "The Evolution of New Industries and the Determinants of Market Structure." *RAND Journal of Economics* 21:24–44. https://www.jstor.org/stable/2555491.

Klepper, S., and K. Simons. 1997. "Technological Extinctions of Industrial Firms: An Inquiry into Their Nature and Causes." *Industrial and Corporate Change* 6:379–460. https://doi.org/10.1093/icc/6.2.379.

Klette, T. J., and S. Kortum. 2004. "Innovating Firms and Aggregate Innovation." *Journal of Political Economy* 112:986–1018. https://doi.org/10.1086/422563.

Krapivsky, P. L., S. Redner, and E. Ben-Naim. 2010. *A Kinetic View of Statistical Physics.* New York, NY: Cambridge University Press.

Levy, M., and S. Solomon. 1996. "Power Laws are Logarithmic Boltzmann Laws." *International Journal of Modern Physics C* 7:595. https://doi.org/10.1142/S0129183196000491.

Lewontin, R., and D. Cohen. 1969. "On Population Growth in a Randomly Varying Environment." *Proceedings of the National Academy of Sciences of the United States of America* 62:1056–1060. https://doi.org/10.1073/pnas.62.4.1056.

Lindley, D. V. 1952. "The Theory of Queues with a Single Server." *Proceedings of the Cambridge Philosophical Society* 48:277–289. https://doi.org/10.1017/S0305004100027638.

Luttmer, E. G. J. 2007. "Selection, Growth, and the Size Distribution of Firms." *Quarterly Journal of Economics* 122:1103–1144. http://www.jstor.org/stable/25098869.

————. 2011. "On the Mechanics of Firm Growth." *Review of Economic Studies* 78:1042–1068. https://doi.org/10.1093/restud/rdq028.

————. 2018. *Slow Convergence in Economies with Organization Capital.* 1–66. Minneapolis, MI: Publisher. https://doi.org/10.21034/wp.748.

Mackenzie, D. 2006. *An Engine, Not a Camera: How Financial Models Shape Markets.* Cambridge, MA: MIT Press.

Makse, H. A., J. S. Andrade, Jr., M. Batty, S. Havlin, and H. E. Stanley. 1998. "Modeling Urban Growth Patterns with Correlated Percolation." *Physical Review E* 58:7054–7062. https://doi.org/10.1103/PhysRevE.58.7054.

Malcai, O., O. Biham, and S. Solomon. 1999. "Power-Law Distributions and Lévy-Stable Intermittent Fluctuations in Stochastic Systems of Many Autocatalytic Elements." *Physical Review E* 60:1299. https://doi.org/10.1103/PhysRevE. 60.1299.

Mandelbrot, B. 1959. "A Note on a Class of Skew Distribution Functions: Analysis and Critique of a Paper by H. A. Simon." *Information and Control* 2:90–99. https://doi.org/10.1016/S0019-9958(59)90098-1.

———. 1961a. "Final Note on a Class of Skew Distribution Functions: Analysis and Critique of a Model Due to H. A. Simon." *Information and Control* 4:198–216. https://doi.org/10.1016/S0019-9958(61)80008-9.

———. 1961b. "Post Scriptum to 'Final Note'." *Information and Control* 4:300–304. https://doi.org/10.1016/S0019-9958(61)80025-9.

———. 1961c. "Stable Paretian Random Functions and the Multiplicative Variation of Income." *Econometrica* 29:517–543. https://doi.org/10.2307/1911802.

———. 1963. "The Variation of Certain Speculative Prices." *Journal of Business* 36:394–419. https://www.jstor.org/stable/2350970.

———. 1964. "The Variation of Some Other Speculative Prices." *Journal of Business* 37:393–413. https://www.jstor.org/stable/2351623.

———. 1965. "A Class of Long-tailed Probability Distributions and the Empirical Distribution of City Sizes." In *Mathematical Explorations in Behavioral Science*, 322–332. Homewood, IL: Richard D. Irwin.

Mansfield, E. 1962. "Entry, Gibrat's Law, Innovation, and the Growth of Firms." *American Economic Review* 52:1023–1051. https://doi.org/http://www. jstor.org/stable/1812180.

Mata, J. 2008. "Gibrat's Law." In *The New Palgrave Dictionary of Economics*, edited by S. N. Durlauf and L. E. Blume. New York, NY: Palgrave Macmillan.

McCloughan, P. 1995. "Simulation of Concentration Development from Modified Gibrat Growth-Entry-Exit Processes." *Journal of Industrial Economics* 43:405–433. https://doi.org/10.2307/2950552.

Metzig, C., and C. Colijn. 2019. "Preferential Attachment in Systems and Networks of Constant Size." arXiv preprint: 10:e22030312.

Mitzenmacher, M. 2004. "A Brief History of Generative Models for Power Law and Lognormal Distributions." *Internet Mathematics* 1 (2): 226–251. https://api.semanticscholar.org/CorpusID:1671059.

Montroll, E. W., and M. F. Shlesinger. 1982. "On 1/f Noise and Other Distributions with Long Tails." *Proceedings of the National Academy of Sciences of the United States of America* 79:3380–3383. https://doi.org/10.1073/pnas.79.10.3380.

Pareto, V. 1896. *Cours d'Economie Politique.* Geneva: Librairie Droz. ~ 407 ~

Perline, R., R. Axtell, and D. Teitelbaum. 2006. *Volatility and Asymmetry of Small Firm Growth Rates Over Increasing Time Frames.* Washington, D.C. https://ideas.repec.org/p/sba/wpaper/06rarpdt.html.

Ramsden, J. J., and G. Kiss-Haypál. 2000. "Company Size Distribution in Different Countries." *Physica A* 277:220–227. https://doi.org/10.1016/S0378-4371(99)00572-5.

Redner, S. 1990. "Random Multiplicative Processes: An Elementary Tutorial." *American Journal of Physics* 58:267–273. https://doi.org/10.1119/1.16497.

Ree, S. 2006. "Power-Law Distributions from Additive Preferential Redistributions." *Physical Review E* 73:026115. https://doi.org/10.1103/PhysRevE.73.026115.

———. 2007. "Generation of Scale-Free Networks Using a Simple Preferential-Rewiring Dynamics." *Physica A* 376:692–698. https://doi.org/10.1016/j.physa.2006.10.011.

Reichstein, T., M. S. Dahl, B. Ebersberger, and M. B. Jensen. 2010. "The Devil Dwells in the Tails: A Quantile Regression Approach to Firm Growth." *Journal of Evolutionary Economics* 20:219–231. https://doi.org/10.1007/s00191-009-0152-x.

Richiardi, M. G. 2004. "Generalizing Gibrat: Reasonable Multiplicative Models of Firm Dynamics." *Journal of Artificial Societies and Social Simulation* 7:34. https://api.semanticscholar.org/CorpusID:9412388.

Saichev, A., Y. Malevergne, and D. Sornette. 2010. *Theory of Zipf's Law and Beyond.* New York, NY: Springer-Verlag.

Samuelson, P. A. 1952. "Economic Theory and Mathematics---An Appraisal." *American Economic Review* 42:56–69. https : / / www . jstor . org / stable / 1910585.

Simon, H. A. 1955. "On a Class of Skew Distribution Functions." *Biometrika* 42:425–440. https://doi.org/10.2307/2333389.

———. 1960. "Some Further Notes on a Class of Skew Distribution Functions." *Information and Control* 3:80–88. https://doi.org/10.1016/S0019-9958(60)90302-8.

———. 1961a. "Reply to 'Final Note' by Benoit Mandelbrot." *Information and Control* 4:217–223.

———. 1961b. "Reply to Dr. Mandelbrot's Post Scriptum." *Information and Control* 4:305–308.

———. 1962. "The Architecture of Complexity." *Proceedings of the American Philosophical Society* 106 (6): 467–482. https://www.jstor.org/stable/985254.

Simon, H. A., and C. Bonini. 1958. "The Size Distribution of Business Firms." *American Economic Review* 48 (4): 607–617. https://www.jstor.org/stable/1808270.

Sornette, D. 2006. *Critical Phenomena in Natural Sciences: Chaos, Fractals, Self-Organization and Disorder—Concepts and Tools.* New York, NY: Springer.

Sornette, D., and R. Cont. 1997. "Convergent Multiplicative Processes Repelled from Zero: Power Laws and Truncated Power Laws." *Journal de Physique I France* 7:431–444. https://doi.org/10.1051/jp1:1997169.

Spitzer, F. 1964. *Principles of Random Walk.* Princeton, NJ: D. Van Nostrand Company, Inc.

Stanley, M. H. R., S. V. Buldyrev, S. Havlin, R. N. Mantegna, M. A. Salinger, and H. E. Stanley. 1995. "Zipf Plots and the Size Distribution of Firms." *Economics Letters* 49:453–457. https://doi.org/10.1016/0165-1765(95)00696-D.

Steindl, J. 1965. *Random Processes and the Growth of Firms.* New York, NY: Hafner Publishing Company.

Sutton, J. 1997. "Gibrat's Legacy." *Journal of Economic Literature* 35:40–59. https://www.jstor.org/stable/2729692.

Teitelbaum, D., and R. L. Axtell. 2005. *Firm Size Dynamics of Industries: Stochastic Growth Processes, Large Fluctuations and the Population of Firms as a Complex System.* Washington, D.C. https://www.govinfo.gov/app/details/GOVPUB-SBA-PURL-LPS95720.

van Kampen, N. G. 2007. *Stochastic Processes in Physics and Chemistry.* Amsterdam, Netherlands: Elsevier.

Winter, S. G. 1993. "On Coase, Competence, and the Corporation." In *The Nature of the Firm: Origins, Evolution, and Development,* edited by O. E. Williamson and S. G. Winter. New York, NY: Oxford University Press.

Yakovenko, V. M., and J. Barkley Rosser. 2009. "Statistical Mechanics of Money, Wealth, and Income." *Reviews of Modern Physics* 81:1703–1725. https://doi.org/10.48550/arXiv.0905.1518.

Yakubu, I. N. 2020. "Testing the Validity of Gibrat's Law in the Context of Profitability and Leverage." *Economics and Finance Letters* 7 (2): 85–91. https://doi.org/10.18488/journal.29.2020.72.85.91.

Zipf, G. K. 1949. *Human Behavior and the Principle of Least Effort.* Cambridge, MA: Addison-Wesley Press.

PART III

Macroeconomic Dynamics & Finance

IMPLICATIONS OF BEHAVIORAL RULES IN AGENT-BASED MACROECONOMICS

Herbert Dawid, Bielefeld University;

Domenico Delli Gatti, Università Cattolica del Sacro Cuore;

Luca Eduardo Fierro, International Institute
for Applied Systems Analysis; and

Sebastian Poledna, Austrian Institute of Economic Research

Abstract

In this paper we examine the role of the design of behavioral rules in agent-based macroeconomic modeling. Based on clear theoretical foundations, we develop a general representation of the behavioral rules governing price and quantity decisions of firms and show how rules used in four main families of agent-based macroeconomic models can be interpreted as special cases of this general representation. We embed the four variations of this rule into a calibrated agent-based macroeconomic framework and show that they all yield qualitatively very similar dynamics in business-as-usual times. However, the impact of demand, cost, and productivity shocks differ substantially depending on which of the four variants of the price and quantity rules are used.

Introduction

Since agent-based models (ABMs) were first developed in the early 2000s,[1] they have become an important part of the toolbox

---

[1] Predecessors closely related in spirit to the agent-based approach, have however been developed substantially earlier. A main example in this respect is the MOSES model (see Eliasson (2023) for an extensive discussion of its development and main results).

for macroeconomic analysis, which features strong behavioral foundations and the potential for versatile policy analysis. Robert Axtell and J. Doyne Farmer (2025), Giovanni Dosi and Andrea Roventini (2019), and Herbert Dawid and Domenico Delli Gatti (2018) provide extensive surveys and discussions of the merit and the potential of the agent-based approach, as well as of different streams of literature that have flourished in this field of research. A noteworthy recent development is the emergence of a substantial stream of ABM literature focusing on the economic impact and potential policy responses for major challenges, such as green transition (e.g., Filatova and Akkerman 2026; Hötte 2020; Lamperti *et al.* 2020; Lamperti, Dosi, and Roventini 2026; Turco *et al.* 2023) or the outbreak of a pandemic like COVID-19 (Delli Gatti and Reissl 2022; Pichler *et al.* 2022; Basurto *et al.* 2023; Poledna *et al.* 2023, e.g.,). Several of these papers highlight the value of integrating macro ABMs with dynamic models developed in other disciplines, such as climate research or epidemiology (see also Savin *et al.* 2023). In terms of policy impact, ABMs are being used increasingly by central banks for analyzing issues such as systemic risk, housing-market dynamics, and inflation (see, e.g., the survey by Borsos *et al.* 2026, ch. 18 in this volume).

On a more conceptual and methodological level, recent work by Sebastian Poledna *et al.* (2023) and Cars H. Hommes *et al.* (2025) demonstrates that ABMs are not only suitable for improving our understanding of the general mechanisms that determine how interactions on the micro level generate macro-level phenomena, but also perform well in terms of short-term forecasting capability. Using calibrations based on Austrian (Poledna *et al.* 2023) and Canadian (Hommes *et al.* 2025) data, Poledna and Hommes *et al.* show that the out-of-sample forecasting performance of the agent-based macroeconomic model is competitive with that of vector auto-regressive (VAR) and DSGE models. The model's rich micro-structure enables sector-specific forecasts, making ABMs fully competitive with forecasts based on VARs.

As the field of agent-based macroeconomics has matured, a relatively small set of *model families* has emerged that can be applied to a large number of topics and policy issues. In this chapter we focus on four families: CANVAS,[2] the framework developed by Poledna *et al.* (2023) and Hommes *et al.* (2025; 2026), the Complex Adaptive Trivial Systems (CATS) introduced by Delli Gatti *et al.* (2011), Luca Riccetti, Alberto Russo and Mauro Gallegati (2013), and Tiziana Assenza, Delli Gatti and Jakob Grazzini (2015); the Eurace@Unibi (EUBI) model by Dawid, Philipp Harting, and Michael Neugart (2018), and Dawid *et al.* (2019); and the Schumpeter meeting Keynes (KS) framework developed by Giovanni Dosi, Giorgio Fagiolo, and Andrea Roventini (2010) and Dosi *et al.* (2015). The focus on these four families is essentially due to the limits of our knowledge base and to the specificity of our competences. We have therefore been forced to leave out a few important frameworks with which we are less familiar, such as the baseline model in Matthias Lengnick (2013), the macroeconomic ABM put forward by Peter Howitt and coauthors (see, e.g., Ashraf, Gershman, and Howitt 2017), the Oxford–INET model developed by Farmer and coauthors (successfully used in Pichler *et al.* 2022), the JAMEL model (Salle and Seppecher 2018), and the recently developed MATRIX model (Ciola *et al.* 2023).

---

[2] CANVAS refers specifically to the version developed by the Bank of Canada. The broader family of models to which CANVAS belongs does not have a formal name, unlike other agent-based model families such as CATS, EUBI, and KS.

These four families of macroeconomic ABMs share the general interpretation of the economy as a system of interacting heterogeneous agents whose behavior is determined by rules rather than equilibrium conditions, in which macro-level dynamic patterns are emergent phenomena due to micro-level interactions. They differ quite substantially, however, with respect to the specifications of the interaction structures in different sectors and markets, as well as of the behavioral rules used by the different agents. To a large extent, these differences are due to the fact that the frameworks have been developed with slightly different research agendas and theoretical foundations in mind. Nevertheless, the heterogeneity of model assumptions and structures makes it challenging to compare results, such as insights about the impact of certain policies, that have been found in different macro-ABM settings. Therefore, it is imperative to systematically analyze the impact of differences in the modeling assumptions for behavioral rules and interaction structures on the dynamics of key economic variables. This analysis not only fosters comparability between ABM-based findings but also generates new insights about the relationship between properties of the micro-level behavior and emergent macro-level phenomena.

~415~

Two of us made a first step in this direction in Dawid and Delli Gatti (2018), which systematically spells out and compares the modeling assumptions underlying eight families of macroeconomic ABMs. The analysis also highlights the common theoretical basis of several of the considered behavioral rules in different macro-ABM families. In this chapter we go a step further and focus on a specific set of important behavioral rules, namely the rules by which firms[3] decide on the quantity to be produced and the price of their product, and make two main contributions.

First, we derive from a clear theoretical micro foundation a unified representation for behavioral rules concerning price and quantity setting, and show that the rules used in each of the macro-ABM families considered (CANVAS, CATS, EUBI, KS) can be interpreted as a special case of that general representation. By so doing we can clearly highlight how the rules differ with respect to the information that is taken into account and also with respect to the underlying economic rationale. Put differently, our unified representation allows us to identify which "channels" are active under the different rules. Second, we systematically compare the macro dynamics across macroeconomic ABMs that differ only with respect to the behavioral rule used for determining quantities and prices of consumption goods. More precisely, we incorporate price–quantity decision rules used in each of the four considered ABM families into a calibrated version of the model, as in Poledna *et al.* (2023), and study how the out-of-sample dynamics of the model under these four types of decision rules compare to each other and to the empirical time series. Furthermore, we study how the different rules react to three types of economic shocks: demand, input prices, and productivity. Based on this exercise, we can isolate the effect of properties of the firm's behavioral rules on the dynamics of key economic variables and thereby gain important insights about the implications of different modeling assumptions. Furthermore, our exercise also provides insights into the performance of different behavioral assumptions. Our exercise also provides insights into the

---

[3] Strictly speaking, for EUBI and KS the rules we consider govern the behavior of consumption goods firms only.

performance of different behavioral assumptions with respect to short-term forecasting.[4]

In the next section, we derive a general representation of price–quantity rules for consumption good firms and show how the rule implemented in each of the four considered families of macro-ABMs can be interpreted as a special case of this representation. Next, we describe the macroeconomic environment in which the price–quantity rules are embedded. We subsequently present the results of our analysis of the implications of the use of different price–quantity rules. We conclude with a discussion of our results and considerations of how to extend our analysis.

A General Treatment of Price–Quantity Rules in Macroeconomic ABMs

THE BASIC FRAMEWORK

Consider the profit maximization problem of a firm (say firm i, with $i = 1, 2, ...F_C$) in an imperfect competition setting. Using notation similar to Dawid and Delli Gatti (2018), we write the demand for the good produced by firm i at time t as

$$Q_{i,t}^D = \chi_t s_{i,t}(P_{i,t}, P_{-i,t}, z_{i,t}, z_{-i,t}), \tag{1}$$

where χ_t is the total demand for consumption goods (which in general depends on macroeconomic dynamics, the household's life cycle, the industry's life cycle, etc.) while $s_{i,t}(.)$ captures

---

[4]Our approach here is to compare the different rules with respect to the implications of their use on the macro level. Hence, we do not compare the strength of the empirical foundations on the micro level. A discussion of different approaches for developing foundations for firms' decision rules in AMBs can be found in Dawid and Harting (2012). They also explicitly discuss the "management-science approach," that is, the use of decision heuristics that are documented in the management literature, as the foundation for the decision rules in the EUBI model.

the market share of firm i. The market share is a function of the firm's own price $P_{i,t}$, the vector of competitors' prices $P_{-i,t}$, the firm's own product characteristics $z_{i,t}$, the vector of product characteristics of competitors $z_{-i,t}$. To be as general as possible, we assume that the functional form of the market share function is firm-specific and time-varying.

A product characteristic is any feature (such as quality, formal appearance, proximity to a given buyer or group of buyers, etc.) that might imply differentiation among goods from the consumers' perspective. This differentiation is the source of imperfect competition: Firms produce *varieties* and therefore have price-setting power on their captive markets.

Notice that in principle χ_t might depend on the average price level and therefore on the price of the individual firm, inasmuch as the latter contributes to the formation of the aggregate price level. Still, we suppose that firm i assumes that changes in its own price do not affect χ_t, but only its market share $s_{i,t}(.)$.

The firm operates in an uncertain environment. For simplicity, we assume that uncertainty concerns (i) the shape of the market share function $s_{i,t}(.)$ and (ii) the vector of competitors' prices $P_{-i,t}$. Hence the firm must estimate the functional form of the market share, which we refer to as the expected market share function and denote with $s_{i,t}^e(.)$ where the superscript e indicates an expectation. Moreover, assuming that the firm's size is negligible, the vector of competitors' prices can be satisfactorily proxied by the aggregate price level P_t, a weighted average of the individual prices. The firm may be not have sufficient information to determine the actual average price level, hence it has to form the expectation, denoted with P_t^e. For simplicity, we assume that the *expected* average price is uniform across firms. Taking these considerations into account, we can write the *expected* demand for the product of firm i as

$$Q_{i,t}^{D,e} = \chi_t s_{i,t}^e(P_{i,t}, P_t^e, z_{i,t}, z_{-i,t}), \tag{2}$$

PRICING DECISION

We assume that technology is linear so that the marginal production cost is constant. We assume moreover that, because the firm does not know with certainty technology and input prices, the firm must form expectations of the marginal cost. We denote the *expected* marginal (production) cost at t by $c_{i,t}^e$. We assume finally that the firm incurs fixed costs $F_{i,t}$. The ith firm sets its own price $P_{i,t}$ in order to maximize expected profits $\Pi_{i,t}^e$:

$$\max_{P_{i,t}} \Pi_{i,t}^e = Q_{i,t}^{D,e}\left(P_{i,t} - c_{i,t}^e\right) - F_{i,t}.$$

Note that for simplicity we abstract from capacity constraints in this formulation. From the first-order condition we obtain

$$P_{i,t} = (1 + \mu_{i,t})c_{i,t}^e \tag{3}$$

with the markup given by

$$\mu_{i,t} = \frac{1}{\epsilon_{i,t}^e - 1} \tag{4}$$

and

$$
\begin{aligned}
\epsilon_{i,t}^e &= -\frac{\partial Q_{i,t}^{D,e}}{\partial P_{i,t}} \frac{P_{i,t}}{Q_{i,t}^{D,e}} \\
&= -\frac{\partial s_{i,t}^e(P_{i,t}, P_t^e, z_{i,t}, z_{-i,t})}{\partial P_{i,t}} \frac{P_{i,t}}{s_{i,t}^e(P_{i,t}, P_t^e, z_{i,t}, z_{-i,t})}
\end{aligned}
$$

denoting the (absolute value of the) *expected price elasticity* of the demand for the product of firm i.

The main problem the firm faces in setting the price is estimating the price elasticity of demand $\epsilon_{i,t}^e$.[5] Different ABMs use different approaches for addressing this problem.

We envision the following protocol for the firm to estimate the elasticity. The firm believes that its market share is essentially determined by the *ratio* of its own price to the average market price $\frac{P_{i,t}}{P_t}$, a proxy of the *relative* price of the ith good. Due to the uncertainty surrounding the average price level, the firm forms an expectation of the relative price: $\frac{P_{i,t}}{P_t^e}$. Moreover, the firm assumes that the relationship between the expected market share and the expected relative price is linear. This means that the slope of the expected market share function (with respect to the expected relative price) is independent of the price levels. This is tantamount to assuming that there exist two parameter values $\zeta_{i,t}^e$ and $\bar{\zeta}_{i,t}^e$, both positive, such that the expected market share can be written as follows:

$$s_{i,t}^e(P_{i,t}, P_t^e, z_{i,t}, z_{-i,t}) = \bar{\zeta}_{i,t}^e - \zeta_{i,t}^e\left(\frac{P_{i,t}}{P_t^e} - 1\right).$$

Product characteristics $z_{i,t}$ and $z_{-i,t}$ are "embodied" in the above mentioned parameter values.

Let's assume finally that all firms have identical market shares if they charge the same price. In this case, the intercept of the expected market share function is uniform across firms and equal to $\bar{\zeta}_{i,t}^e = \frac{1}{FC}$. In the end, therefore, we can write the expected market share as follows:

$$s_{i,t}^e = \frac{1}{FC} - \zeta_{i,t}^e\left(\frac{P_{i,t}}{P_t^e} - 1\right), \tag{5}$$

---

[5] In principle, the firm also faces potential capacity constraints (or inventory targets), which might imply that an expansion of output induced by a decrease in price leads to an increase in marginal costs. For simplicity, this case is not considered in the present setting.

where $\zeta^e_{i,t}$ measures the *sensitivity* of the firm's market share to the relative price of its product.[6]

Using this equation, the expected elasticity of the demand for the product of firm i becomes

$$\epsilon^e_{i,t} = \frac{\zeta^e_{i,t}}{P^e_t} \frac{P_{i,t}}{s^e_{i,t}}.$$

Inserting this expression into equation (4) and solving for $P_{i,t}$ in equation (3) gives the following representation of the firm's optimal price

$$P_{i,t} = c^e_{i,t} + \frac{s^e_{i,t} P^e_t}{\zeta^e_{i,t}}. \tag{6}$$

From the pricing rule we retrieve the mark-up:

$$\mu_{i,t} = \frac{s^e_{i,t} P^e_t}{\zeta^e_{i,t} c^e_{i,t}}. \tag{7}$$

Based on equation (7), we can formulate the following recursive representation of the markup:

$$\mu_{i,t} = \mu_{i,t-1} \times \underbrace{\frac{s^e_{i,t}}{s^e_{i,t-1}}}_{\substack{\text{exp. change} \\ \text{in market share}}} \times \underbrace{\frac{P^e_t}{P^e_{t-1}}}_{\substack{\text{exp. change} \\ \text{in av. price}}}$$

$$\Big/ \underbrace{\frac{\zeta^e_{i,t}}{\zeta^e_{i,t-1}}}_{\substack{\text{exp. change} \\ \text{in sensitivity}}} \times \underbrace{\frac{c^e_{i,t}}{c^e_{i,t-1}}}_{\substack{\text{exp. change} \\ \text{in marg. cost}}}. \tag{8}$$

It is worth noting that the change of the reciprocal of expected sensitivity, $\frac{1/\zeta^e_{i,t}}{1/\zeta^e_{i,t-1}}$, is closely related to unexpected demand. In the absence of rationing and capacity constraints, actual output would coincide with expected demand: $Y_{i,t} =$

~421~

---

[6]A simple microfoundation for this demand structure is presented in Appendix A.

$Q_{i,t}^{D,e}$. Actual demand $Q_{i,t}^{D}$, in turn, is increasing with the current value of $\zeta_{i,t}$. A lower (higher) realization of $\zeta_{i,t}$ relative to $\zeta_{i,t}^{e}$ (for given prices), therefore, yields a higher (lower) actual demand relative to expected demand. This results in an increase (reduction) of unfulfilled demand. Based on these considerations, we interpret the ratio $\frac{1/\zeta_{i,t}^{e}}{1/\zeta_{i,t-1}^{e}}$ as a proxy for unfulfilled demand expressed as a percentage of output. Hence, we can rewrite and reinterpret the representation of the markup as follows:

$$\mu_{i,t} = \mu_{i,t-1} \times \underbrace{\frac{s_{i,t}^{e}}{s_{i,t-1}^{e}}}_{\substack{\text{exp. change} \\ \text{in market share}}} \times \underbrace{\frac{P_{t}^{e}}{P_{t-1}^{e}}}_{\substack{\text{exp. change} \\ \text{in av. price}}}$$

$$\times \underbrace{\frac{1/\zeta_{i,t}^{e}}{1/\zeta_{i,t-1}^{e}}}_{\substack{\text{rate of} \\ \text{excess demand}}} \Bigg/ \underbrace{\frac{c_{i,t}^{e}}{c_{i,t-1}^{e}}}_{\substack{\text{exp. change} \\ \text{in marg. cost}}} . \tag{9}$$

Equation (9) shows that different *channels* might drive a change in the markup, each one identified by a change in a variable: market share, average price, rate of excess demand, and marginal cost. This representation can be used as an encompassing basis that may generate different heuristics. The main families of macroeconomic ABMs, in fact, differ with respect to the channels which are embedded in the pricing heuristics they adopt. To compare these heuristics, we introduce a dummy for each of the variables showing up in equation (9). These dummies will be denoted with δ^{x}, where x is an element in the set $X = \{s_{i}^{e}, P^{e}, Q_{i}^{D}, c_{i}^{e}\}$. We therefore will write each of the four growth factors appearing in equation (9) in the form $(1+\delta^{x}\pi_{t}^{x})$ or $(1+\delta^{x}\gamma_{t}^{x})$, where δ^{x} is set to 1 (0) if the channel corresponding to variable x is active (inactive) in any specific ABM, π_{t}^{xN} is the growth rate of nominal variables (price level and marginal cost) and γ_t^{xR} denotes the growth rate of real variables (market share and the rate of excess demand). Hence we can write the markup as follows:

$$\mu_{i,t} = (\delta^{rec} + \mu_{i,t-1}) \frac{(1 + \delta^{s_i^e} \gamma_{i,t}^{s_i^e})(1 + \delta^{P^e} \pi_t^{P^e})(1 + \delta^{Q_i^D} \gamma_{i,t}^{Q_i^D})}{(1 + \delta^{c_i^e} \pi_{i,t}^{c_i^e})} - \delta^{rec}$$

(10)

with $\gamma_{i,t}^{s_i^e}$ denoting the growth rate of the firm's expected market share, $\pi_t^{P^e}$ the growth rate of the expected price level, $\gamma_{i,t}^{Q_i^D}$ the unfulfilled demand expressed as a percentage of output, and $\pi_t^{c_i^e}$ the expected growth rate of firm i's marginal cost. In the CATS and CANVAS model families, the recursive representation given in equation (10) is applied to $(1 + \mu_{i,t})$ rather than $\mu_{i,t}$. Formally, we represent this by introducing the dummy δ^{rec}, which is set to one for CATS and CANVAS and to zero for EUBI and KS.

In the implementation of the model, the expected growth rates are determined as follows:

$$\gamma_{i,t}^{s_i^e} = \frac{s_{i,t}^e}{s_{i,t-1}^e} - 1 = \frac{s_{i,t-1}}{s_{i,t-2}} - 1 = \gamma_{t-1}^{s_i}$$

$$\pi_t^{P^e} = \frac{P_t^e}{P_{t-1}^e} - 1 = \frac{P_{t-1}}{P_{t-2}} \frac{(1 + \pi_t^e)}{(1 + \pi_{t-1}^e)} - 1$$

$$\pi_{i,t}^{c_i^e} = \frac{c_{i,t}^e}{c_{i,t-1}^e} - 1 = \frac{c_{i,t-1}}{c_{i,t-2}} \frac{(1 + \pi_t^e)}{(1 + \pi_{t-1}^e)} - 1$$

$$\gamma_{i,t}^{Q_i^D} = \frac{Q_{i,t-1}^D}{Y_{i,t-1}} - 1.$$

We assume that firms have naive expectations about the growth rate of their market share, so that the expected growth rate of the market share in t is equal to the actual growth rate of the market share in $t - 1$: $\gamma_t^{s_i^e} = \gamma_{t-1}^{s_i}$. As to expectations of the average market price, we assume that $P_t^e = P_{t-1}(1 + \pi_t^e)$, with π_t^e the expected overall inflation rate estimated using an AR(1) model applied to the time series of the producer price index (PPI), which is re-estimated every period.[7] Similarly, we assume that expectations of the nominal marginal costs are updated using the expected inflation rate:

$$c_{i,t}^e = c_{i,t-1}(1 + \pi_t^e). \qquad (11)$$

Finally, as discussed above, $\gamma_{i,t}^{Q_i^D}$ proxies the ratio $\frac{1/\zeta_{i,t}^e}{1/\zeta_{i,t-1}^e}$ under the assumption of naive expectations about $\zeta_{i,t}$, that is, $\zeta_{i,t}^e = \zeta_{i,t-1}$.

Using this notation, we can write the general pricing rule as in equation (3) with $\mu_{i,t}$ given by equation (10) and $c_{i,t}^e$ given by equation (11).

QUANTITY DECISION

Generally speaking, in macroeconomic ABMs, firms are assumed to set the desired scale of production $Y_{i,t}^*$ in order to satisfy the demand they expect to receive:

$$Y_{i,t}^* = Q_{i,t}^{D,e}. \qquad (12)$$

When inventories of final goods can be stored, desired production must take into account that the stock of accumulated inventories up to the previous period $\Delta_{i,t}$ can be used to satisfy current demand.

In this case desired production is $\tilde{Y}_{i,t}^* = \max[0, Q_{i,t}^{D,e} - \Delta_{i,t}].$[8]

---

[7] Inflation expectations are computed in every period as $\log(1 + \pi_t^e) = \alpha_{t-1}^\pi \pi_{t-1} + \beta_{t-1}^\pi + \epsilon_{t-1}^\pi$. Where α_{t-1}^π and β_{t-1}^π are re-estimated every period on the time series of inflation $\pi_{t'}$ where $t' = -T', -T' + 1, -T' + 2, \ldots, 0, 1, 2, \ldots, t-1$. ϵ_{t-1}^π is a random shock with zero mean and variance re-estimated every period from past observations over the last $T' + t - 1$ periods.

[8] In some of the considered macroeconomic ABMs, firms include positive inventory targets in their production planning. Since this seems to have little relevance for our analysis we abstract from such inventory planning here.

Using $Q_{i,t-1}^{D} = (1 + \gamma_{i,t}^{Q_i^D})Y_{i,t-1}$, we can write expected demand in t as

$$Q_{i,t}^{D,e} = (1 + \gamma_{i,t}^{Q_i^D})(1 + \gamma_{i,t}^{Y_i})Y_{i,t-1}, \qquad (13)$$

where $\gamma_{i,t}^{Y_i}$ is the expected growth rate of demand of firm i. In line with the existing literature, we assume that firms use the estimated growth rate of *aggregate* demand as a proxy for this rate: $\gamma_{i,t}^{Y_i} = \gamma_t^{Y}$ for all i, with γ_t^{Y} estimated on the basis of an AR(1) model applied to past demand data (which might be proxied by sales). Since supply is set to satisfy expected demand (see eq. (12)) desired output changes (relative to output in the previous period) if the firm adjusts (i) the quantity produced following a market disequilibrium (excess demand or supply) and/or (ii) its expectation of aggregate demand. In symbols:

$$Y_{i,t}^{*} = \left(1 + (1 - \delta^{Q_i^D})\gamma_{i,t}^{Q_i^D}\right)(1 + \delta^{Y_i}\gamma_t^{Y})Y_{i,t-1}. \qquad (14)$$

where $\delta^{Q_i^D}$ is the dummy already used in the markup rule (eq. (10)). Quantity adjustment is not active if $\delta^{Q_i^D} = 1$ (hence $1 - \delta^{Q_i^D} = 0$). In this case, in fact, the firm reacts to excess demand or supply with a price change and does not adjust quantities. On the contrary, for $\delta^{Q_i^D} = 0$ (hence $1 - \delta^{Q_i^D} = 1$) only the quantity is adjusted in response to a difference between demand and supply and the price does not change. Furthermore, the firm takes into account the expected change in aggregate demand in its production planning if $\delta^{Y_i} = 1$. If $\delta^{Y_i} = 0$, desired output does not react to changes in aggregate demand.

APPLICATION TO KEY ABM FAMILIES

In this subsection we show how the price–quantity heuristics in the CANVAS, CATS, EUBI and KS families of macroeconomic ABMs can be interpreted as special cases of our general model formulation.

Table 1. Price–quantity rules in four main families of macroeconomic ABMs.

MODEL	$\delta^{s_i^e}$	δ^{P^e}	$\delta^{Q_i^D}$	$\delta^{c_i^e}$	δ^{rec}	δ^{Y_i}
CANVAS	0	0	cond.	0	1	1
CATS	0	0	cond.	1	1	0
EUBI	1	1	0	1	0	1
KS	1	0	0	0	0	0

The relationship between these models and our general formulation is summarized in table 1. Whereas the first five columns refer to dummies describing different (strategic) aspects of firm's pricing strategy, the last one indicates in how far firms have naive expectations about future demand, or anticipate future changes based on past data. We briefly go through the different dummy variables governing which of the different channels present in our general formulation are active in each model.

The *market power channel* (captured by $\delta^{s_i^e}$), is most explicitly present in KS, where markups are adjusted in parallel to changes in the market share. In EUBI the channel is present implicitly, due to the fact that firms choose profit-maximizing prices based on the estimated[9] sensitivity of its demand with respect to price changes. As can also be seen in (7), this procedure implies a positive relationship between price and market share. The price adjustment rules in CATS and CANVAS do not incorporate any dependence of price on market share, and therefore $\delta^{s_i^e} = 0$ for these models.

The *competitors' prices channel* (δ^{P^e}), is directly present in the EUBI model, due to the fact that firms communicate the (last period) prices of the competitors to the consumers participating in the surveys they use to estimate demand. Hence, the derived optimal price is an approximation of a best

---

[9]Firms use consumer surveys as the basis of this estimation in EUBI.

reply to competitors' price setting decisions, which points to a positive correlation between the individual price and the prices charged by the firm's competitors. In KS the competitors' prices have no direct influence on the firm's price. In the CATS and CANVAS models the average market price does not directly influence the level of the firm's price, but determines whether the latter is adjusted upwards (in case of excess demand and underpricing in the previous period) or downwards (in case of excess supply and overpricing).[10] The level of the average market price in CATS and CANVAS determines whether the excess demand channel (captured by $\delta^{Q_i^D}$) is active or not. In particular, $\delta^{Q_i^D} = 1$ if either the firm's price in $t - 1$ was below the average market price in case of excess demand or $P_{i,t-1} > \bar{P}_t$ in case of excess supply. In KS no direct relationship between excess supply/demand and pricing can be established, although there is an indirect effect through a potential change in the firm's market share. In EUBI excess supply has no impact on future prices, and excess demand would affect the price only if capacity expansion is too costly for the firm to be carried out. In light of this we set $\delta^{Q_i^D} = 0$ for KS and EUBI, although some effect of excess demand on pricing is possible in EUBI.

The *cost absorption channel* ($\delta^{c_i^e}$) describes to which extent firms reduce (increase) their markup if their (marginal) production costs go up (down). When $\delta^{c_i^e} = 1$ the markup decreases at the same rate with which the marginal cost increases, so that cost increases are fully absorbed and the firm's profit margin is independent of the size of the cost. This is equivalent to assuming zero pass-through from cost to price. This is the case in CATS, where pricing does not directly depend on the size of the production costs. On the contrary,

~427~

---

[10]Underpricing (overpricing) occurs when the firm's price is lower (higher) than the average price.

both in CANVAS and KS the markup does not depend on the cost level, such that cost increases are fully passed through to consumers. In EUBI firms take into account the estimated production costs in t when determining their optimal price, such that, in accordance with (7), the markup is inversely related to the cost level, such that $\delta^{c_i^e} = 1$.

As discussed above, the dummy δ^{rec} expresses whether changes induced by the four channels apply only to the mark-up $\mu_{i,t}$, or to the entire price-cost ratio $(1 + \mu_{i,t})$. For EUBI and KS the former holds ($\delta^{rec} = 0$), while the latter applies for CANVAS and CATS ($\delta^{rec} = 1$). Finally, the last column of table 1 refers to the formation of demand expectation of the firms. In CATS and KS firms have naive expectations about the size of total demand in t ($\delta^{Y_i} = 0$), whereas in CANVAS and EUBI firms choose their production quantity using an estimate of the change in demand from $t - 1$ to t.

The Macroeconomic Environment

In this section we give a brief overview over the macroeconomic model in which these price–quantity rules are embedded. Similarly to most MABMs, our framework comprises five classes of agents: Firms, households, banks, a government and a central bank, whose relations are summarized in figure 1.

Each firm belongs to a sector g ($g = 1, 2, \ldots, G$), sectors are organized in an input–output (IO) network, with each sector having size I_g. Firms belonging to sector g produce a single homogeneous good also indexed by g, and the economy as a whole produces G heterogeneous goods. We index firms using i ($i = 1, 2, \ldots, I = \sum_g I_g$) and use the notation $g(i)$ to indicate i's sector. Firms produce using labor, capital, and intermediate goods.

The household sector comprises H ($h = 1, 2, \ldots, H$) persons. Every individual has an *activity status*, that is, a type of economic activity from which she receives an income. The activity status is categorized into H^{act} economically active and H^{inact} economically inactive persons. Economically active persons are H^W workers and I investors. Workers are further divided in H_t^{E} employed and H_t^{U} unemployed persons. Each person also buys goods in the consumption market.

We also assume (i) a representative bank taking deposits from firms and households, extending loans to firms, and receiving advances from the central bank; (ii) a government consuming on the retail market (government consumption), levying taxes, and providing social contributions and benefits to households; (iii) a central bank setting the policy rate, providing liquidity to the banking system, holding reserves for the bank, and purchasing government bonds.

Finally, the foreign sector is modeled as an exogenous, aggregate entity fully calibrated on data, meaning that import prices, import supply, and export demand are directly taken from data.

In the remainder of this section, we will provide further details only concerning firms' and households' behavior, since these parts are essential to understand the results presented in the next section. The interested reader should refer to Poledna *et al.* 2023, and in particular its appendix, from which this section borrows extensively, for a more complete description of the model.

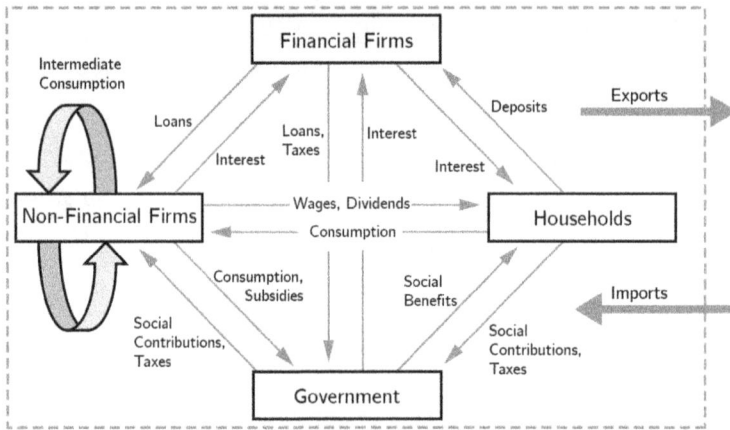

Figure 1. Model flow chart. Rectangles represent agent classes, arrows represent money flows, and the dashed frame contains the domestic economy.

FIRMS

Production

Firm i produces a single good of type $g(i)$ using labor $(N_{i,t})$, intermediate goods $(M_{i,t-1})$, and physical capital $(K_{i,t-1})$, which are combined in a Leontief production function:

$$Y_{i,t} = \min\left(\tilde{Y}_{i,t}^{*}, \beta_{g(i)} M_{i,t-1}, \alpha_{i,t} N_{i,t}, \kappa_{g(i)} K_{i,t-1}\right) \qquad (15)$$

where $\tilde{Y}_{i,t}^{*}$ is the desired production; $\beta_{g(i)}$ and $\kappa_{g(i)}$ are the productivities of intermediate goods and capital for any firms in sector $g(i)$; $\alpha_{i,t}$ is the *effective* labor productivity of firm i at time t. We assume labor productivity $(\bar{\alpha}_{g(i)})$ to be time-invariant and sector-specific. However, since $N_{i,t}$ denotes the number of workers employed by firm i at time t, and workers can work part-time or overtime, $N_{i,t}$ may differ from the effective amount of labor employed in production. Therefore, $\alpha_{i,t}$ adjusts in such a way to ensure that the effective amount of labor enters the production function correctly (see eq. (19) below).

Investment

In each period firm i adjusts its real investment demand $\left(I^{\mathrm{d}}_{i,t}\right)$ to the expected wear and tear of the capital stock $\left(K_{i,t}\right)$, where we assume that only the capital actually employed in production depreciates at the sector-specific rate $\delta_{g(i)}$:

$$I^{\mathrm{d}}_{i,t} = \frac{\delta_{g(i)}}{\kappa_{g(i)}} \min\left(\tilde{Y}^{*}_{i,t}, \kappa_{g(i)} K_{i,t-1}\right) \qquad (16)$$

The capital stock $K_{i,t}$ can be seen as a bundle of G goods where each g-good is associated with a weight b^{K}_{g}. These weights are assumed homogeneous across firms and sectors, therefore each firm demands a quantity of g-good $b^{\mathrm{K}}_{g} I^{\mathrm{d}}_{i,t}$ for the purpose of investment.

Intermediate Goods Demand

Each firm i holds a stock of input goods $M_{i,t}$. From this stock of intermediate input goods, firm i takes out materials for production as needed, and it keeps these goods in positive supply to avoid shortfalls impeding production. In each period, firm i has to decide on the desired amount of intermediate goods $\left(\Delta M^{\mathrm{d}}_{i,t}\right)$ that it intends to purchase to keep its stock in positive supply:

$$\Delta M^{\mathrm{d}}_{i,t} = \frac{\min\left(\tilde{Y}^{*}_{i,t}, \kappa_{g(i)} K_{i,t-1}\right)}{\beta_{g(i)}}. \qquad (17)$$

Firms thus try to keep their stock of material input goods within a certain relationship to $\tilde{Y}^{*}_{i,t}$ by accounting for planned material input use in this period. As for physical capital, the stock of intermediate goods $M_{i,t}$ can be seen as a bundle of G goods. In the case of intermediate goods, we assume the weights $b^{M}_{g(i),g}$ to be sector-specific, so that each firm demands a quantity of intermediate good g $\left(\Delta M^{\mathrm{d}}_{i,g,t}\right)$ equal to $b^{M}_{g(i),g} \Delta M^{\mathrm{d}}_{i,t}$.

Labor Demand

The labor requirement of firm i $\left(N_{i,t}^{\mathrm{d}}\right)$ is defined in accordance to desired scale of production $(\tilde{Y}_{i,t}^{*})$ and the sector-specific labor productivity $(\bar{\alpha}_{g(i)})$:

$$N_{i,t}^{\mathrm{d}} = \max\left(1, \frac{\min\left(\tilde{Y}_{i,t}^{*}, \kappa_{g(i)}K_{i,t-1}\right)}{\bar{\alpha}_{g(i)}}\right) \qquad (18)$$

As firms can be constrained on the labor, intermediate goods, and physical capital markets, they might need to adjust their effective labor input by requiring overtime work in case of labor shortages, or part-time work in case of intermediate inputs and physical capital shortages. We deal with this issue by adjusting the labor productivity of firm $i(\alpha_{i,t})$ as follows:

$$\alpha_{i,t} = \bar{\alpha}_{g(i)} \min\left(1.5, \frac{\min\left(\tilde{Y}_{i,t}^{*}, \beta_{g(i)}M_{i,t-1}, \kappa_{g(i)}K_{i,t-1}\right)}{N_{i,t}\bar{\alpha}_{g(i)}}\right) \\ (19)$$

Where the maximum overtime allowed is 50% of a full-time position.

To remunerate part-time and overtime labor as compared to a full-time position, the average wage $(\bar{w}_i)$ paid by firm i is adapted accordingly:

$$w_{i,t} = \bar{w}_i \min\left(1.5, \frac{\min\left(\tilde{Y}_{i,t}^{*}, \beta_{g(i)}M_{i,t-1}, \kappa_{g(i)}K_{i,t-1}\right)}{N_{i,t}\bar{\alpha}_{g(i)}}\right)$$

Where $w_{i,t}$ is the real wage paid by firm i and nominal hourly wages are pegged to inflation expectations.

HOUSEHOLDS

Activity Status

An employed worker h of firm i in period t receives wage $w_{h,t} = w_{i,t}$. Unemployed workers supply labor to firms with open vacancies through a labor market modeled as a random *search-and-matching* process. All unemployed persons receive unemployment benefits, which are a fraction θ^{UB} of the wage received in the last period of employment.

We assume that each firm is owned by an investor, that is, the number of investors matches the number of firms. Each investor receives income in the form of dividends whenever net profits are positive. We assume limited liability, that is, in the case of bankruptcy, the associated losses are borne by the creditor and not the investor household.

An economically inactive person h receives social benefits sb_t^{inact} and does not look for a job. Additionally, each household receives additional social transfers sb_t^{other} from the government.

Consumption

Households consume a fraction of their expected disposable net income $(Y_{h,t}^{e})$. Expected disposable net income inclusive of social transfers is determined according to the household's activity status and the associated income from labor, expected profits or social benefits, as well as tax payments, the consumer price index of the last period, and inflation expectations (π^{et}). The consumption budget (net of VAT) of household $h(C_{h,t}^{d})$ is thus given by:

$$C_{h,t}^{d} = \frac{\psi Y_{h,t}^{e}}{1 + \tau^{\text{VAT}}} \tag{20}$$

where $\psi \in (0,1)$ is the propensity to consume out of expected income and τ^{VAT} is a value added tax rate on consumption. Consumers then allocate their consumption budget to purchase different goods from firms. The consumption budget of the hth household to purchase the gth good is

$$C_{h,g,t}^{\mathrm{d}} = b_g^{\mathrm{HH}} C_{h,t}^{\mathrm{d}} \qquad (21)$$

Where b_g^{HH} is the homogeneous and time-invariant consumption coefficient for the good g.

How Do Different Types of Price–Quantity Rules Affect Economic Dynamics?

To investigate differences across price–quantity rules, we conduct two sets of exercises: forecasting under business-as-usual scenarios and analyzing model reactions to predefined, exogenous shocks. For these exercises, we calibrate the framework using Eurostat data for the Austrian economy, following the calibration approach outlined in Poledna *et al.* 2023. Starting from a reference quarter, we run simulations with different price–quantity rules over a 12-quarter horizon, allowing for a systematic comparison under varying economic conditions.

Before presenting the results, it is important to clarify that they do not necessarily reflect the exact predictions of the models under consideration. To be precise, our results pertain to the rules discussed in "Application to Key ABM Families" within the Poledna *et al.* 2023 framework. The difference may seem subtle, and often comes down to nuances, but it is important to remember that these rules are applied somewhat outside their original context. As a result, the outcomes may differ from those produced within their original models. Whenever we identified such potential differences, we made an effort to alert the reader.

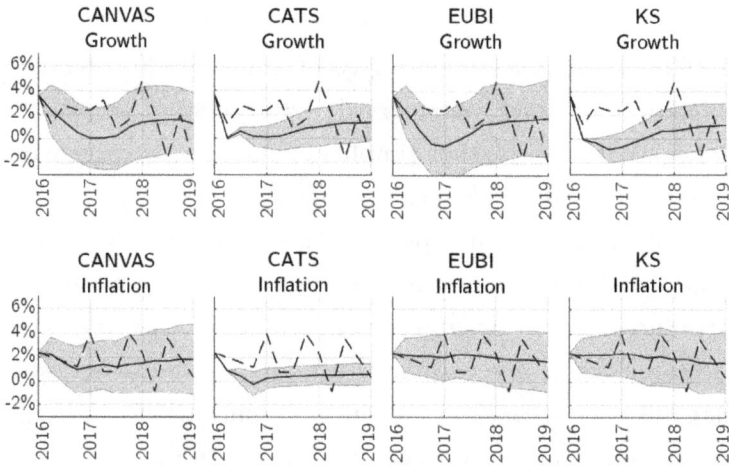

Figure 2. Out-of-sample forecasts with the different price–quantity rules. Forecasts for GDP growth and inflation measured by the CPI are 12 quarters ahead from Q1 2017 until Q4 2019 for the Austrian economy. The black lines show the respective forecasts with the different price–quantity rules and observed Eurostat data for Austria is shown with the dashed line. For each forecast, one standard deviation is plotted around the mean trajectory. Model results are obtained as an average of 500 Monte Carlo simulations.

FORECASTING

To illustrate the differences among the respective price–quantity rules under a business-as-usual scenario, we present representative 12 quarters ahead out-of-sample forecasts in figure 2. The figure displays GDP growth and inflation forecasts, measured by the Consumer Price Index (CPI), generated by the CANVAS, CATS, EUBI, and KS price–quantity rules, with observed Eurostat data for Austria provided as a benchmark. Overall, in the business-as-usual scenario, the price–quantity rules produce qualitatively similar forecasts for GDP growth and inflation. As seen in figure 2, inflation forecasts are particularly consistent across the CANVAS, EUBI, and KS price-setting rules, with these rules generating closely aligned projections over the forecast horizon.

Similarly, the growth forecasts of the CANVAS and EUBI rules show a strong degree of comparability. Notable differences, however, are observed in the CATS and KS quantity choices. Both of these rules simplify by assuming naive demand expectations, where the expected demand is set equal to the demand of the previous period. This assumption results in somewhat lower forecast performance in the short run and introduces a bias in the projections, as these models are slower to adjust to changes in economic conditions. For the same reason, the CATS price-setting rule also shows a somewhat lower forecast performance and an overall bias in its projections.

These observations from figure 2 are further corroborated by the quantitative evaluation of forecast performance along the lines of Poledna *et al.* 2023 shown in Appendix B. Table 2 in this appendix provides detailed out-of-sample forecast performance, showing root mean square error (RMSE) statistics for different forecast horizons. The results indicate that the CANVAS and EUBI price–quantity rules tend to outperform the CATS and KS rules over short forecast horizons, where the latter models show a tendency for increased forecast errors. Table 3 in Appendix B further highlights the mean biases of the models across different forecast horizons, showing that the CATS and KS models tend to produce more biased forecasts for short forecast horizons.

SHOCKS

We experiment with three different shocks: a demand shock, a cost-push shock, and a non-idiosyncratic technological shock. Each shock is imposed in isolation and involves a permanent 10% increase in one of the following variables: government spending (public consumption), import prices, or labor productivity. Figure 3 summarizes adjustments to shocks

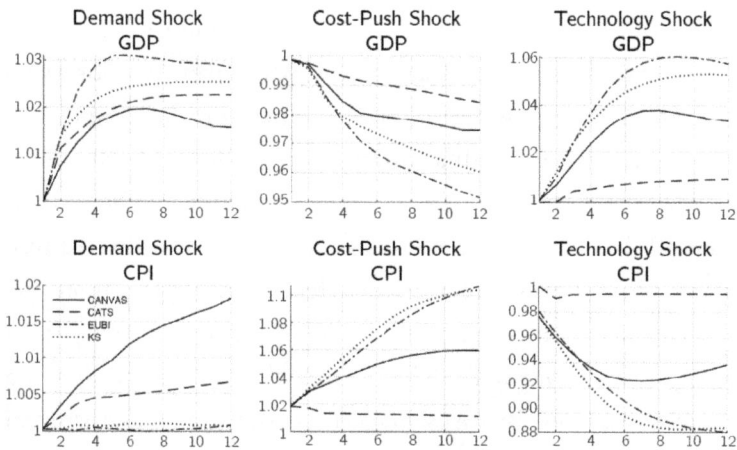

Figure 3. Responses of GDP and the Consumer Price Index (CPI) to three types of exogenous shocks---demand shock, cost-push shock, and technology shock---across four different price–quantity rules: CANVAS, CATS, EUBI, and KS. The simulations are run over a 12-quarter horizon, starting from a reference quarter calibrated with Eurostat data for the Austrian economy. The top panels show the relative changes in GDP, while the bottom panels display the corresponding effects on CPI.

across different rules, focusing on GDP and CPI expressed as deviations from their baseline values.

The *demand shock* triggers inflation only in the CANVAS and CATS rules, while the CPI remains unaffected in EUBI and KS.[11] This difference is due to the absence of the demand-pull channel in the price equations of EUBI and KS, where the entire demand shock is passed into quantities. When demand-pull is not passed on to prices, the real effects of demand shocks are

---

[11]It is important to emphasize once again that our results do not necessarily or perfectly reflect the predictions of the models under consideration. Indeed, although the implemented price and quantity rules closely align with those of the models, they are applied somewhat outside their original context. For instance, in EUBI and KS, nominal wages grow when the labor market is tight. It follows that in the original EUBI and KS frameworks, a positive demand shock would likely lower unemployment and therefore trigger wage inflation, which would then be passed on to prices, resulting in some level of positive inflation.

more pronounced, resulting in larger multipliers for EUBI and KS compared to CANVAS and CATS. Moreover, we observe a stronger GDP response in EUBI as compared to KS. The reason is that in EUBI, individual quantity decisions are also linked to expected aggregate demand growth, which in the case of demand shocks, boils down to an additional, coordinated pull factor. Finally, the inflationary effect is stronger under the CANVAS rule compared to CATS. This difference stems from how nominal wages feed back into the price equation via the cost channel. In CATS, wage increases are fully absorbed in the markup, unlike under CANVAS, where we have full pass-through. Therefore, nominal wages being pegged to the CPI, demand-pull inflation leads to nominal wage growth, which further drives prices under rules assuming positive pass-through.[12]

The *cost-push shock* impacts only models featuring cost channel in the price equation, specifically CANVAS, EUBI, and KS.[13] However, responses are heterogeneous across rules, with CANVAS exhibiting moderate inflation as compared to EUBI and KS. This difference stems from the demand-pull channel in the CANVAS price rule. Cost-push shocks typically lead to stagflation, where the recessionary aspect of the shock causes individual firms to experience lower-than-expected demand. In CANVAS, this prompts firms to reduce prices, counteracting the inflationary pressures from the cost-push shock. We also observe a slight difference in CPI under the EUBI and KS rules, with the latter featuring more inflation. This is due to the

---

[12] It should be noted that in this particular case, the cost channel constitutes a second-order effect, and therefore it is not sufficient to generate inflation. Since nominal wages are tied to inflation (but not to labor-market tightness), the cost channel only activates if the shock initially triggers inflation through demand-pull, which is not the case for EUBI and KS.

[13] The slight increase in CPI observed for CATS is a mechanical effect seen across all models, as import prices are included in the CPI calculation.

different recursions adopted by the two models in the markup rule (eq. 9), implying perfect pass-through (i.e., more inflation) for KS and only partial pass-through (i.e., less inflation) for EUBI. Finally, the magnitude of the recession is tightly linked to the extent of the pass-through, with EUBI and KS showing the strongest GDP loss, CANVAS showing moderate GDP loss, and CATS being only slightly impacted.

The *nonidiosyncratic technological shock* is symmetric to the cost-push shock. While CANVAS, EUBI, and KS see reductions in inflation, the magnitude differs. CANVAS shows a moderate deflationary effect due to its demand-pull channel, which diminishes the price effect of productivity growth. In contrast, EUBI and KS show stronger deflationary effects, with KS exhibiting the most significant price reduction due to its perfect pass-through mechanism. The productivity shock also leads to an increase in GDP across all models, with the strongest growth observed in EUBI and KS, and a moderate increase in CANVAS. CATS, as expected, is only slightly affected.

Conclusions

The design of behavioral rules is a key ingredient of agent-based macroeconomic models and an important determinant of differences between these models. In this paper we make two main contributions to foster a systematic analysis of the impact of the design of behavioral rules on model output. First, focusing on a particular set of firm decisions, namely price and quantity choice, we derive a general representation of rules that allows to systematically categorize rules according to which channels are active in determining the action under the rule. We exploit this approach by considering price and quantity rules in four families of agent-based macroeconomic models, but the approach is much more generally applicable as a tool to describe and compare different rules systematically.

Second, we embed the different rules in an identical macroeconomic framework, which allows us to isolate their effect on the economic dynamics emerging in the model. Our forecasting exercise reveals only minor differences across rules, indicating that in a business-as-usual scenario, the choice of a specific rule has little impact on the results. However, we find that the model's response to macroeconomic shocks differs significantly across the rules, both quantitatively and qualitatively. This variability underscores the need for a careful analysis of these rules to understand which ones perform best under specific conditions. In particular, using data for time windows including severe disruptions, such as supply respectively demand shocks triggered by COVID-19, or cost-push shocks due to energy price hikes, would allow one to examine performance of the models in terms of matching empirical observations. Such an analysis is beyond the current chapter and left for future work.

The results presented here should be seen only as a first step. The analysis should be extended to other important classes of decision rules and a broader spectrum of agent-based macroeconomic models. Work along these lines will improve our understanding of the drivers of differences in behavior across different models and play an important role in interpreting results, such as insights about policy effects, that have been obtained in different model frameworks. Finally, this type of analysis might also provide guidance concerning which types of behavioral rules perform best in terms of short- and medium-term forecasting. ✤

Acknowledgments

Luca Eduardo Fierro and Sebastian Poledna acknowledge funding from the Austrian Central Bank (OeNB) Anniversary Fund (Jubiläumsfonds) grant number 18790.

REFERENCES

Ashraf, Q., B. Gershman, and P. Howitt. 2017. "Banks, Market Organization, and Macroeconomic Performance: An Agent-Based Computational Analysis." *Journal of Economic Behavior and Organization* 135:143–180. https://doi.org/10.1016/j.jebo.2016.12.023.

Assenza, T., D. Delli Gatti, and J. Grazzini. 2015. "Emergent Dynamics of a Macroeconomic Agent Based Model with Capital and Credit." *Journal of Economic Dynamics and Control* 50:5–28. https://doi.org/10.1016/j.jedc.2014.07.001.

Axtell, R. L., and J. D. Farmer. 2025. "Agent-Based Modeling in Economics and Finance: Past, Present, and Future." *Journal of Economic Literature* 63 (1): 197–287. https://doi.org/10.1257/jel.20221319.

Basurto, A., H. Dawid, P. Harting, J. Hepp, and D. Kohlweyer. 2023. "How to Design Virus Containment Policies? A Joint Analysis of Economic and Epidemic Dynamics Under the COVID-19 Pandemic." *Journal of Economic Interaction and Coordination* 18 (2): 311–370. https://doi.org/10.1007/s11403-022-00369-2.

Borsos, A., A. Carro, A. Glielmo, M. Hinterschweiger, J. Kaszowska-Mojsa, and A. Uluc. 2026. "Agent-Based Modeling at Central Banks: Recent Developments and New Challenges." In *The Economy as an Evolving Complex System IV*, edited by R. M. del Rio-Chanona, M. Pangallo, J. Bednar, E. D. Beinhocker, J. Kaszowska-Mojsa, F. Lafond, P. Mealy, A. Pichler, and J. D. Farmer. Santa Fe, NM: SFI Press.

Ciola, E., E. Turco, A. Gurgone, D. Bazzana, S. Vergalli, and F. Menoncin. 2023. "Enter the MATRIX Model: A Multi-Agent Model for Transition Risks with Application to Energy Shocks." *Journal of Economic Dynamics and Control* 146:104589. https://doi.org/10.1016/j.jedc.2022.104589.

Dawid, H., and D. Delli Gatti. 2018. "Agent-Based Macroeconomics." *Handbook of Computational Economics* 4:63–156. https://doi.org/10.1016/bs.hescom.2018.02.006.

Dawid, H., and P. Harting. 2012. "Capturing Firm Behavior in Agent-Based Models of Industry Evolution and Macroeconomic Dynamics," edited by G. Bünsdorf, 103–130. Cheltenham, UK: Edward-Elgar Publishing. https://ideas.repec.org/h/elg/eechap/14183_6.html.

Dawid, H., P. Harting, and M. Neugart. 2018. "Cohesion Policy and Inequality Dynamics: Insights from a Heterogeneous Agents Macroeconomic Model." *Journal of Economic Behavior and Organization* 150:220–255. https://doi.org/10.1016/j.jebo.2018.03.015.

Dawid, H., P. Harting, S. van der Hoog, and M. Neugart. 2019. "Macroeconomics with Heterogeneous Agent Models: Fostering Transparency, Reproducibility and Replication." *Journal of Evolutionary Economics* 29:467–538. https://doi.org/10.1007/s00191-018-0594-0.

Delli Gatti, D., S. Desiderio, E. Gaffeo, P. Cirillo, and M. Gallegati. 2011. *Macroeconomics from the Bottom-Up.* Vol. 1. Milan, Italy: Springer.

Delli Gatti, D., and S. Reissl. 2022. "Agent-Based Covid Economics (ABC): Assessing Non-pharmaceutical Interventions and Macro-Stabilization Policies." *Industrial and Corporate Change* 31, no. 2 (March): 410–447. https://doi.org/10.1093/icc/dtac002.

Dosi, G., G. Fagiolo, M. Napoletano, A. Roventini, and T. Treibich. 2015. "Fiscal and Monetary Policies in Complex Evolving Economies." *Journal of Economic Dynamics and Control* 52:166–189. https://doi.org/10.1016/j.jedc.2014.11.014.

Dosi, G., G. Fagiolo, and A. Roventini. 2010. "Schumpeter Meeting Keynes: A Policy-Friendly Model of Endogenous Growth and Business Cycles." *Journal of Economic Dynamics and Control* 34 (9): 1748–1767. https://doi.org/10.1016/j.jedc.2010.06.018.

Dosi, G., and A. Roventini. 2019. "More is Different... and Complex! The Case for Agent-Based Macroeconomics." *Journal of Evolutionary Economics* 29:1–37. https://doi.org/10.1007/s00191-019-00609-y.

Eliasson, G. 2023. "Bringing Markets Back into Economics: On Economy Wide Self-Coordination by Boundedly Rational Market Agents." *Journal of Economic Behavior and Organization* 216:686–710. https://doi.org/10.1016/j.jebo.2023.10.001.

Filatova, T., and J. Akkerman. 2026. "Complexity Economics' View on Physical Climate Change Risks and Adaptation." In *The Economy as an Evolving Complex System IV,* edited by R. M. del Rio-Chanona, M. Pangallo, J. Bednar, E. D. Beinhocker, J. Kaszowska-Mojsa, F. Lafond, P. Mealy, A. Pichler, and J. D. Farmer. Santa Fe, NM: SFI Press.

Harvey, D., S. Leybourne, and P. Newbold. 1997. "Testing the Equality of Prediction Mean Squared Errors." *International Journal of Forecasting* 13 (2): 281–291. https://doi.org/10.1016/S0169-2070(96)00719-4.

Hommes, C., M. He, S. Poledna, M. Siqueira, and Y. Zhang. 2025. "CANVAS: A Canadian Behavioral Agent-Based Model for Monetary Policy." *Journal of Economic Dynamics and Control* 172:104986. https://doi.org/10.1016/j.jedc.2024.104986.

Hommes, C., S. Kozicki, S. Poledna, and Y. Zhang. 2026. "How an Agent-Based Model Can Support Monetary Policy in a Complex Evolving Economy." In *The Economy as an Evolving Complex System IV,* edited by R. M. del Rio-Chanona, M. Pangallo, J. Bednar, E. D. Beinhocker, J. Kaszowska-Mojsa, F. Lafond, P. Mealy, A. Pichler, and J. D. Farmer. Santa Fe, NM: SFI Press.

Hötte, K. 2020. "How to Accelerate Green Technology Diffusion? Directed Technological Change in the Presence of Coevolving Absorptive Capacity." *Energy Economics* 85:104565. https://doi.org/10.1016/j.eneco.2019.104565.

Lamperti, F., G. Dosi, M. Napoletano, A. Roventini, and A. Sapio. 2020. "Climate Change and Green Transitions in an Agent-Based Integrated Assessment Model." *Technological Forecasting and Social Change* 153:119806. https://doi.org/10.1016/j.techfore.2019.119806.

Lamperti, F., G. Dosi, and A. Roventini. 2026. "A Complex System Perspective on the Economics of Climate Change, Boundless Risk, and Rapid Decarbonization." In *The Economy as an Evolving Complex System IV,* edited by R. M. del Rio-Chanona, M. Pangallo, J. Bednar, E. D. Beinhocker, J. Kaszowska-Mojsa, F. Lafond, P. Mealy, A. Pichler, and J. D. Farmer. Santa Fe, NM: SFI Press.

Lengnick, M. 2013. "Agent-Based Macroeconomics: A Baseline Model." *Journal of Economic Behavior and Organization* 86:102–120. https://doi.org/10.1016/j.jebo.2012.12.021.

Mincer, J. A., and V. Zarnowitz. 1969. "The Evaluation of Economic Forecasts." In *Economic Forecasts and Expectations: Analysis of Forecasting Behavior and Performance,* 3–46. NBER.

Pichler, A., M. Pangallo, R. M. del Rio-Chanona, F. Lafond, and J. D. Farmer. 2022. "Forecasting the Propagation of Pandemic Shocks with a Dynamic Input-Output Model." *Journal of Economic Dynamics and Control* 144:104527. https://doi.org/10.1016/j.jedc.2022.104527.

Poledna, S., M. G. Miess, C. Hommes, and K. Rabitsch. 2023. "Economic Forecasting with an Agent-Based Model." *European Economic Review* 151:104306. https://doi.org/10.1016/j.euroecorev.2022.104306.

Riccetti, L., A. Russo, and M. Gallegati. 2013. "Leveraged Network-Based Financial Accelerator." *Journal of Economic Dynamics and Control* 37 (8): 1626–1640. https://doi.org/10.1016/j.jedc.2013.02.008.

Salle, I., and P. Seppecher. 2018. "Stabilizing An Unstable Complex Economy on the Limitations of Simple Rules." *Journal of Economic Dynamics and Control* 91:289–317. https://doi.org/10.1016/j.jedc.2018.02.014.

Savin, I., F. Creutzig, T. Filatova, J. Foramitti, T. Konc, L. Niamir, K. Safarzynska, and J. van den Bergh. 2023. "Agent-Based Modeling to Integrate Elements from Different Disciplines for Ambitious Climate Policy." *Wiley Interdisciplinary Reviews: Climate Change* 14 (2): e811. https://doi.org/10.1002/wcc.811.

Turco, E., D. Bazzana, M. Rizzati, E. Ciola, and S. Vergalli. 2023. "Energy Price Shocks and Stabilization Policies in the MATRIX Model." *Energy Policy* 177:113567. https://doi.org/10.1016/j.enpol.2023.113567.

Appendix

A   Microfoundations for the Demand Model

The simple linear demand function (eq. 5) used by the firms can be derived from simple interaction structures with heterogeneous consumers. An example of a market structure giving rise to this form of expected demand is a market setup in which each consumer c intending to buy the product in t visits two producers and chooses between these producers based on a utility function $u_{i,t}^c = \bar{u} - \tilde{\zeta}_t \frac{P_{i,t}}{\bar{P}_t}$ and a (stochastic) idiosyncratic preference between the producers. The fact that the disutility of paying $P_{i,t}$ is normalized by $\bar{P}_t$ captures that prices are evaluated in terms of purchasing power and pure inflationary effects are neutralized. More precisely, consumer visiting firms i_1 and i_2 purchases from producer i_1 if and only if

$$u_{i_1,t}^c - u_{i_2,t}^c \geq \epsilon_{i_1,i_2}^c,$$

where ϵ_{i_1,i_2}^c captures the idiosyncratic preference of consumer c between the producers and is assumed to be uniformly distributed in $[-k, k]$ and iid across consumers and firm pairs. Assuming that k is sufficiently large such that $\tilde{\zeta}_t \frac{P_{i_1,t} - P_{i_2,t}}{\bar{P}_t} \in [-k, k]$, then the probability that consumer c buys from firm i_1 is given by

$$
\begin{aligned}
q_{i_1}(i_2) &= (\epsilon_{i_1,i_2}^c \leq u_{i_1,t}^c - u_{i_2,t}^c) \\
&= \left(\epsilon_{i_1,i_2}^c \leq \tilde{\zeta}_t \frac{P_{i_2,t} - P_{i_1,t}}{\bar{P}_t^e} \right) \\
&= \frac{1}{2k} \left(\tilde{\zeta}_t \frac{P_{i_2,t} - P_{i_1,t}}{\bar{P}_t} - (-k) \right) \\
&= \frac{1}{2} - \frac{\tilde{\zeta}_t}{2k} \frac{P_{i_1,t} - P_{i_2,t}}{\bar{P}_t}.
\end{aligned}
$$

Based on this, the probability that an arbitrary consumer c buys from firm i can be calculated as

$$\text{IP}(c \text{ buys from } i)$$

$$= \text{IP}(c \text{ visits } i \text{ in } t) \sum_{\tilde{i} \neq i} (c \text{ visits } \tilde{i} \text{ in } t | c \text{visits } i \text{ in } t) q_i(\tilde{i})$$

$$= \frac{2}{F_C} \frac{1}{F_C - 1} \sum_{\tilde{i} \neq i} \left(\frac{1}{2} - \frac{\tilde{\zeta}_t}{2k} \frac{P_{i,t} - P_{\tilde{i},t}}{\bar{P}_t} \right)$$

$$= \frac{1}{F_C} - \frac{\tilde{\zeta}_t}{k} \sum_{\tilde{i}=1}^{F_C} \frac{P_{i,t} - P_{\tilde{i},t}}{F_C \bar{P}_t}$$

$$= \frac{1}{F_C} - \frac{\tilde{\zeta}_t}{k} \left(\frac{P_{i,t}}{\bar{P}_t} - 1 \right)$$

$$= \frac{1}{F_C} - \zeta_t \left(\frac{P_{i,t}}{\bar{P}_t} - 1 \right),$$

with $\zeta_t = \frac{\tilde{\zeta}_t}{k}$. The corresponding expectation of firm i at time t about this parameter is denoted by $\zeta_{i,t}^e$. This yields an expected market share of firm i given by equation 5.

B   Forecast Performance

Table 2. Out-of-Sample forecast performance

	GDP	INFLATION	HOUSEHOLD CONSUMPTION	INVESTMENT
AR(1)	*RMSE-statistic for different forecast horizons*			
1q	0.48	0.36	0.76	1.42
2q	0.69	0.37	0.92	1.98
4q	1.17	0.37	1.24	3.16
8q	2.01	0.37	2.1	4.49
12q	2.8	0.37	2.87	6.09
CANVAS	*Percentage improvements (+) or losses (-) relative to AR (1) model*			
1q	0.2 (0.83)	-11.7 (0.04)	-3.1 (0.79)	3.9 (0.14)
2q	0.1 (0.93)	-29.8 (0.04)	-23.2 (0.30)	6.9 (0.06*)
4q	-0.4 (0.95)	-8.6 (0.61)	-8.2 (0.80)	8.9 (0.11)
8q	7.9 (0.72)	-18.3 (0.00)	30.1 (0.45)	12.8 (0.04**)
12q	25.4 (0.49)	-30.7 (0.23)	38.7 (0.45)	14.1 (0***)
CATS	*Percentage improvements (+) or losses (-) relative to AR (1) model*			
1q	-24.1 (0.08*)	-50.6 (0.00***)	-13.4 (0.39)	-8.4 (0.04**)
2q	-22.1 (0.25)	-50 (0.00***)	-30.8 (0.23)	-13 (0.04**)
4q	-2 (0.93)	-38.9 (0.00***)	-28.5 (0.49)	-10.4 (0.06*)
8q	17.1 (0.69)	-16 (0.03**)	2.3 (0.96)	-4.5 (0.66)
12q	21.1 (0.65)	-7.1 (0.08*)	23.1 (0.74)	3.5 (0.59)
EUBI	*Percentage improvements (+) or losses (-) relative to AR (1) model*			
1q	0.2 (0.83)	-11.7 (0.04**)	-3.1 (0.79)	3.9 (0.14)
2q	-1.1 (0.54)	-15.5 (0.08*)	-26.2 (0.27)	4.6 (0.20)
4q	-9.6 (0.44)	-6.3 (0.56)	-17.1 (0.64)	5.4 (0.25)
8q	-7.7 (0.77)	-11.2 (0.28)	27.1 (0.50)	12.2 (0.05*)
12q	11.8 (0.66)	5.1 (0.00***)	39.7 (0.51)	19.1 (0.01***)
KS	*Percentage improvements (+) or losses (-) relative to AR (1) model*			
1q	-24.4 (0.08*)	-11.7 (0.04**)	-14.3 (0.37)	-8.2 (0.04**)
2q	-40.9 (0.08*)	-17 (0.09*)	-42.3 (0.14)	-17.9 (0.02**)
4q	-29 (0.36)	-8.6 (0.43)	-42.4 (0.35)	-19.7 (0.01**)
8q	8.6 (0.87)	-12.4 (0.29)	6.2 (0.90)	-12.3 (0.33)
12q	26.1 (0.68)	6.4 (0.17)	27.1 (0.70)	-0.8 (0.86)

Table 3. Mean biases of models.

	GDP	INFLATION	HOUSEHOLD CONSUMPTION	INVESTMENT
AR(1)	*Mean biases for different forecast horizons*			
1q	0.0004 (0.34)	-0.0007 (0.23)	0.0019 (0.26)	-0.0035 (0.20)
2q	0.0011 (0.39)	-0.008 (0.17)	0.0032 (0.07)	-0.0084 (0.02**)
4q	0.0031 (0.27)	-0.0009 (0.26)	0.0065 (0.00)	-0.0183 (0.00***)
8q	0.0085 (0.02)	-0.001 (0.18)	0.0133 (0.00)	-0.0322 (0.00***)
12q	0.0149 (0.00)	-0.0007 (0.47)	0.0202 (0.00)	-0.0469 (0.00***)
CANVAS	*Mean biases for different forecast horizons*			
1q	0.0003 (0.29)	-0.0007 (0.00)	-0.0017 (0.14)	-0.0034 (0.23)
2q	0.001 (0.35)	-0.0019 (0.00)	-0.004 (0.01***)	-0.0075 (0.03**)
4q	0.0031 (0.28)	-0.0016 (0.00)	-0.0062 (0.01***)	-0.015 (0.00***)
8q	0.0077 (0.03**)	0.0013 (0.00)	-0.0044 (0.25)	-0.0229 (0.00***)
12q	0.0064 (0.03**)	0.0034 (0.00)	-0.0004 (0.62)	-0.0361 (0.00***)
CATS	*Mean biases for different forecast horizons*			
1q	-0.0039 (0.00***)	-0.0038 (0.00***)	-0.0033 (0.02***)	-0.0076 (0.00***)
2q	-0.0049 (0.00***)	-0.004 (0.00***)	-0.0066 (0.00***)	-0.0138 (0.00***)
4q	-0.005 (0.04**)	-0.0038 (0.00***)	-0.0109 (0.00***)	-0.0245 (0.00***)
8q	-0.0012 (0.22)	-0.0025 (0.00***)	-0.015 (0.00***)	-0.0348 (0.00***)
12q	0.0025 (0.05**)	-0.0019 (0.03**)	-0.0155 (0.00***)	-0.0444 (0.00***)
EUBI	*Mean biases for different forecast horizons*			
1q	0.0003 (0.29)	-0.0007 (0.00***)	-0.0017 (0.14)	-0.0034 (0.23)
2q	0.0004 (0.48)	-0.0009 (0.00***)	-0.0049 (0.00***)	-0.0082 (0.02**)
4q	0.002 (0.61)	-0.0011 (0.02**)	-0.0077 (0.00***)	-0.0164 (0.00***)
8q	0.0106 (0.02**)	-0.0009 (0.01**)	-0.0057 (0.09***)	-0.0208 (0.00***)
12q	0.0131 (0.01***)	-0.0005 (0.60)	-0.0051 (0.23)	-0.0313 (0.00***)
KS	*Mean biases for different forecast horizons*			
1q	-0.004 (0.00***)	-0.0007 (0.00***)	-0.0036 (0.01**)	-0.0075 (0.01***)
2q	-0.0067 (0.00***)	-0.0009 (0.00***)	-0.0079 (0.00***)	-0.0151 (0.00***)
4q	-0.0092 (0.00***)	-0.0009 (0.01**)	-0.0126 (0.00***)	-0.0281 (0.00***)
8q	-0.0055 (0.22)	-0.0011 (0.01***)	-0.0139 (0.00***)	-0.0386 (0.00***)
12q	-0.0023 (0.49)	-0.0009 (0.44)	-0.0137 (0.00***)	-0.0488 (0.00***)

HOW AN AGENT-BASED MODEL CAN SUPPORT MONETARY POLICY IN A COMPLEX EVOLVING ECONOMY

Cars Hommes, Bank of Canada,
University of Amsterdam, and Tinbergen Institute;
Sharon Kozicki, Bank of Canada;
Sebastian Poledna, Austrian Institute of Economic Research; and
Yang Zhang, Bank of Canada

Abstract

This chapter sets out a central banker's perspective on the dynamics of inflation in Canada since the onset of the COVID-19 pandemic using a behavioral agent-based model developed at the Bank of Canada. The model features two crucial assumptions based on empirical evidence: first, firms use simple rules to make price-quantity decisions, and second, inflation expectations are adapted through experience. While the course of the economy in recent years has posed some new challenges about many traditional macroeconomic models, an agent-based model can explain inflation by using simple heuristics and closely tracking survey evidence. We also include an example from policymakers on how an agent-based model can be used to develop insights to support monetary policy in Canada. Finally, we suggest some future avenues for research.[1]

The Advantages of an Agent-Based Model

The COVID-19 pandemic was a watershed event for economists, especially those who work for central banks and other policy

---

[1] Opinions expressed in this chapter are those of the authors and do not necessarily reflect those of the Bank of Canada or its staff. Any remaining errors are ours.

institutions. The consequences of the pandemic and related public-health policies on economies were massive and global. It was one of those rare episodes that economists describe as Knightian uncertainty—a time when risks cannot be quantified or measured because of a lack of historical data or understanding. In this chapter we discuss how an alternative agent-based model (ABM) can benefit monetary policy, and thus the management of the economy, during just such a period of turmoil. András Borsos *et al.* (2026, ch. 18 in this volume) provide an extensive overview of the recent developments and new challenges of using agent-based models in central banks.

At the Bank of Canada, the main macroeconomic policy models used to inform monetary policy prior to the pandemic took little account of individual sectors of the economy and covered just a small cross section of consumers. Firms' pricing behavior was assumed not to be state contingent. The existing models assumed that expectations were fully rational and agents had perfect knowledge of the underlying model. They were estimated using data collected during the inflation targeting period. While these years were marked by cyclical fluctuations, inflation rarely fell outside the Bank of Canada's control range of 1–3%, and when it did, it did not do so for long. Inflation expectations were in line with the target range, and the target itself served as a reliable forecast for inflation a year ahead and even beyond. While the economy was subject to domestic- and foreign-sourced supply-and-demand shocks, swings in demand tended to be more important than swings in supply in determining excess demand.

All this changed with the onset of the pandemic. Supply shocks became at least as important as demand shocks, and each tended to have a different degree of persistence. For the first time, global supply disruptions became a critical issue, adding to firms' costs and taking an unusually long time to subside. Because of the size and expected persistence of these cost shocks, firms began to raise prices more often than in the past. Inflation accelerated rapidly beyond the Bank of Canada's target range, and stubbornly stayed there. The result was that expectations of inflation over the short and medium term followed prices as they went up but lagged on the way down.

Models with embedded supply chains, regulated pricing behavior, and nonrational expectations proved to be valuable components of an expanded toolkit. In this chapter, we discuss the development and adoption of an agent-based model called CANVAS (Canadian Behavioural Agent-Based Model) at the Bank of Canada.

By way of background, agent-based models explain the behavior of a system by simulating the behavior of each individual "agent" within it. These agents and the systems they inhabit could be the consumers in an economy, fish within a shoal, particles in a gas, or even galaxies in the universe. The strength of agent-based models is that they show how even simple behaviors can combine from the 'bottom up' to recreate the more complex behaviors observed in the real world. An example would be how the decisions of each individual fish create the seemingly organized and unpredictable movements of the shoal.

This bottom-up approach is in contrast to top-down models, which assume how agents' behaviors will combine together, sometimes by assuming that all agents are identical or that the behavior of a single agent can represent the aggregate of the behavior of all. The different approaches have different strengths. The agent-based approach to problem-solving began in the physical sciences but has now spread to many other disciplines including biology, ecology, computer science and epidemiology. In recent years, agent-based models have become more common in economics and several macroeconomic ABMs have been developed featuring a clear common core as discussed in Herbert Dawid and Domenico Delli Gatti (2018). One example could be that individual consumer decisions to accumulate mortgage debt could collectively cause financial market distress through too much aggregate leverage.

CANVAS is one of the first ABMs used by a central bank to improve its understanding of inflation targeting and related macroeconomic forecasting and policy analysis. It builds upon the framework of Sebastian Poledna *et al.* (2023), who pioneered the development of a small open-economy ABM for forecasting and policy analysis. CANVAS offers an alternative approach by simulating at a scale of 1:100 the behavior of a diverse group of individuals to provide a comprehensive macro-level view of the economy. See Cars Hommes *et al.* (2025) for a detailed description of the CANVAS model.

A behavioral agent-based model offers both empirical and practical advantages. Our first objective was to design a comprehensive framework that integrates reliable household and business data in a production network for the Canadian economy. We also wanted to use a common sample to evaluate the forecasting performance of CANVAS in comparison with the more established vector autoregressive (VAR) and dynamic stochastic general equilibrium (DSGE) models. We have found that CANVAS outperforms DSGE models in forecasting gross domestic product (GDP) growth and key components like consumption. This head-to-head performance is encouraging and suggests that a model based on diverse agents interacting in imperfect markets has the potential to enrich macroeconomic policy analysis. Our third goal was to improve our understanding of the interplay between evolving expectations and cost-push shocks in the context of an inflation-targeting policy regime. Finally and perhaps most importantly, we have actively applied a macroeconomic agent-based model on the front lines of monetary policy and integrated it as a new component of a central bank's toolkit.

In contrast to the challenges faced by traditional models during the pandemic, we are confident that an agent-based model like CANVAS is well-suited to interpreting micro- and macroeconomic data and thus a useful tool as part of the suite of models used by central banks for forecasting key variables as well as for scenario-based policy analysis. We share some experiences below where the Bank of Canada used CANVAS to provide a variety of perspectives to support monetary policy.

Why Pricing Behavior Matters to Central Banks

Firms' pricing behavior is relevant for understanding how various supply-and-demand shocks may affect inflation, including the persistence of their impacts. Because monetary policy usually takes time to work through the macroeconomy, the likely persistence of shock impacts matters. Often, monetary policy tends to "look through" high-frequency movements. However, when impacts of shocks on inflation are expected to be both large and persistent, it may be appropriate for monetary policy responses to be scaled up more than proportionately in order to lean against the potential for inflation expectations to drift and slow down the return of inflation to target.

The pandemic period also revealed that it is not enough to look at measures of excess demand for understanding firm pricing.

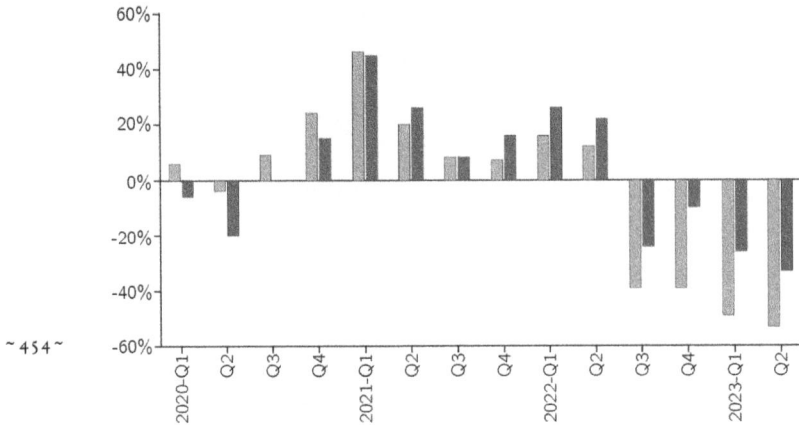

Figure 1. Asymmetry in cost pass-through. The figure plots the evolution of shares of Canadian firms' experiencing higher input prices (left bars, in light gray), along with share of firms' expressing intention to adjust output prices (right bars, in dark gray). There is a significant pass-through from higher input prices to output prices as inflation surges to its peak in 2022. However, after experiencing the decline in input costs in 2023, fewer Canadian firms indicating intention to lower their output prices *(Source: Business Outlook Survey—Second Quarter of 2023, Bank of Canada)*

While Canadian data shows a correlation between inflation measures and estimates of excess demand, during the pandemic, cost drivers became of great importance. The most important of the supply-side drivers was global supply-chain disruptions that had large impacts on production, costs, and availability of many products. Such details are outside the scope of the standard macroeconomic policy models.

The Calvo assumption governing the frequency of price adjustment also proved to be unrealistic. When faced with larger-than-average persistent increases in costs, firms have an incentive to change prices more frequently. This shortens the lag between developments in macro drivers (including those on the demand side) and subsequent movements in inflation. This, in turn, may influence the appropriate size and timing of policy responses to shocks.

Why does a central bank need an agent-based model? The best motivation is to look at our recent experience over the pandemic. Our traditional structural and semistructural models showed their limitations.

We have learned three important lessons from this experience:

- **We need to better understand how firms set their prices.** Our survey conducted during the pandemic suggests that the most relevant factors that affected Canadian firms' pricing behavior are cost pressures, followed by labor shortages and supply chain considerations.

~ 455 ~

- **There is asymmetry in cost pass-through.** Figure 1 plots the evolution of firms' response to input price growth (in dark gray), along with firms' intention to adjust output prices (in light gray). It is interesting to see that there is a timely and significant pass-through of higher input price to output prices as inflation surges to its peak in 2022. However, despite the decline in input costs in 2023, Canadian firms appeared to have no intention and were not taking any actions to lower their output prices.

- **Agents form expectations based on experience.** It has long been accepted that, in practical terms, nominal interest rates cannot fall below an effective level (usually zero) because investors can always earn a zero nominal return simply by holding cash. This concept has been termed the "effective (or zero) lower bound" on nominal interest rates. Theoretically, the existence of this effective lower bound (thereafter ELB) limits a central bank's ability to provide further stimulus to the economy through conventional decreases in policy rates below such level. In face of large shocks and when policy rate is lowered to the ELB, monetary policy needs to work through expectations to help lower real interest rates to stimulate the economy.

The pandemic experience suggests that better modeling the formation of expectations is not just relevant near the ELB; it is something we need to understand at all times. The theory of rational inattention suggests that firms and households typically pay little attention to inflation when it is low and stable (Sims 2003). Informational frictions make paying attention to inflation costly and simply not worth the effort. But when inflation is high and volatile, paying attention to inflation and monetary policy becomes more important— the marginal benefit is greater than when inflation is low and stable. In such situations, households and businesses benefit from incorporating views on inflation into their economic decisions and expectations. Similarly, survey and lab experiments have also shown us that people tend to learn about inflation based on their own experiences. It is more of an adaptive process.

Adapting Models to Behavior

A large body of literature documents how the formation of expectations often deviates from rational expectations. However, the relevant question for policy is whether these deviations are large enough to matter for monetary policy decisions. From the time when the inflation target was set at 2% in 1995 until the start of the pandemic in the first quarter of 2020, inflation in Canada tended to only fall outside the inflation control range of 1–3% for short periods of time. Inflation expectations were well-anchored on the inflation target, and assumptions on rational versus learning behavior in the formation of expectations had only small impacts on the outlook for inflation. Thus, adjusting models to include nonrational assumptions regarding expectations formation was not a high priority. The pandemic provided an exceptional deviation from the prior twenty-five years. Assumptions on expectations formation became critical for economic projections and evaluations of risks.

RATIONAL EXPECTATIONS ARE NOT THE FULL STORY

In a complex environment of fundamental uncertainty and imperfect information, obtaining accurate forecasts is nearly impossible. Firms in such a complex environment may choose simple forecasting methods that closely monitor the relevant variables, even if such methods fail to understand the complete model of the economy. This approach fits within the concept of procedural rationality (Gigerenzer 2015). Lab experiments with human subjects show that in an environment that is too complex to fully understand, subjects use simple forecasting rules, such as adaptive expectations, low-order autoregressive (AR) processes, and trend-following rules (Hommes 2021). A forecasting method that meets these requirements is the AR(1) forecasting rule: this is a simple procedure, based on the observed mean and persistence, for projecting past trends into the future with relatively high forecasting capabilities.

~457~

Under adaptive learning—that is, recursive updating of parameters—the gaps between expected and realized values of state variables will close gradually. Learning ensures that the unconditional mean and autocorrelations of the unknown nonlinear stochastic process—which describes the actual law of motion of the model economy—concur with the unconditional mean and autocorrelations of the AR(1) process in the long run. In fact, adaptive learning with the AR(1) rule leads to convergence to a behavioral learning equilibrium where agents learn to use an optimal AR(1) rule in the complex ABM economy, one of the simplest types of misspecification equilibrium put forth in the adaptive learning literature (Hommes and Zhu 2014).

A BEHAVIORAL AGENT-BASED MODEL BASED ON LAB INSIGHTS

In this chapter we only present the structure and main assumptions underlying the CANVAS model. For details of the CANVAS agent-based model, see Hommes *et al.* (2025). The CANVAS model economy is structured into six sectors: (1) nonfinancial corporations (firms); (2) households; (3) the general government; and (4) financial corporations (banks), including (5) the central bank. These four sectors interact with (6) the rest of the world through imports and exports. Each sector is populated by heterogeneous agents whose balance sheets and economic flows are set according to data from national accounts.

The firm sector is made up of nineteen industries (the North American Industry Classification System (NAICS) by Canada's national statistical agency), where each industry (firm) produces a perfectly substitutable good using labor, capital, and intermediate inputs with Leontief technology. Firms, subject to fundamental uncertainty, use a simple AR(1) rule to form expectations of the output and (producer price) inflation. Given these, they set prices and quantities. Output is sold to households or to other firms, or is exported. Firm investment is conducted according to the expected wear and tear on capital.

Households earn income and consume in markets characterized by search and matching processes. Employed households supply labor and earn wages; unemployed households receive unemployment benefits; investor households obtain dividend income; and inactive households receive social benefits. Additional social transfers are distributed equally to all households. Similar to firms, households also form AR(1) expectations about the expected growth rate and expected inflation.

The government collects taxes, distributes social as well as other transfers, and engages in government consumption. The banking sector obtains deposits from households as well as from firms and provides loans to firms. Bank profits are calculated as the difference between interest payments received on firm loans and deposit interest paid to holders of bank deposits, as well as write-offs due to credit defaults. The central bank sets the policy rate according to a generalized Taylor rule, provides liquidity to the banking system, and takes deposits from the bank in the form of reserves. Furthermore, the central bank purchases government bonds, acting as a creditor to the government. To model interactions with the rest of the world, a segment of the firm sector is engaged in import–export activities. As we model a small open economy whose limited volume of trade does not affect world prices, we obtain trends of exports and imports from exogenous projections based on national accounts.

Interactions between agents in the model take place on decentralized markets, characterized by search and matching. These interactions are governed by explicit behavioral rules that depict the micro behavior and institutional design of the considered economic system. This search and matching mechanism depends on the probability of a firm being chosen by a customer, which is determined by (1) the offering price of the firm and (2) the size of the firm. The purchased amount then depends on the consumer's consumption budget and the seller's supply. Markets do not necessarily clear; however, the ABM constantly tends toward an approximate equilibrium state, and markets, in general, tend to be close to an equilibrium between supply and demand.

A Fresh Understanding of Growth and Inflation

In our model, there are two variables that agents need to form expectations over the expected output and producer price inflation. We assume agents form expectations in a homogeneous way: They are boundedly rational and use AR(1) rules to forecast variables in the model economy.[2] This rule is misspecified as it ignores cross-correlations and nonlinearities. However, agents continuously learn and update the parameters of their AR(1) forecasting rule and, in the long run, converge to a so-called misspecification equilibrium where agents have learned an optimal AR(1) rule consistent with the mean and first-order autocorrelation (a behavioral learning equilibrium as introduced by Hommes and Mei Zhu 2014 and Hommes *et al.* 2023).

Equation (1) summarizes the firm's expectations of the real GDP growth rate ($\gamma^e(t)$) and inflation ($\pi^e(t)$), measured by the log first difference of the GDP deflator:

$$\gamma^e(t) = e^{\alpha^\gamma(t-1)\gamma(t-1)+\beta^\gamma(t-1)+\epsilon^\gamma(t-1)} - 1$$
$$\pi^e(t) = e^{\alpha^\pi(t-1)\pi(t-1)+\beta^\pi(t-1)+\epsilon^\pi(t-1)} - 1 \tag{1}$$

where parameters $\alpha^\gamma(t-1)$, $\alpha^\pi(t-1)$, $\beta^\gamma(t-1)$, and $\beta^\pi(t-1)$ are re-estimated every period with the time series of output growth $\gamma(t')$ and inflation $\pi(t')$ where $t' = -T', -T'+1, -T'+2, \ldots, 0, 1, 2, \ldots, t-1$. $\epsilon^\gamma(t-1)$, and $\epsilon^\pi(t-1)$ are random shocks with zero mean and variance re-estimated every period from past observations over the last $T' + t - 1$ periods.

---

[2] This modeling choice is comparable to other adaptive mechanisms, such as VAR expectations as used in the US Federal Reserve's FRB/US macroeconomic model (a large-scale estimated general equilibrium model of the US economy that was developed at the Federal Reserve Board) (Brayton *et al.* 1997), or expectations according to an exponential moving average (EMA) model, as in Assenza, Delli Gatti, and Grazzini (2015).

The Production Network

The production network in traditional DSGE models is often simplified and does not have sufficient granularity to analyze the impact of shocks in specific sectors. The introduction of a detailed Canadian production network follows the lines of Frank Smets, Joris Tielens, and Jan van Hove (2019) and Hassan Afrouzi and Saroj Bhattarai (2023) and considers how shocks to individual sectors could permeate throughout the economy. Naturally, the effect of shock propagation depends on the input–output linkage, the degree of substitutability of intermediate inputs, and the complex price and quantity adjustments of individual firms when responding to shocks.

We use a detailed Canadian input–output database to calibrate the production network based on Canada's classification as a small open economy. Figure 2 suggests that there are important domestic production linkages through supply and demand of intermediate inputs across the nineteen sectors. The energy commodity sector (including mining, quarrying, and oil and gas extraction), for example, is a key sector of the Canadian economy at the aggregate level. Canada was the world's fourth largest oil producer in 2018, and investment in the oil and gas sector accounts for about 30% of total business investment. Canada is also a net oil exporter, with about 75% of commodity production being exported. As a result, the sharp energy price increase had a significant impact on the Canadian economy through terms of trade effects.

What's worth highlighting is that both the energy commodity sector and the nonenergy commodity sector play important roles in the Canadian production network because they are critical inputs for the manufacturing goods sector. Since the manufacturing sector supplies nearly all parts of

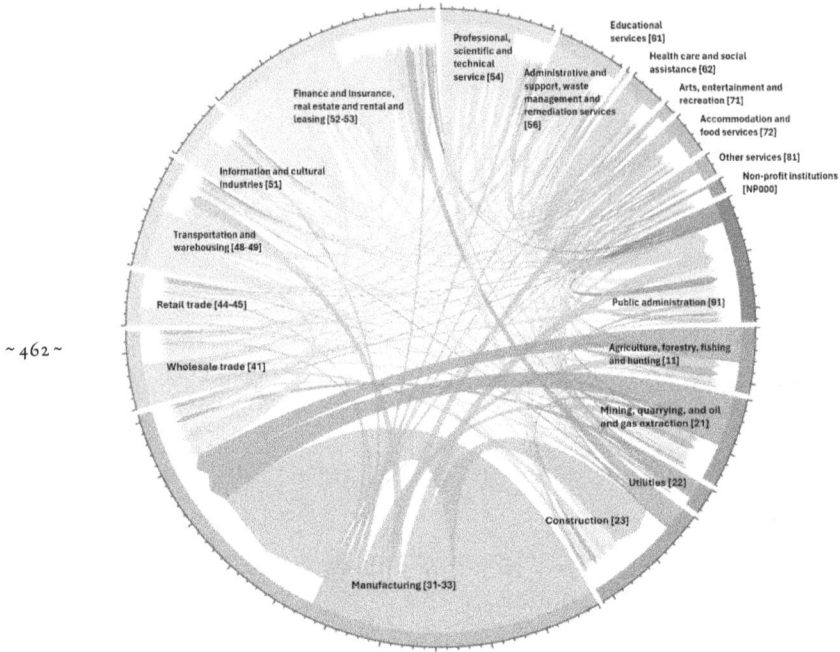

Figure 2. Visualization of the Canadian production network. The intersector input–output linkage is shown for nineteen sectors with corresponding two-digit North American Industry Classification System code for reference: Agriculture, forestry, fishing, and hunting [11]; Mining, quarrying, and oil and gas extraction [21]; Utilities [22]; Construction [23]; Manufacturing [31–33]; Wholesale trade [41]; Retail trade [44–45]; Transportation and warehousing [48–49]; Information and cultural industries [51]; Finance and insurance, real estate, and rental and leasing [52–53]; Professional, scientific and technical services [54]; Administrative and support, waste management, and remediation services [56]; Educational services [61]; Health care and social assistance [62]; Arts, entertainment, and recreation [71]; Accommodation and food services [72]; Other services (except public administration) [81]; Non-profit institutions serving households [NP000]; Public administration [91]. *(Source: Statistics Canada, authors' calculation)*

the economy, any fluctuation in commodity prices would naturally have effects that propagate throughout the Canadian economy.

Each firm uses labor input for production measured by the number of persons employed. Demand for labor during each period is determined according to the firm's desired scale of activity and its average labor productivity.

Each period, each firm chooses real investment, which adjusts the real capital stock. As in standard DSGE models, capital adjustment in CANVAS is not immediate and is time-consuming. The firms' investment decision process is modeled by a simple heuristic that accounts for both expectations of the aggregate economy and firm-specific conditions. Every period, each firm observes realized demand and makes a forecast of future demand according to the expected rate of economic growth. Conditional on demand for business investment, the firm's capital stock adjusts accordingly when accounting for depreciation. Under adaptive learning, should realized growth rates surpass growth expectations, investment in subsequent periods will adapt to the approximate trend equilibrium level, and vice versa. As a result, the resulting trend of business investment tends to approximate the trend equilibrium path of this model economy.

~ 463 ~

Each firm also needs intermediate inputs for production. We assume that each firm holds an inventory stock of each type of input goods. Each period, the firm follows a heuristic of maintaining its inventory stock in positive supply and choosing the desired amount of intermediate goods and raw materials to avoid shortfalls of material input that would impede production. The realized demand for intermediate goods depends on a search-and-matching process. If this firm does not succeed in acquiring the materials it intends to purchase, it will be limited in its production possibilities.

We define the cost-push inflation $\pi_i^c(t)$ for each sector as:

$$
\pi_i^c(t) = \underbrace{\frac{(1 + \tau^{SIF})\bar{w}_i}{\bar{\alpha}_i}\left(\frac{\bar{P}^{HH}(t-1)}{\bar{P}_{g=s}(t-1)} - 1\right)}_{\text{Unit labor costs}}
$$

$$
+ \underbrace{\frac{1}{\beta_i}\left(\frac{\sum_g a_{sg}\bar{P}_g(t-1)}{\bar{P}_{g=s}(t-1)} - 1\right)}_{\text{Unit production material costs}} \quad (2)
$$

$$
+ \underbrace{\frac{\delta_i}{\kappa_i}\left(\frac{\bar{P}^{CF}(t-1)}{\bar{P}_{g=s}(t-1)} - 1\right)}_{\text{Unit capital costs}}
$$

$$
\forall i \in I_s
$$

where $\bar{P}_{g=s}(t)$ is the average price among competitors of firm i, $\bar{\alpha}_i$ indicates the average labor productivity, and $\bar{w}_i$ is the average real wage, defined as gross wages, which include both salary costs and employers' contributions to social insurance charged with a rate τ^{SIF}. $\frac{1}{\beta_i}\sum_g a_{sg}$ is real unit expenditures on intermediate production input by industry s on good g, weighted by the average product price index $\bar{P}_g(t)$. δ_i/κ_i are unit capital costs, conditional on the average price of capital goods ($\bar{P}^{CF}(t)$) and capital depreciation relative to productivity growth (where δ_i is the firm-specific capital depreciation rate and κ_i is the productivity coefficient for capital).

Price-Quantity Decisions Based on a Simple Rule

As discussed above, we assume each firm is boundedly rational and uses a simple AR(1) rule to forecast output growth and (producer price) inflation. Every period, the firm sets both the supply of its product and related price based on its expectation for the aggregate economy, firm-specific cost structure, and market demand conditions.

In each period, a firm from an industry produces real output of a principal product with Leontief technology that combines intermediate inputs, labor, and capital. A firm's supply choice depends on its expectations of the aggregate economy and individual supply/demand versus pricing power. It is driven by three sources: (1) its product supply from the previous period; (2) the expected economy-wide economic growth rate; and (3) the realized firm-specific growth rate of demand from the previous period.[3]

Firms set product prices and determine supply based on both aggregate expectations for overall economic conditions as well as firm-specific conditions for cost pressures and market demand.

In addition to setting price and quantity based on aggregate expectations and cost pressures from the production network, each firm also chooses to alter its previous period's quantity or price based on its perceived market conditions. We assume that, due to different information and search costs, each firm has a certain degree of pricing power in its local market, so that the law of one price does not apply. We also assume that firms cannot change their quantity and price at the same time.[4]

---

[3] The assumption of Leontief production technology is consistent with the data and is in line with the literature (Assenza, Delli Gatti, and Grazzini 2015). Moreover, as our explicit aim was to derive the simplest possible ABM that has the features we desire, we relegate all further extensions of the model, such as assumptions on technological progress that change technology coefficients, to further research.

[4] This modeling choice is adapted from Delli Gatti *et al.* (2011), who use a similar price–quantity heuristic and is motivated by empirical surveys of managers' pricing and quantity decisions; see, e.g., Seiichi Kawasaki, John McMillan, and Klaus Zimmermann (1982) and Venkataraman Bhaskar, Stephen Machin, and Gavin C. Reid (1993). Dawid *et al.* (2026, ch. 13 in this volume) also provide a detailed discussion of the price-quantity rules in macroeconomic agent-based models. Moreover, there is empirical evidence from laboratory studies showing subjects use similar price–quantity settings heuristics in monopolistic price–quantity settings (see, e.g., Assenza *et al.* 2015).

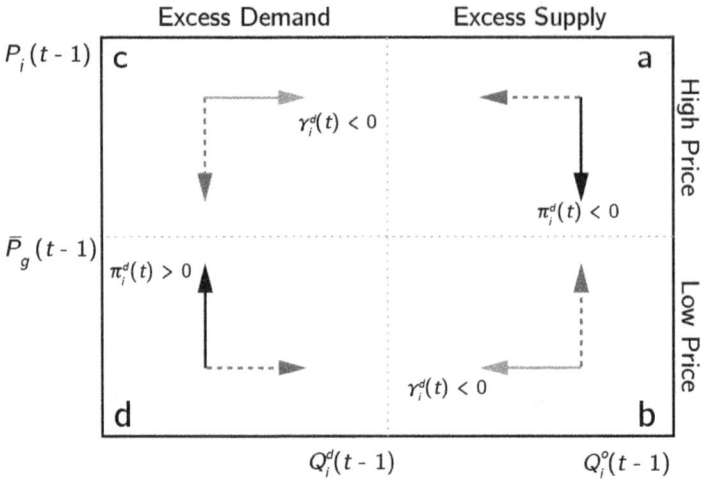

Figure 3. The four scenarios for the firm's quantity and price setting. In scenarios (b) and (c), firms adjust their quantities according to equation 4; in scenarios (a) and (d), firms adjust their prices as in equation 5.

As a reflection of firms' expectation error concerning demand, "excess supply" is defined as the difference between production and sales. Firms form their expectations of market conditions using two indicators that both rely on observed information from the previous period $t - 1$: (1) the level of excess supply, which is the difference between the previous period's supply $Q_i^o(t - 1)$ and realized demand $Q_i^d(t - 1)$; and (2) the deviation of the firm's price $P_i(t - 1)$ from the average price among competitors $\bar{P}_g(t - 1)$ for product g. Each firm considers four possible price–quantity setting scenarios (depicted in fig. 3).

Effectively, the firm-specific growth rate of demand only differs from zero in scenarios (b) and (c):

$$
\gamma_i^d(t) = \begin{cases}
\dfrac{Q_i^d(t-1)}{Q_i^o(t-1)} - 1 & \text{if } Q_i^o(t-1) \leq Q_i^d(t-1) \\
& \text{and } P_i(t-1) \geq \bar{P}_g(t-1) \\
& \text{or if } Q_i^o(t-1) > Q_i^d(t-1) \quad (3) \\
& \text{and } P_i(t-1) < \bar{P}_g(t-1) \\
0 & \text{otherwise.}
\end{cases}
$$

Each firm's supply choice depends on its expectations of the aggregate economy and individual supply/demand versus pricing power. In our model, it is driven by three sources:

- product supply from the previous period $Q_i^o(t-1)$;

- the expected, economy-wide economic growth rate $(\gamma^e(t))$; and

- the realized, firm-specific growth rate of demand from the previous period, $\gamma_i^d(t)$.

$$
Q_i^s(t) = Q_i^o(t-1)(1 + \gamma^e(t))(1 + \gamma_i^d(t)) \qquad (4)
$$

Similar to equation 3, in scenarios (a) and (d) firms adjust their prices in the following manner:

$$
\pi_i^d(t) = \begin{cases}
\dfrac{Q_i^d(t-1)}{Q_i^o(t-1)} - 1 & \text{if } Q_i^o(t-1) \leq Q_i^d(t-1) \\
& \text{and } P_i(t-1) < \bar{P}_g(t-1) \\
& \text{or if } Q_i^o(t-1) > Q_i^d(t-1) \quad (5) \\
& \text{and } P_i(t-1) \geq \bar{P}_g(t-1) \\
0 & \text{otherwise.}
\end{cases}
$$

Thus, depending on the scenario, firms change either prices or quantities (at the same rate) to ensure that supply (in nominal terms) matches expected (nominal) demand. Notice that $Q_i^d(t-1)/Q_i^o(t-1)-1$ is a dimensionless quantity and is equal to the nominal relative change $(P_i(t-1)Q_i^d(t-1))/(P_i(t-1)Q_i^o(t-1))-1$, which can be reduced to $Q_i^d(t-1)/Q_i^o(t-1)-1$.

Inflation Determination

With some aggregation and simplification, firm i's nominal price $P_i(t)$ in our model can be decomposed into three components: (1) expectations of economy-wide inflation $\pi^e(t)$ (aggregate inflation expectation); (2) the cost structure of the firm $\pi_i^c(t)$ (cost-push inflation); and (3) the firm-specific growth rate of demand $\pi_i^d(t)$ (demand-pull inflation):

$$P_i(t) = P_i(t-1) \cdot \underbrace{(1+\pi_i^d(t))}_{\substack{\text{Demand-pull}\\\text{inflation}}} \cdot \underbrace{(1+\pi_i^c(t))}_{\substack{\text{Cost-push}\\\text{inflation}}} \cdot \underbrace{(1+\pi^e(t))}_{\substack{\text{Aggregate}\\\text{inflation expectation}}}$$

$$(6)$$

This specification is broadly in line with business survey evidence in Canada found in Asghar, Fudurich, and Voll (2023). In normal times, firms' input costs are the key driver of their output prices. When there is broader input cost pressure, the pass-through of input costs to output prices could increase. During such times, inflation expectations could also further influence the way a firm sets its prices.

Using CANVAS to Manage the Inflation Challenge

Before introducing CANVAS for policy analysis, we conducted formal evaluation of the out-of-sample forecasting performance of CANVAS in comparison with standard macroeconomic models such as a VAR and a DSGE model (see Hommes *et al.* 2025). The DSGE model is ToTEM, an estimated large-scale multisectoral model that is used for projection and policy analysis at the

Bank of Canada (see Corrigan *et al.* 2021). Relative to both alternative models, CANVAS has richer heterogeneity in both households and firms. On the demand side, CANVAS models households of varying income levels, and persons in the labor force are associated with additional characteristics such as their age, sex, and occupation. On the supply side, each of the nineteen industries is populated by firms that differ in market share, balance sheet conditions, and ownership structures.

CANVAS shows great strength and outperforms the DSGE model in forecasting GDP growth and key components like consumption. This competitive forecasting performance is encouraging, suggesting that a macroeconomic ABM that encompasses rich household and firm heterogeneity where agents interact in incomplete markets has the potential to enrich macroeconomic forecasting.

The learning mechanism helps align theoretical work on behavior with new survey evidence on Canadian firms' pricing behavior observed through the pandemic (Asghar, Fudurich, and Voll 2023). These enhanced features provide us confidence in using CANVAS to help understand the economic impact from the COVID-related lockdown measures within particular industries and specific populations.

CANVAS HAS RICHER FRAMEWORK TO UNDERSTAND INFLATION

Conditional on the state of the economy in 2019Q-4, we use the model to conduct a counterfactual exercise of eight-quarters-ahead out-of-sample projections until 2021Q-4. The light gray dashed line in figure 4 shows inflation data and the dotted line shows the mean model-implied inflation projections under this scenario. In contrast to the benchmark cases from the Bank

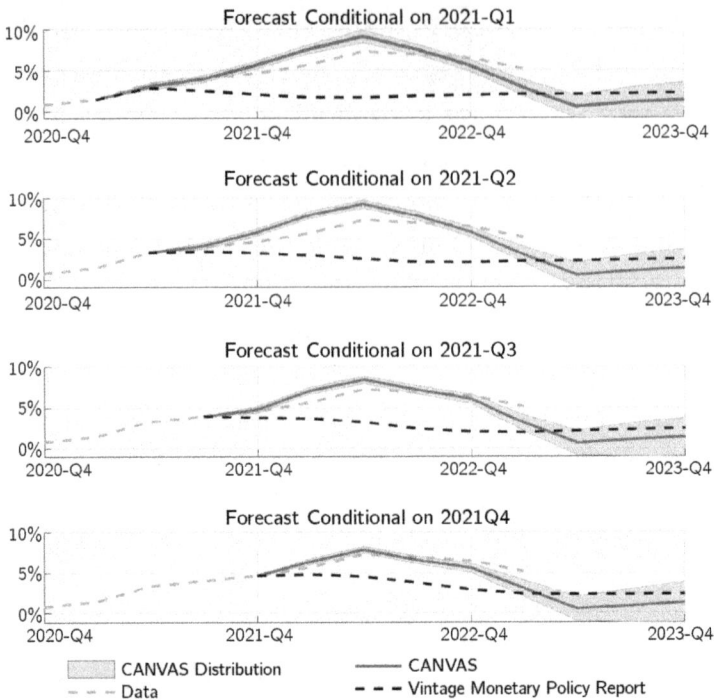

Figure 4. The conditional forecast of the CANVAS model would have pointed to emerging upside risk to an inflation surge. The light gray dashed line shows inflation data and the dark gray dotted line shows the mean forecast based on model simulation. In contrast to the projection from the Bank of Canada's Monetary Policy Reports, conditional forecast with CANVAS in 2021 would have predicted a surge in inflation in 2022. *(Source: Statistics Canada, authors' calculation)*

of Canada's Monetary Policy Reports (dark gray dotted line), CANVAS predicted a surge in inflation in 2021 and 2022.

Key drivers behind inflation surge. On the one hand, the cost-push mechanism characterizing inflation in 2021 is intuitive. An increase in production inputs costs leads firms to increase their prices. Such a mechanism can be found in most macroeconomic models. On the other hand, the demand-pull mechanism characterizing inflation in the postrestrictions period in 2020 is less obvious. Firms increase prices, not

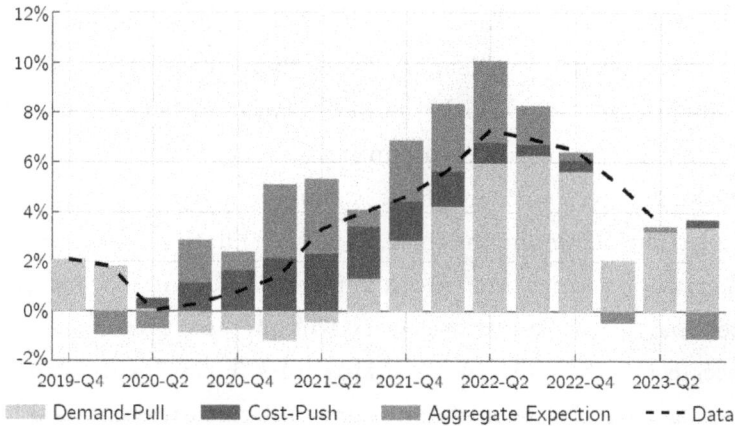

Figure 5. Inflation surge explained by three key drivers *(Source: Statistics Canada, authors' calculation)*

because the *level* of demand is high—in fact, the level of real GDP in 2020Q-3 in Canada is still below its prepandemic value—but because the *growth* of demand is high. This mechanism is specific to the heuristic assumed in equation (6) and has a well-defined implication.

Figure 5 provides a decomposition of the GDP deflator inflation in Canada from a CANVAS simulation. The model suggests that higher costs are the primary driver of inflation during 2020–2021 as shown by the dark gray bars. As soon as inflation surpasses the 2% target, inflation expectations start contributing to higher inflation as illustrated by the top light gray bars. While both expectations and cost considerations remain relevant, since the second half of 2022 when the Canadian economy gradually reopens, demand has become an important driving force to the inflation surge. To this end, the notion that firms could have exploited their market power to raise prices by more than increased input costs could be the mechanism at play, similar to the recent discussion (e.g., Glover, Mustre-del-Río, and von Ende-Becker 2023;

Hansen, Toscani, and Zhou 2023). When consumer demand is less sensitive to prices, or when there is reduced market competition, firms' market power increases which could lead them to set higher prices to gain greater markups. In our model with adaptive expectations, it is clear that there is a rotating contribution among the three drivers of inflation. When costs increase, firms may increase their prices by allowing for greater cost pass-through but not necessarily increase their markups because of the concern for competitors. Firms' expectations play a crucial role in determining inflation dynamics since firms pay greater attention to the inflation data to assess how long the higher cost situation may persist. When inflation perception starts to rise, even current costs decline; since mid-2023, markups may continue to increase as firms are concerned about their future costs. While market demand remains strong, there is more incentive to maintain keep prices high to maintain profit margin.

Granular level of inflation dynamics. Since CANVAS models firm heterogeneity, it helps expand the capacity of macroeconomic analysis with its ability to model dynamics at the firm and industry levels. We construct a counterfactual scenario and illustrate the firm-level inflation distribution within each sector in figure 6.

In this scenario, firms react to the sharp increase in individually perceived demand observed in the postlockdown period by increasing their prices. As a result, demand-pull inflation explains most of the inflationary pressure in Canada over 2020–2021. As global activity rebounds since mid-2021, an increasingly high share of firms expects demand to grow quickly. As they observe higher inflation following the global supply chain disruption, a greater share of firms would also expect increased inflation persistence. For an individual firm

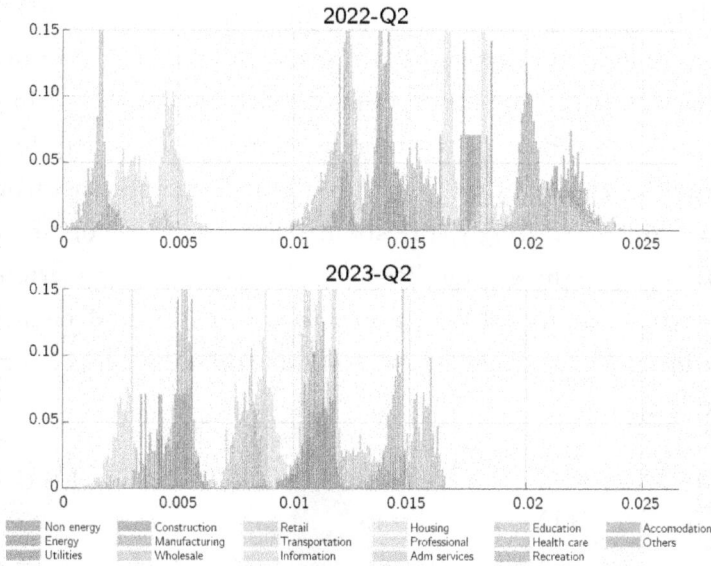

Figure 6. CANVAS provides granular insights beyond traditional models. The figure shows snapshots of simulated sectoral inflation distribution when inflation was at its peak in summer 2022 and a year into the future (in 2023-Q2). The simulation illustrates the heterogeneity across sectors, with the means of inflation of the commodity, manufacturing, and transportation sectors relatively higher than other sectors. There is also notable difference of inflation dispersion across sectors. *(Source: Statistics Canada, authors' calculation)*

that relies on utilizing labor, capital, and intermediate inputs from other suppliers, the rebound of global activities implies more firms would experience rising input costs pressure, which directly affects production costs and price setting.

We show a snapshot of sectoral inflation distribution when inflation was at its peak in summer 2022. In the simulation, the means of inflation of the commodity, manufacturing and transportation sectors are relatively higher than other sectors. Since the commodity sector supplies the critical inputs for the manufacturing goods sector, when there is large and persistent surge in commodity prices, coupled with port strikes that hinder functioning in the transportation sector, sectoral inflation surges and the production and distribution of manufacturing goods is jammed. Since the manufacturing sector supplies nearly all parts of the economy, such disruption would have effects that propagate throughout the Canadian economy. Interestingly, the simulated distribution of firm-level inflation shows large dispersion. This type of information is particularly useful for policymakers to better understand the implications of the oil price movement and its impact on Canada.

MONETARY POLICY DECISION-MAKING IN TIMES OF UNCERTAINTY

The inflation mechanism in CANVAS is closely related to new lab experimental evidence. Isabelle Salle, Yuriy Gorodnichenko, and Olivier Coibion (2024) show that recalled experiences of inflation by individuals affect their forward-looking expectations about inflation. In particular, experience with high inflation leads to more attentive behavior and less central bank credibility, while experience with low inflation leads to more credibility. Furthermore, experience with rising inflation leads to higher inflation expectations. The authors also find that there is asymmetric behavior as individuals tend to overreact to their forecast errors in the case of rising inflation treatment.

Making monetary policy decisions amid such a high degree of uncertainty is a challenge. The baseline economic projection models used in the Bank of Canada's Monetary Policy Report generally assume that inflation expectations are well anchored on the 2% inflation target. However, in July 2022, inflation was climbing and broadening, inflation expectations moved upward, and it became important to consider the upside risk to inflation. If these rising inflation expectations occur while the labor market is tight and wage growth is strong, there is an increased risk that a self-reinforcing wage–price spiral could raise the inflation outlook.

We developed a wage–price spiral scenario in CANVAS to illustrate the uncertainty when inflation was approaching its peak level in July 2022. In this risk scenario, when inflation remains persistently high, more households and firms base their inflation expectations only on the most recent inflation data. As a result, longer-term inflation expectations become de-anchored and stay above the target.

De-anchored inflation expectations lead firms to set prices even higher. Similarly, in response to higher expected inflation, workers bargain for persistently higher wage growth to protect against anticipated losses in purchasing power. The resulting stronger wage growth feeds into production costs and prompts firms to raise prices even further. This process boosts inflation expectations, perpetuating the spiral.

While the likelihood of a wage–price spiral was assessed to be negligible, policymakers are keenly aware of the risks associated with such a spiral. Concerns about a worst-case scenario like this can lead to amplification in the optimal monetary policy response (e.g., Giannoni 2002; Onatski and Stock 2002; Tetlow and von zur Muehlen 2009). To break this vicious circle, monetary policy needs to tighten more than in the base case to create additional excess supply so that long-run inflation expectations remain well-anchored. This is illustrated in figure 7, where in the wage–price spiral scenario, considerably weaker real GDP growth is required to bring inflation down, and even so the decline in inflation is delayed and more gradual than in the base case. This scenario was included in the Bank of Canada's Monetary Policy Report

a. CPI Inflation, Year-Over-Year
Percentage Change

b. Real GDP, Quarterly Growth at
Annual Rates

——— Base-Case Projection   ——— Wage-Price Spiral Scenario

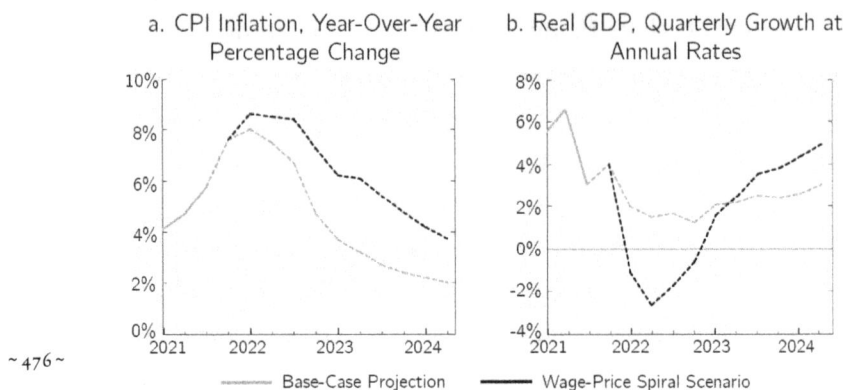

Figure 7. In the wage–price spiral scenario (dark gray dashed line), inflation is higher (left plot) and growth is slower (right plot) than corresponding levels in the base-case projection (light gray dotted line). *(Source: Monetary Policy Report, July 2022, Bank of Canada)*

of July 2022 accompanying the policy announcement at the time.

The Way Forward

Macroeconomic models play an essential role in the conduct of monetary policy, and the Bank of Canada has a long history of using them for projections and policy analysis (Poloz 2017). The models have grown vastly more sophisticated through the years and have many benefits. But while they tend to feature a rich demand side, they have performed less strongly in analyzing supply—a shortcoming that became clear during the COVID-19 pandemic.

For policy models used to support real-world decision-making for an inflation-targeting central bank, it is important to have inflation equations that are effective at explaining and forecasting inflation. Given challenges before and during the pandemic, a better understanding of inflation determination is a focus for research and development by an inflation-targeting central bank. CANVAS helps bridge modeling gaps at the Bank of Canada in three main areas:

- The individual firm's cost structures of integrate critical domestic and global input-output linkages in Canadian production networks. During the nearly thirty years of the "great moderation" era prior to the pandemic, supply shocks (such as energy price fluctuations) typically had temporary effects on inflation. As demand rebounded following the reopening from pandemic lockdowns, the economy experienced a series of large negative supply shocks such as a negative labor supply, the disruption of global supply chains, and the subsequent surge of commodity prices. Traditional models that only focus on demand as the main driver of aggregate imbalances between supply and demand in the economy were insufficient in gauging inflationary pressures to guide policy actions. These experiences triggered a critical need for a more granular, multisector model to help better understand inflation dynamics when the costs of production depend crucially on the economic structure and linkages.

- Inflation expectations depart from rational expectation assumptions. Agents are boundedly rational. In this chapter, we have emphasized adaptive learning of simple AR(1) rules, but the models can easily be extended to switching between heterogeneous forecasting rules. As a result, the persistence of inflation and central bank credibility become endogenous, and inflation expectations could become de-anchored in the absence of an aggressive monetary policy response. The experience of applying CANVAS to help understand the formation and importance of inflation expectations during a high-inflation episode has taught us to be more agile in considering alternative ways to understand how the complex economic system is evolving. Our household and corporate surveys have provided new evidence on how people and firms form expectations during stable inflation time as well as when inflation fluctuates. New lab experimental evidence shows that experience with high inflation leads to higher inflation expectations. Inflation memory could potentially explain the persistence of inflation expectations and the perception gap not explained by the structural models. These are areas future macroeconomic agent-based models could explore to allow us to fill important gaps in our understanding of the inflation psychology.

- While many central banks have emphasized the joint role played by both persistent input cost pressure and excess demand in contributing to an inflation surge, there is little direct role in traditional DSGE models in capturing excess demand and how a central bank uses its instrument and communication in managing it. Similarly, there is little emphasis in acknowledging the central bank's role in providing confidence and managing inflation expectations under enormous uncertainty. Detailed macroeconomic agent-based models that encompass heterogeneity, a realistic supply side of the economy, and behavior into one framework can offer some promising performance in the Canadian experience. Yet, the journey of effectively integrating ABM into our policy toolkit is not smooth. It is complicated by the level of investment policymakers would need to make to fully understand the ins and outs of the model and clearly communicate model uncertainties to the public. When inflation expectations are high and central banks are under tight scrutiny in effectively restoring price stability, knowing how to use the intuitive narrative from the models to connect with a broader audience has become critical. We have resorted to using effective data visualization to get to the story behind model simulations, introducing interactive "what-if" type tools to allow the model insights to be more accessible, and explain the models in plain and simple languages to make sure the key messages get through.

The pandemic presented an excellent opportunity to better understand the dynamics of inflation with the help of an agent-based model. While we have learned a great deal, we are still learning. Our work will not be done without more experiments and greater clarity. However, one conclusion is already clear: Policy scenario analysis using agent-based models based on microeconomic data will become essential tools for central banks and many other policymaking institutions around the world. We hope our work will enhance the collective knowledge of these models, helping us to get a firmer grip on inflation— whatever it may throw at us. ✸

Acknowledgments
We would like to thank Doyne Farmer, John Geanakoplos and participants of the workshop "Complex-System Approaches to Twenty-First Century Challenges: Inequality, Climate Change and New Technologies" for stimulating discussions and helpful feedback. Sebastian Poledna acknowledges funding from the Austrian Academy of Sciences (ÖAW), grant number DATA 2023-23.

REFERENCES

Afrouzi, H., and S. Bhattarai. 2023. *Inflation and GDP Dynamics in Production Networks: A Sufficient Statistics Approach.* Working Paper 31218. National Bureau of Economic Research. https://doi.org/10.3386/w31218.

Asghar, R., J. Fudurich, and J. Voll. 2023. *Firms' Inflation Expectations and Price-Setting Behaviour in Canada: Evidence from a Business Survey.* Staff Analytical Note 2023-3. Bank of Canada. https://publications.gc.ca/pub?id=9.921978&sl=0.

Assenza, T., D. Delli Gatti, and J. Grazzini. 2015. "Emergent Dynamics of a Macroeconomic Agent-Based Model with Capital and Credit." *Journal of Economic Dynamics and Control* 50:5–28. https://doi.org/10.1016/j.jedc.2014.07.001.

Assenza, T., J. Grazzini, C. Hommes, and D. Massaro. 2015. "PQ Strategies in Monopolistic Competition: Some Insights from the Lab." *Journal of Economic Dynamics and Control* 50:62–77. https://doi.org/10.1016/j.jedc.2014.08.017.

Bhaskar, V., S. Machin, and G.C. Reid. 1993. "Price and Quantity Adjustment Over the Business Cycle: Evidence from Survey Data." *Oxford Economic Papers* 45 (2): 257–268. https://doi.org/10.1093/oxfordjournals.oep.a042091.

Borsos, A., A. Carro, A. Glielmo, M. Hinterschweiger, J. Kaszowska-Mojsa, and A. Uluc. 2026. "Agent-Based Modeling at Central Banks: Recent Developments and New Challenges." In *The Economy as an Evolving Complex System IV,* edited by R. M. del Rio-Chanona, M. Pangallo, J. Bednar, E. D. Beinhocker, J. Kaszowska-Mojsa, F. Lafond, P. Mealy, A. Pichler, and J. D. Farmer. Santa Fe, NM: SFI Press.

Brayton, F., E. Mauskopf, D. Reifschneider, P. Tinsley, J. Williams, B. Doyle, and S. Sumner. 1997. "The Role of Expectations in the FRB/US Macroeconomic Model." *Federal Reserve Bulletin,* https://doi.org/10.17016/bulletin.1997.83-4.

Champagne, J., and R. Sekkel. 2018. "Changes in Monetary Regimes and the Identification of Monetary Policy Shocks: Narrative Evidence from Canada." *Journal of Monetary Economics* 99:72–87. https://doi.org/10.1016/j.jmoneco.2018.06.002.

Corrigan, P., H. Desgagnés, J. Dorich, V. Lepetyuk, W. Miyamoto, and Y. Zhang. 2021. *ToTEM III: The Bank of Canada's Main DSGE Model for Projection and Policy Analysis.* Technical Report 119. Bank of Canada. https://publications. gc.ca/pub?id=9.902581&sl=0.

Dawid, H., and D. Delli Gatti. 2018. "Agent-Based Macroeconomics." In *Handbook of Computational Economics,* edited by C. Hommes and B. LeBaron, 4:63–156. Amsterdam, Netherlands: Elsevier. https://doi.org/10.1016/bs.hescom.2018. 02.006.

Dawid, H., D. Delli Gatti, L. E. Fierro, and S. Poledna. 2026. "Implications of Behavioral Rules in Agent-Based Macroeconomics." In *The Economy as an Evolving Complex System IV,* edited by R. M. del Rio-Chanona, M. Pangallo, J. Bednar, E. D. Beinhocker, J. Kaszowska-Mojsa, F. Lafond, P. Mealy, A. Pichler, and J. D. Farmer. Santa Fe, NM: SFI Press.

Delli Gatti, D., S. Desiderio, E. Gaffeo, P. Cirillo, and M. Gallegati. 2011. *Macroeconomics from the Bottom-Up.* 54–57. Milan, Italy: Springer Milan. https://doi.org/10.1007/978-88-470-1971-3.

Giannoni, M. P. 2002. "Does Model Uncertainty Justify Caution? Robust Optimal Monetary Policy in a Forward-Looking Model." *Macroeconomic Dynamics* 6 (1): 111–144. https://doi.org/10.1017/S1365100502027062.

Gigerenzer, G. 2015. *Risk Savvy: How to Make Good Decisions.* London, UK: Allen Lane.

Glover, A., J. Mustre-del-Río, and A. von Ende-Becker. 2023. "How Much Have Record Corporate Profits Contributed to Recent Inflation?" *Federal Reserve Bank of Kansas City Economic Review* 108 (1): 1–13. https://doi.org/10. 18651/ER/v108n1GloverMustredelRiovonEndeBecker.

Hansen, L. P., and T. J. Sargent. 2007. "Recursive Robust Estimation and Control Without Commitment." *Journal of Economic Theory* 136 (1): 1–27. https:// doi.org/10.1016/j.jet.2006.06.010.

Hansen, N.-J., F. Toscani, and J. Zhou. 2023. *Euro Area Inflation After the Pandemic and Energy Shock: Import Prices, Profits and Wages.* Working Paper 2023/131. IMF. https://doi.org/10.5089/9798400245473.001.

Hommes, C. 2021. "Behavioral and Experimental Macroeconomics and Policy Analysis: A Complex Systems Approach." *Journal of Economic Literature* 59 (1): 149–219. https://doi.org/10.1257/jel.20191434.

Hommes, C., M. He, S. Poledna, M. Siqueira, and Y. Zhang. 2025. "CANVAS: A Canadian Behavioral Agent-Based Model for Monetary Policy." Special Issue in Honor of Jasmina Arifovic, *Journal of Economic Dynamics and Control* 172:104986. https://doi.org/10.1016/j.jedc.2024.104986.

Hommes, C., K. Mavromatis, T. Özden, and M. Zhu. 2023. "Behavioral Learning Equilibria in New Keynesian Models." *Quantitative Economics* 14 (4): 1401–1445. https://doi.org/10.3982/QE1533.

Hommes, C., and M. Zhu. 2014. "Behavioral Learning Equilibria." *Journal of Economic Theory* 150:778–814. https://doi.org/10.1016/j.jet.2013.09.002.

Kawasaki, S., J. McMillan, and K. F. Zimmermann. 1982. "Disequilibrium Dynamics: An Empirical Study." *American Economic Review* 72 (5): 992–1004. https://www.jstor.org/stable/1812018.

Onatski, A., and J. H. Stock. 2002. "Robust Monetary Policy Under Model Uncertainty in a Small Model of the US Economy." *Macroeconomic Dynamics* 6 (1): 85–110. https://doi.org/10.1017/S1365100502027050.

Poledna, S., M. G. Miess, C. H. Hommes, and K. Rabitsch. 2023. "Economic Forecasting with an Agent-Based Model." *European Economic Review* 151:104306. https://doi.org/10.1016/j.euroecorev.2022.104306.

Poloz, S. S. 2017. *Models and the Art and Science of Making Monetary Policy.* Bank of Canada. https://www.bankofcanada.ca/2017/01/models-art-science-making-monetary-policy/.

Salle, I., Y. Gorodnichenko, and O. Coibion. 2024. *Lifetime Memories of Inflation: Evidence from Surveys and the Lab.* Working Paper 31996. NBER. https://doi.org/10.3386/w31996.

Sims, C. A. 2003. "Implications of Rational Inattention." *Journal of Monetary Economics* 50 (3): 665–690. https://doi.org/10.1016/S0304-3932(03)00029-1.

Smets, F., J. Tielens, and J. van Hove. 2019. *Pipeline Pressures and Sectoral Inflation Dynamics.* Working Paper 351. National Bank of Belgium. https://doi.org/10.2139/ssrn.3346371.

Tetlow, R. J., and P. von zur Muehlen. 2009. "Robustifying Learnability." *Journal of Economic Dynamics and Control* 33 (2): 296–316. https://doi.org/10.1016/j.jedc.2008.06.005.

MACROECONOMIC FLUCTUATIONS AS EMERGENT BEHAVIOR WHEN AGENTS INTERACT AND ACCUMULATE

Paul Beaudry, University of British Columbia and NBER;

Dana Galizia, Carleton University; and

Franck Portier, University College London and CEPR

Abstract

In most economic environments, agents interact with each other and accumulate diverse stocks such as capital or durable goods. This paper uses an agent-based-type setup to clarify how different forms of interactions translate to different types of aggregate dynamics. In particular, we focus on the aggregate dynamic properties that can emerge when agents' actions act as complements. The resulting dynamics depends on the strength of complementarities, how past accumulation decisions affect current actions, and the extent to which decisions are sluggish. Depending on the relative strength of such forces, we show how hysteresis, two-cycles, or smooth-limit cycles can emerge as aggregate outcomes.

Introduction

In standard models of the business cycle, the agents that populate the model are typically assumed to dislike fluctuations. Notwithstanding this preference for stability, aggregate fluctuations exist, both in the real world, and in the models built to try to understand it. An important question is: Why?

The most common approach to addressing this issue—the one used in most dynamic stochastic general equilibrium

(DSGE) models[1]—is to model the economy as smoothly converging to a stable steady state in the absence of any shocks, with fluctuations only emerging as a result of exogenous shocks to fundamentals. In that sense, in these models there is no fundamental conflict between what the individual agents want and what the endogenous equilibrium forces favor.[2] Rather, fluctuations occur because of external *exogenous* forces acting upon this economy.

In this paper, we discuss an alternative explanation for the conflict between the individual desire for stability and the fact that fluctuations may nonetheless emerge: that the endogenous equilibrium forces themselves may act as a destabilizing force. That is, could macroeconomic fluctuations such as boom–bust episodes be mostly the consequence of interactions between agents, in the sense that a collection of noninteracting one-agent economies would display much less variability?

The phenomenon of there being a qualitative difference between individual-level behavior and aggregate outcomes has been addressed in various scientific fields. As the physicist Philip W. Anderson (1972) wrote, "More is different," meaning that multicomponent physical systems can exhibit macroscopic behavior that cannot be understood from the laws that govern their microscopic parts, a feature known as emergent behavior. Emergent structures can be found in many natural phenomena: hurricanes, complex crystals, and the patterns of sand dunes are well-known examples. Swarm behavior of marching locusts, schooling fish, and flocking birds are famous examples of emergent phenomena in biology.

---

[1] See Finn Kydland and Edward Prescott (1982) for pioneering work, and Frank Smets and Rafael Wouters (2007) for the seminal New-Keynesian incarnation.

[2] Counterexamples are the DSGE with limit cycles of Beaudry, Galizia, and Portier (2015) and (2020).

There is a large literature studying emergent phenomena across many scientific fields. However, while the existence of a two-way feedback between microstructure and macrostructure has been recognized within economics for as long as the field has existed—from the invisible hand of Adam Smith (1776) to the work of Friedrich August von Hayek (1948) and Thomas C. Schelling (2006)—there has been little effort in systematically understanding emergent phenomena in the context of equilibrium aggregate economic fluctuations. This chapter aims to help fill this gap. In particular, we study what we believe is the minimal model that one can write down to generate emergent aggregate outcomes. It features accumulation, interactions between agents, and a simple decision rule. In that framework, we study how the strength of agents' interactions affect aggregate outcomes, and make explicit the conditions under which the aggregate economy exhibits behavior qualitatively similar to what would occur in a one-agent version, and when it instead displays emergent phenomena such as hysteresis and endogenous fluctuations.

Our approach is closely related to the agent-based-model (ABM) literature.[3] Similar to the approach in this chapter, in ABMs individual agents follow heuristic decision rules and interact with each other, and the aggregate properties of the model cannot generally be directly inferred from examining only the individual or micro behavior. However, in contrast to large parts of the ABM literature, within-period interactions are modeled as equilibrium outcomes; that is, the spillover and

---

[3]See Schelling (1971) for one of the very first applications to social sciences, Domenico Delli Gatti *et al.* (2018) for an account of agent-based modeling in economics, and Robert L. Axtell and J. Doyne Farmer (2022) for a recent survey. See also Delli Gatti *et al.* (2008) for an application of ABMs to business fluctuations. In the field of macroeconomics, see also András Borsos *et al.* (2026, ch. 18 in this volume), Herbert Dawid *et al.* (2026, ch. 13 in this volume) and Cars Hommes *et al.* (2026, ch. 14 in this volume).

spill-back effects of one agent's actions on the actions of others is assumed to be played out within the period. Furthermore, the model dynamics will come from accumulation, not from learning (Sargent 1993; Evans and Honkapohja 2001) nor the from dynamic selection of different heuristics via some fitness criterion (De Grauwe and Ji 2019).

The rest of the paper is organized as follows. The first section presents the model without interactions; the next introduces interactions; the third studies the local dynamics; and the fourth section analyses the global dynamics. The final section focuses on the limit cycle case and introduces stochastic shocks.

The Baseline Model Without Interactions

The environment we consider is one with a large number N of agents indexed by i. They live on separate islands and do not interact. Each agent makes effort e_{it}, which accumulates into a capital stock of some kind X_{it} (e.g., physical capital, a durable consumption good, etc.). The accumulation equation is given by

$$X_{it+1} = (1 - \delta)X_{it} + e_{it}, \qquad 0 < \delta < 1. \qquad (1)$$

Suppose initially that there is no interaction between agents and that the decision rule for agent i's effort is given by

$$e_{it} = \max\{\alpha_0 - \alpha_1 X_{it} + \alpha_2 e_{it-1}, 0\}, \qquad (2)$$

where $\alpha_0 > 0$, $-1 < \alpha_1 < 1$, and $0 < \alpha_2 < 1$. In this decision rule, the effect of X_{it} on effort can be negative ($\alpha_1 > 0$) so as to reflect some underlying decreasing returns to capital accumulation, or positive ($\alpha_1 < 0$) so as to reflect that a higher asset level can increase the productivity of effort. Finally, the effect of the previous period's action is positive so as to reflect sluggishness in the response of effort to current conditions, due for example to adjustment costs of some kind.[4] Since any steady state (SS) of this model must be nonzero, and since we will largely be concerned with local dynamics around a particular SS, in practice we will often ignore the non-negativity constraint captured by the max operator in this decision rule wherever doing so is justified.

We will always restrict the analysis to symmetric allocations, so that $e_{it} = e_t$ and $X_{it} = X_t$. Ignoring the non-negativity constraint, the aggregate dynamics of the economy are then given by the linear system:

$$\begin{pmatrix} e_t \\ X_t \end{pmatrix} = \underbrace{\begin{pmatrix} \alpha_2 - \alpha_1 & -\alpha_1(1-\delta) \\ 1 & 1-\delta \end{pmatrix}}_{M_L} \begin{pmatrix} e_{t-1} \\ X_{t-1} \end{pmatrix} + \begin{pmatrix} \alpha_0 \\ 0 \end{pmatrix}. \quad (3)$$

We make the following assumption.

Assumption A1. *The parameters satisfy:*

(i) $\alpha_1 > -(1-\alpha_2)\delta$,

(ii) $\alpha_1 < \left(\sqrt{\alpha_2} - \sqrt{1-\delta}\right)^2$.

Assumption A1(i) is a necessary and sufficient condition for there to exist a SS. In particular, letting $\rho^* \equiv 1 - \alpha_2 + \alpha_1/\delta$, if a SS exists then it is unique and given by $SS_0 = (e^s, X^s)$, where $e^s = \alpha_0/\rho^*$ and $X^s = e^s/\delta$. Clearly this is only a valid SS if $e^s, X^s \geq 0$, which is equivalent to assumption A1(i) holding. Assumption A1(ii), meanwhile, ensures that, in this model with no interaction, individual decisions favor stable monotonic dynamics, as stated formally in the following proposition.

---

[4] This decision rule can be thought of as a heuristic as in an ABM model, or as derived from a microfounded set up. In Beaudry, Galizia, and Portier (2024), we present a large game-theoretical structure from which qualitatively similar decision rules can be derived.

Proposition 1. *Under assumption A1, the eigenvalues of the matrix M_L are both real, positive, and inside the unit circle. Therefore, the unique steady state is stable, and the system converges to it without cycles.*

All proofs are in the appendix. According to proposition 1, the dynamics are extremely simple, with the system converging to SS_0 for any starting values of $X_{it} = X_t$ and $e_{it-1} = e_{t-1}$.

Adding Interactions

Now we allow for bridges between the different islands, so that agents are interacting with each other.

DECISION RULE

We now extend the previous setup to allow for interactions between individuals by having the effort rule be given instead by

$$e_{it} = \max\left\{\alpha_0 - \alpha_1 X_{it} + \alpha_2 e_{it-1} + F\left(e_t\right), 0\right\} \qquad (4)$$

where $e_t = N^{-1}\sum e_{jt}$ is the economy-wide average level of effort, and F is a function that controls the nature of interactions between agents (which we discuss further shortly). The law of motion for X is unchanged (i.e., eq. (1)). As before, since the SSs will be strictly positive and we are largely interested in the local dynamics around a SS, we will ignore the non-negativity constraint captured by the max operator wherever doing so is justified. In a symmetric equilibrium, the system can then be written

$$\begin{cases} e_t - F\left(e_t\right) = \alpha_0 - \alpha_1 X_t + \alpha_2 e_{t-1}, \\ \qquad X_{t+1} = \left(1 - \delta\right) X_t + e_t. \end{cases} \qquad (5)$$

THE INTERACTION FUNCTION

The function F captures the type of interactions that exist between agents. In particular, we assume that agents take the average action in the economy as given, so that, for a given X_{it} and e_{it-1}, equation (4) can be interpreted as the best-response rule of an individual to the average action.

In a fully specified model, interactions between agents could be mediated through prices, market thickness, or externalities, among other things. For example, an increase in aggregate investment could raise the price of investment goods, causing individual agents to want to reduce their own investment. In this case F would be a decreasing function, since an increase in e_t has a negative effect on e_{it}. Alternatively, in the model we consider in Beaudry, Galizia, and Portier (2020), an increase in aggregate investment raises the employment rate, which lowers the risk premium faced by households on their borrowing, causing individual agents to want to increase their own investment. In this case F would be an increasing function.

In order to maintain generality in our analysis here, we will not be specific about exactly where F comes from. However, in order to allow for a cleaner characterization of the model dynamics, we do make a number of assumptions about the basic functional form of F, which are combined here into assumption A2.

Assumption A2. *The interaction function F is given by $F(e) = \rho \widetilde{F}(e)$, where $\widetilde{F}$ belongs to the following family of functions:*

(i) *$\widetilde{F}$ is three times continuously differentiable;*

(ii) *$\widetilde{F}(e^s) = 0$, where e^s is the steady state of the model without interactions;*

(iii) *$\widetilde{F}(0) > -\alpha_0$;*

(iv) $0 < \widetilde{F}'(e) \leq 1$, *with* $\widetilde{F}'(e^s) = 1$;

(v) $\lim_{e \to 0} \widetilde{F}'(e) = \lim_{e \to \infty} \widetilde{F}'(e) = 0$;

(vi) $\widetilde{F}''(e) > 0$ *on* $e < e^s$ *and* $\widetilde{F}''(e) < 0$ *on* $e > e^s$, *with* $\widetilde{F}'''(e^s) < 0$.

The function $\widetilde{F}$ here is a strictly increasing (possibly unbounded) sigmoid function whose slope attains a unique maximum of 1 at $e = e^s$. For $\rho \neq 0$, the interaction function F inherits these properties, except (a) F is decreasing instead of increasing if $\rho < 0$ (i.e., F is an inverted sigmoid), and (b) the maximum value of $|F'(e)|$ is $|\rho|$ instead of 1. Thus, the type of strategic interaction is governed by the sign of ρ (complementarity if $\rho > 0$, substitutability of $\rho < 0$), while the strength of those interactions is assumed to be strongest at $e = e^s$, becoming monotonically weaker as e moves away from e^s in either direction.

The parameter ρ simultaneously controls both the type of strategic interaction (substitutability vs. complementarity), as well as the maximum strength of those interactions. The following assumption imposes the additional condition that if there are complementarities ($\rho > 0$ case) then those complementarities are always weak, in the sense that $F'(e) < 1$ everywhere.

Assumption A3. $\rho < 1$.

PERIOD-T EQUILIBRIUM

In a given period t, the (symmetric) equilibrium is given by $e_t = \alpha_0 - \alpha_1 X_t + \alpha_2 e_{t-1} + F(e_t)$, which is indexed by the history as summarized by $\alpha_t = \alpha_0 - \alpha_1 X_t + \alpha_2 e_{t-1}$.

Proposition 2. *Under assumptions* A1–A3, *there exists a unique period-t equilibrium.*

In figure 1, for the case where $\alpha_t + F(0) > 0$, we illustrate graphically the determination of the period-t equilibrium with, respectively, substitutabilities, strong complementarities ($F'(e) > 1$ for some values of e), and weak complementarities. In particular, we plot the decision rule $e_{it} = \alpha_t + F(e_t)$ for each of these configurations. A symmetric equilibrium occurs at the intersection of this decision rule with the 45-degree line $e_{it} = e_t$. As the figure makes clear, the only case that can possibly result in multiple such equilibria is the case of strong complementarities, but this is ruled out by assumption A3.

Figure 2 shows agent i's best response function for two values of the intercept—that is, for two different histories $\alpha_t = \alpha_0 - \alpha_1 X_t + \alpha_2 e_{t-1}$—so as to make clear its dependence on past effort decisions and on the stock variable X_t. The dynamics of the system are induced by the fact that past effort decisions determine the location of the best response function, which in turn determines the period-t equilibrium level of effort. Thus, today's decision feeds into the determination of next period's intercept.[5]

STEADY STATE(S)

Under assumption A2(ii), the SS of the model without interactions, SS_0, is also a SS of the model with interactions (regardless of the value of ρ). Recalling the definition $\rho^* \equiv 1 - \alpha_2 + \alpha_1/\delta$ from the no-interactions model, which is strictly positive by assumption A1(i), we have the following proposition.

---

[5]Under assumption A3, the period-t equilibrium depicted in figure 2 is stable under a *tâtonnement*-type adjustment process. This stability property is not the focus of the current paper. Instead we are interested in the explicit dynamics induced by the system (eq. (5)).

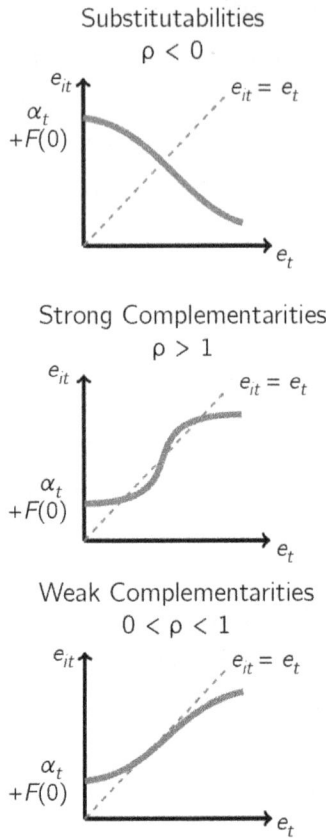

Figure 1. Period-t Equilibrium. This figure plots agent i's policy rule (equation (4)): $e_{it} = {}_t + F(e_t)$, with ${}_t = {}_0 - {}_1 X_t + {}_2 e_{t-1}$. Period-$t$ equilibria are at intersections of the policy rule and the 45-degree line $e_{it} = e_t$.

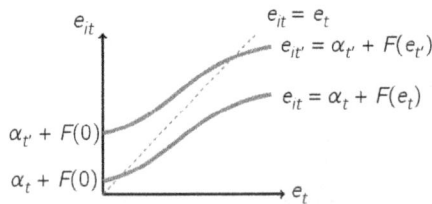

Figure 2. Policy Rule for Two Different Histories with Weak Complementarities. This figure plots the best response rule (equation (4)): $e_{it} = \alpha_t + F(e_t)$, with $\alpha_t = \alpha_0 - \alpha_1 X_{it} + \alpha_2 e_{it-1}$. The intercepts α_t and $\alpha_{t'}$ correspond to two different histories of the model.

Proposition 3. *Suppose assumptions* A1–A3 *hold. Then:*

(i) *If $\rho \leq \rho^*$—which is necessarily the case if either $\rho \leq 0$ (substitutabilities) or $\alpha_1 \geq \delta\alpha_2$—then SS_0 is the unique steady state.*

(ii) *If $\rho > \rho^*$—which is only possible if both $\rho > 0$ (complementarities) and $\alpha_1 < \delta\alpha_2$—then there are two steady states in addition to SS_0 (one with $e > e^s$ and one with $e < e^s$).*

Proposition 3 establishes that, for the case of substitutabilities ($\rho < 0$), the unique SS of the system is the no-interaction one SS_0, and that, in order for multiple SS_s to emerge (if they ever do), there must be sufficiently strong complementarities.

Our goal is now to use this framework to examine how the dynamics of this system are affected by the properties of the interaction function F, and especially what conditions on F will give rise to emergent equilibria.

Local Dynamics

We now consider the dynamics given by equation (5). The first-order approximation of the system around SS_0 is given by

$$\begin{pmatrix} e_t \\ X_t \end{pmatrix} = \underbrace{\begin{pmatrix} \frac{\alpha_2 - \alpha_1}{1-\rho} & -\frac{\alpha_1(1-\delta)}{1-\rho} \\ 1 & 1-\delta \end{pmatrix}}_{M} \begin{pmatrix} e_{t-1} \\ X_{t-1} \end{pmatrix}$$

$$+ \begin{pmatrix} \left(1 - \frac{\alpha_2-\alpha_1}{1-\rho}\right) e^s + \frac{\alpha_1(1-\delta)}{1-\rho} X^s \\ 0 \end{pmatrix}. \tag{6}$$

In order to understand the dynamics of the system, it is informative to first look at local dynamics in the neighborhood of this SS, the qualitative nature of which is determined by the eigenvalues of the matrix M. In particular, SS_0 can be locally stable or unstable,[6] with either a pair of complex conjugate eigenvalues, or a pair of real eigenvalues each of which could be either positive or negative. When both eigenvalues are (weakly) positive we will say that the system exhibits monotonic dynamics since, given any initial position at date o, for t large enough e_t and X_t will be monotone functions of t. On the other hand, if the system has any negative or complex eigenvalues we will say that it exhibits cycles or oscillations.

Recall that $\rho = F'(e^s)$ parameterizes the type and degree of the strategic interactions at SS_0. When there are no strategic interactions ($\rho = 0$), we have $M = M_L$, and therefore, as noted above, the eigenvalues of M will be real, positive and smaller than one, so that we have stable monotonic dynamics.

We next consider what happens as substitutability ($\rho < 0$) is introduced and allowed to become arbitrarily large. The following proposition characterizes the outcome.

Proposition 4. *Under assumptions* A1 *and* A2, *as* ρ *decreases from* 0 *to* $-\infty$:

(i) *Both eigenvalues of* M *remain inside the complex unit circle, so that the system remains locally stable.*

(ii) *The eigenvalues are never negative.*

(iii) *If either* $\alpha_2 \leq 1 - \delta$, $\alpha_2 \leq \alpha_1$, *or* $\alpha_1 \leq 0$, *then the eigenvalues are always real and positive, so that there are never cycles.*

---

[6] Note that this system is purely backward-looking, and therefore the presence of any unstable eigenvalues implies that the system will generally diverge from the SS. With some abuse of terminology, for the sake of brevity we use "unstable" to refer to any situation with at least one unstable eigenvalue, even if there is also a stable eigenvalue (i.e., the SS is properly a saddle).

(iv) *If all of the conditions from part (iii) fail to hold, then there are values ρ_1, ρ_2 with $-\infty < \rho_1 < \rho_2 < 0$ such that the eigenvalues are complex if $\rho \in (\rho_1, \rho_2)$, and real and positive for all other values of $\rho \leq 0$.*

Proposition 4 indicates that, when the actions of others play the role of strategic substitutes with one's own action, this favors stability of the system. Strategic substitutability is typical of Walrasian settings, where the price mechanism mediates all strategic interactions, which we believe explains why Walrasian models typically feature stable dynamics.

Next, the following proposition characterizes how the presence of weak strategic complementarities ($\rho \in (0, 1)$) affects the local stability of the system.

Proposition 5. *Under assumptions A1 and A2, as ρ increases from zero towards one, the system will at some point necessarily become locally unstable, and will thereafter remain unstable as ρ continues to increase.*

Proposition 5 establishes that, if ρ is large enough (but smaller than 1), the system will be unstable, even though it would be stable in the absence of complementarity. Intuitively, the instability arises here because agents have an incentive to bunch their actions together. As a result, any initial individual desire to choose a high/low effort level because the date-t state vector (X_t, e_{t-1}) is different from the SS is then amplified in equilibrium. If the amplification is strong enough (ρ sufficiently large), e_t will be such that the resulting date-$(t + 1)$ state vector (X_{t+1}, e_t) will be even more extreme than (X_t, e_{t-1}); that is, the system will be unstable. The dynamics of the economy as a whole is therefore qualitatively different from the dynamics of an individual taken in isolation.

Global Dynamics

A change of the local stability properties when a parameter varies, such as the ones described above for the case where ρ increases toward 1, is referred to as a bifurcation in the theory of dynamical systems. For a system with the structure we consider here, there are essentially three possible types of bifurcations that can occur as ρ changes. The first, known as a flip bifurcation, occurs as an eigenvalue passes through -1. The second, known as a Hopf bifurcation (more accurately referred to as a Neimark–Sacker bifurcation in discrete-time systems), occurs when a pair of complex conjugate eigenvalues crosses the unit circle together. The last, which we refer to as a $+1$ bifurcation, occurs as an eigenvalue passes through $+1$.[7] We already established in proposition 5 that one of these bifurcations must occur as ρ increases toward 1. We now formally state the conditions that determine which particular type of bifurcation will occur.

Proposition 6. *Under assumptions* A1 *and* A2, *the loss of stability that necessarily occurs as ρ increases from zero towards one will take the form of:*

(i) *a flip bifurcation if* $\alpha_2 < \frac{\alpha_1}{(2-\delta)^2}$,

(ii) *a Hopf bifurcation if* $\frac{\alpha_1}{(2-\delta)^2} < \alpha_2 < \frac{\alpha_1}{\delta^2}$, *or*

(iii) *a $+1$ bifurcation if* $\alpha_2 > \frac{\alpha_1}{\delta^2}$.

The conditions in proposition 6 can be understood as follows. Recall that α_2 parameterizes the degree of sluggishness, while α_1 parameterizes the degree of decreasing returns to

---

[7]Bifurcations involving an eigenvalue passing through $+1$ are typically subcategorized according to their nonlinear dynamics (e.g., as a fold, cusp, pitchfork, etc.). Since we have not yet determined which of these bifurcations will occur in this environment, for the moment we will use the more generic term "$+1$ bifurcation." We return to this issue shortly.

capital (or increasing returns when $\alpha_1 < 0$). When α_2 is sufficiently large relative to α_1, the sluggishness channel dominates, and therefore the "momentum" generated by the dynamics is positive, in the sense that a higher e_t favors a higher e_{t+1}, which in turn favors a higher e_{t+2}, etc. As a result, instability in this case emerges via a positive eigenvalue, that is, a +1 bifurcation.

On the other hand, when α_2 is sufficiently small relative to α_1, the decreasing returns channel dominates, causing "negative" momentum: a higher e_t favors a *lower* e_{t+1}, which in turn favors a *higher* e_{t+2}, etc. This produces "jagged" oscillations in e, wherein e flips back and forth over e^s each period. In this case, instability emerges via a negative eigenvalue, that is, a flip bifurcation.

Finally, if α_2 is of an intermediate magnitude relative to α_1, the sluggishness and decreasing returns channels play off each other, causing more complicated dynamics. Specifically, the sluggishness channel dominates for a while, which favors a sustained rise or fall in e, until the accumulated effect of this rise or fall (as embodied in X) becomes extreme enough that the decreasing returns channel takes over, causing e to then reverse direction. The end result is "smooth" oscillations in e, wherein e may spend more than one consecutive period on one side of e^s before switching to the other. In this case, instability emerges via a pair of complex eigenvalues, that is, a Hopf bifurcation.

Letting $\bar{\rho} \in (0, 1)$ denote the value of ρ at which the bifurcation occurs, for ρ in a neighborhood of $\bar{\rho}$ the qualitative dynamics of the linearized system are straightforward to characterize: for $\rho < \bar{\rho}$, the system converges to the no-interaction SS, SS_0, as $t \to 0$, while for $\rho > \bar{\rho}$ the system becomes unbounded as $t \to 0$. Convergence or divergence happens either monotonically (+1 bifurcation), with smooth oscillations (Hopf bifurcation), or with jagged oscillations (flip bifurcation).

When considering the full nonlinear version of the model, however, the local qualitative dynamics near the bifurcation point $\bar{\rho}$ will generally be more complicated and will depend on the nature of the nonlinearities embodied in the function F. The following proposition establishes some key properties of the nonlinear dynamic system near the bifurcation point.

Proposition 7. *Under assumptions A1 and A2, as ρ increases through $\bar{\rho}$:*

(i) *If the system experiences a flip bifurcation, then the steady state remains unique, and a two-period cycle[8] either emerges or disappears.*

(ii) *If the system experiences a Hopf bifurcation, then the steady state remains unique, and an additional isolated closed invariant curve[9] either emerges or disappears.*

(iii) *If the system experiences a $+1$ bifurcation, then it is a pitchfork bifurcation, that is, characterized by the emergence of two additional steady states, one with $e < e^s$ and another with $e > e^s$.*

---

[8] Letting $x \equiv (X, e)$, and writing the evolution of the equilibrium system as $x_t = G(x_{t-1})$, a k-period cycle is defined as a fixed point of the system $x' = G^k(x)$ that is not also a fixed point of the system $x' = G^j(x)$ for some $j = 1, \ldots, k-1$. Here, G^k denotes the k-th iterate of G, i.e., $G^1(x) = G(x)$, $G^k = G(G^{k-1}(x))$ for $k \geq 2$.

[9] A set $A \subset \mathbb{R}^n$ is a *curve* if it can be written as the image $A = f(I)$, where f is a continuous function and I is some interval in $\mathbb{R}$. A curve A is *closed* if we can, in this definition, use $I = [a, b]$ and a function f that satisfies $f(a) = f(b)$ (so that the curve starts and ends at the same place). A set A is an *invariant set* of the map G (i.e., of the system $x_t = G(x_{t-1})$) if $x \in A$ implies $G(x) \in A$. A closed invariant curve A of the map G is *isolated* if there is a sufficiently narrow band around A that excludes every part of every other closed invariant curve of G.

Several points are worth emphasizing/clarifying about proposition 7. First, note that extra SSs appear only in the $+1$ bifurcation case. Second, the isolated closed invariant curve referred to in the Hopf case can be visualized as a closed loop surrounding SS_0 in (X, e)-space. Third, if the system begins on the two-cycle in the flip case or on the closed invariant curve in the Hopf case, then the system will remain on that cycle/curve—and therefore experience continual fluctuations—indefinitely. Fourth, while the cycle in the flip case is necessarily two periods long, the Hopf case can in principle be associated with cycles of any length longer than two. Thus, we view the latter case as more likely to be relevant for studying macroeconomic fluctuations.

~ 499 ~

In each case, the bifurcation is associated with a qualitative change in the global dynamics of the system, as captured by the emergence or disappearance of a special set of points:[10] the two-cycle in the flip case, the closed invariant curve in the Hopf case, and the two additional SSs in the $+1$ (pitchfork) case. These sets are illustrated in figure 3. For the flip case shown in panel (a), the two-cycle is indicated by the points LC_H and LC_L. Starting from any point on this two-cycle, the system jumps back and forth each period, exhibiting jagged oscillations.

For the Hopf case shown in panel (b) of figure 3, the closed invariant curve is indicated by the solid black loop surrounding SS_0. If the system starts from any point on this loop, it will traverse it indefinitely, exhibiting smooth cycles (i.e., it takes more than two periods for the system to complete a full revolution around the loop).

---

[10]Note that "disappearance" here would involve a situation where the special set first emerges at some $\rho < \bar{\rho}$ (i.e., while the system was still stable), and then subsequently disappears as ρ increases through $\bar{\rho}$.

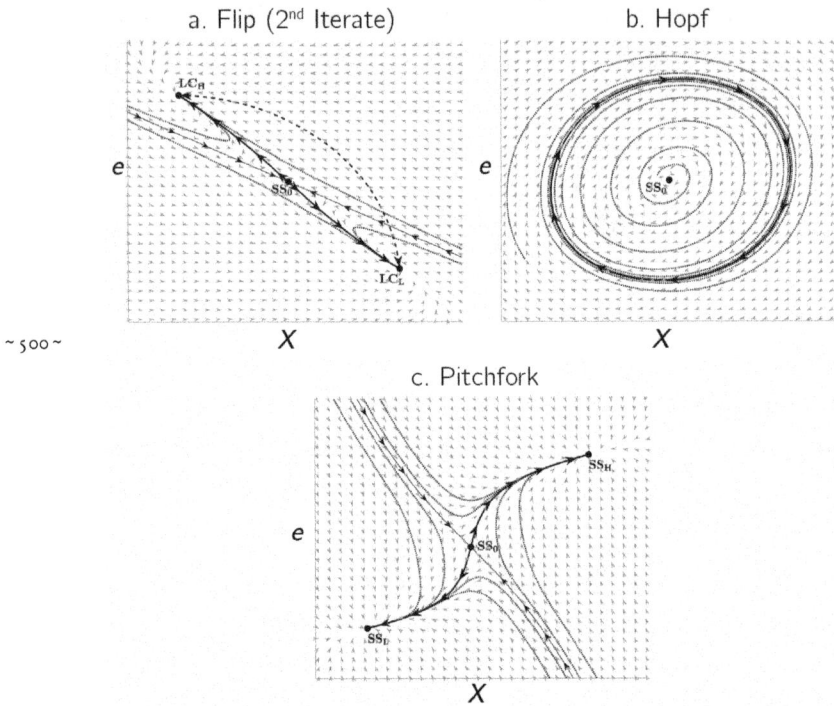

Figure 3. Global Dynamics After the Bifurcation. This figure shows phase diagrams illustrating the dynamics following a bifurcation.

Finally, for the $+1$ (i.e., pitchfork) case in panel (c), we simply have two additional SSs, a "low" one $SS_L \equiv (X_L, e_L)$, and a "high" one $SS_H \equiv (X_H, e_H)$.

Proposition 7 established that a particular special set of points either emerges or disappears as ρ increases through $\bar{\rho}$, and we have described what the dynamics look like when the system begins exactly from a point in one of these sets. This immediately raises two additional questions: (1) Does the special set of points actually emerge, or does it disappear? (2) What do the dynamics look like when the system does *not* begin exactly from one of the special points? The following proposition answers these questions.

Proposition 8. *Suppose assumptions* A1 *and* A2 *hold. Then the special sets described in proposition 7 always emerge as ρ increases through $\bar{\rho}$, and further in all cases these special sets attract all nearby trajectories.*

For the flip case, proposition 8 states that the two-cycle ((LC_L, LC_H) in figure 3(a)) in fact emerges as ρ increases through $\bar{\rho}$, and further that this two-cycle attracts all nearby trajectories. Similarly, for the Hopf case, the isolated closed invariant curve (the solid black loop in fig. 3(b)) emerges as ρ increases through $\bar{\rho}$, and it again attracts all nearby trajectories. Thus, the two-cycle in the flip case and the isolated closed invariant curve in the Hopf case are both attractive limit cycles. Finally, in the pitchfork bifurcation case, proposition 8 states that the two additional SSs that emerge as ρ increases through $\bar{\rho}$ ((SS_L, SS_H) in fig. 3(c)) are each locally stable.

The post-bifurcation dynamics for the Hopf case is illustrated by the various arrows in panel (b) of figure 3. The light gray arrows indicate the local direction of motion conditional on currently being at that point in the phase space, with examples of typical trajectories given by the dotted curves. We see from the diagram that, regardless of whether a trajectory begins inside or outside the solid black limit cycle, as long as it does not start exactly at SS_0 it will always converge to that limit cycle over time.

Panel (a) of figure 3 similarly illustrate the dynamics for the flip case. Note that, because of the violent nature of the oscillations in the flip case, a standard phase diagram would be difficult to make sense of. Thus, letting $x_t \equiv (e_t, X_t)$ and writing our system as $x_t = G(x_{t-1})$, panel (a) instead presents the phase diagram for the second iterate of the system, that, for the map $G^2(x) \equiv G(G(x))$, which gives the position of the system two periods later when it is currently at x. The limit cycle points LC_L and LC_H are stable SSs of G^2, so that nearby trajectories converge to them. In terms of the original map G, this convergence corresponds to jagged oscillations in both e and X, each typically jumping back and forth across its SS_0 value each period, with the system converging eventually to the two-cycle. Note that, because SS_0 is actually a saddle in this case (when the first eigenvalue becomes unstable, the other is still stable), it has both a stable arm (i.e., a saddle path) and an unstable arm. The saddle path is indicated by the thin solid curve with arrows pointing toward SS_0, indicating that if the system begins exactly on this path it will eventually converge to SS_0. The unstable arm, meanwhile, is indicated by the thick solid curve with arrows pointing away from SS_0 that connect it to the limit cycle points.

Finally, panel (c) of figure 3 indicates the dynamics following a pitchfork bifurcation. In this case, the two new SS_s, SS_L and SS_H, are locally stable, attracting all nearby trajectories. As in the flip case, SS_0 is a saddle, so that it has both a saddle path (thin solid curve with arrows pointing towards SS_0) and an unstable arm (thick solid curve with arrows pointing away from SS_0 that connects it to the two new SS_s). If the system begins exactly on the saddle path, it will converge to SS_0. Trajectories beginning anywhere in the region northeast of this saddle path will converge to SS_H, while those beginning anywhere southwest of it will converge to SS_L. In this way, we see that a small perturbation from SS_0 can permanently affect the system, that is, the dynamics display hysteresis.

To understand these dynamics, recall that, regardless of the type of bifurcation, when SS_0 becomes unstable, it does so because the complementarities—which cause agents to want to bunch their actions together—have become strong enough locally that any initial desire by agents to deviate from the SS action e^s gets successively larger over time through the dynamic mechanisms of accumulation and sluggishness. However, because the slope of F becomes flatter as e moves away from e^s in either direction, the complementarities become weaker as the system diverges from SS_0. Thus, as the dynamics push the economy further and further from the unstable SS_0, the complementarity that was responsible for that instability eventually weakens to the point that the system no longer favors divergence. Put differently, the weaker complementarities away from SS_0 reduce the bunching incentive enough so that the private incentives—that is, those present in the no-interactions baseline model, which favor stability—take over and contain the system, preventing it from exploding.

As a final point, recall that a necessary condition for a $+1$ bifurcation is $\alpha_2 > \alpha_1/\delta^2$. Given that, quantitatively, δ is typically taken to be quite small (e.g., on the order of 0.05 or less for a quarterly model), this is only possible for some $\alpha_2 < 1$ if capital exhibits increasing returns ($\alpha_1 < 0$) or at most very weak decreasing returns ($\alpha_1 \in (0, \delta^2)$). Thus, the general insight we take away is that limit cycles (whether smooth or jagged) are likely to emerge in our setting if capital exhibits even modest decreasing returns, and if complementarities are (i) strong enough to create instability near the SS, but (ii) tend to die out as the economy moves away from the SS. In an economic environment, it is quite reasonable to expect that complementarities are likely to die out if activity gets very large. For example, if investment demand gets sufficiently large, some resource constraints are likely to become binding, causing incentives akin to strategic substitutes to emerge instead of complementarities. Similarly, physical constraints, such as non-negativity restrictions on investment or capital, would prevent the economic system from diverging to zero. Such forces will generally favor the emergence of attractive limit cycles in the presence of complementarities.

Stochastic Limit Cycles

AN EXAMPLE

~504~   As noted in the introduction, standard business-cycle models typically generate fluctuations entirely via exogenous shocks, in the sense that in the absence of such shocks the system would converge monotonically to a SS. We have shown, however, that in dynamic environments with accumulation, strategic complementarities between agents' actions can readily create limit cycles, which presents a potential alternative explanation for business-cycle fluctuations. We showed that this can arise even when individual-level behavior favors stability, in that the system would converge to the SS in the absence of agent interactions. Moreover, in our environment, the complementarities are modest, in that they imply interaction elasticities less than one. However, the cyclical dynamics generated by any deterministic limit cycle model will be far too regular to match the patterns observed in macroeconomic data. For example, with some abuse of notation let e now indicate the deviation of effort from its no-interaction SS level e^s, and suppose that F is (locally) given by the simple cubic function $F(e_t) = \rho e_t - \xi e_t^3$, with $\xi > 0$. In figure 4 we plot an example of the path of e_t along a limit cycle in this setup. The observed pattern is not a perfect sine wave, but it is clearly far more regular than observed real-world fluctuations, which tend to be quite irregular.

Thus, limit cycles cannot, on their own, explain observed macroeconomic fluctuations. As we show now, however, if

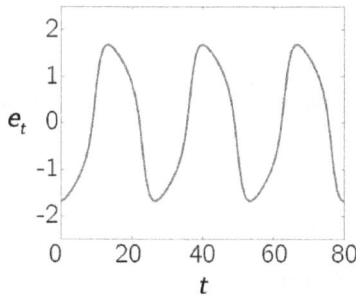

Figure 4. Deterministic Simulation of e_t. Figure shows deterministic simulation of equations (1)–(7) when $F(I) = \rho I - \xi I^3$, $\alpha_1 = 0.04$, $\alpha_2 = 0.6$, $\delta = 0.05$, $\rho = 0.65$, and $\xi = 0.1$.

we take a deterministic limit cycle model, and then (as in standard business-cycle models) add an exogenous shock process to it, then the resulting "stochastic limit cycle" model is capable of producing much more realistic-looking fluctuations. Importantly, as we will see, it only takes a relatively small shock process—that is, one that has only a small effect on the magnitude of the fluctuations—to create a high degree of irregularity.[11] In this sense, the endogenous limit cycle forces are still doing the bulk of the heavy lifting in generating the economic fluctuations. This is in stark contrast to standard business-cycle models, where it is the shocks that drive all fluctuations.

Suppose we modify our agents' decision rule (eq. (4)) to

$$e_{it} = \alpha_0 - \alpha_1 X_{it} + \alpha_2 e_{it-1} + F(e_t) + \mu_t \qquad (7)$$

where μ_t is a stationary exogenous stochastic force.[12] Assuming

[11] An alternative route, which we do not pursue here, would be to consider chaotic dynamics, which would create entirely endogenous fluctuations that also exhibit irregularity.

[12] Note that the form of equation (7) reflects the fact that the stochastic driving force is common across agents, as well as our focus on symmetric equilibria.

the parameters are such that in the absence of any shocks the model would feature a limit cycle, how does the addition of this exogenous stochastic force affect equilibrium dynamics? Importantly, the presence of μ_t does not simply add noise around an otherwise-deterministic cycle, as for example would be the case in panel (a) of figure 5, which is what we get if we simply add the AR(1) exogenous process $\mu_t = \gamma \mu_{t-1} + \sigma \epsilon_t$ to the value of e_t taken from the deterministic cycle in figure 4 (where ϵ_t is i.i.d. $N(0,1)$, and we set $\gamma = 0.9$ and $\sigma = 0.15$). Rather, the exogenous forces in equation (7) not only produce random "amplitude" shifts that temporarily perturb the system from the limit cycle, but also create random "phase" shifts that accelerate or delay the cycle itself. Furthermore, even though μ_t is stationary, and while the amplitude displacement caused by a shock is temporary, the *phase* displacement will actually have a permanent component. Thus, following a one-time shock, the system will eventually converge back to the limit cycle, but will be permanently either ahead or behind where it would have been in the absence of the shock.

To illustrate this effect, consider the stochastic model given by equations (1) and (7) with the same parameter values used in figure 4 and panel (a) of figure 5. Figure 6 shows the effect, beginning from a point on the deterministic limit cycle, of a one-time one-standard-deviation temporary shock to μ. The light gray line shows the path that would have occurred in absence of the one-time shock, while the dark gray shows the perturbed path. As the figure makes clear, the perturbed path eventually returns to the limit cycle, but is permanently out of sync with (behind, in this case) the unperturbed path; that is, the temporary shock induces a random-walk in the phase

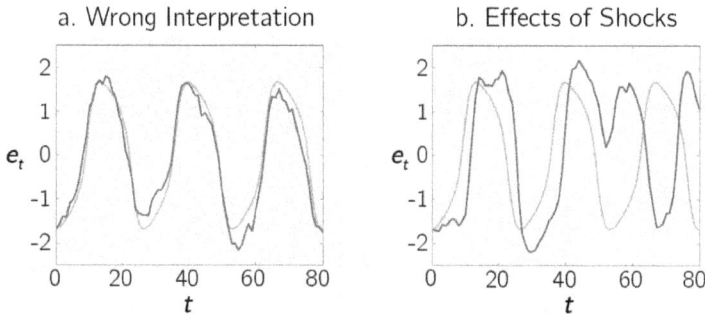

a. Wrong Interpretation   b. Effects of Shocks

Figure 5. Stochastic Limit Cycles. Light gray lines in both panels of the figure plot the deterministic limit cycle from figure 4. The dark gray line in panel (a) is obtained by adding a simulation of the AR(1) stochastic process $\mu_t = \gamma\mu_{t-1} + \sigma\epsilon_t$ to the deterministic limit cycle (i.e., to the light gray line). The dark gray line in panel (b) is obtain by simulating equations (1) and (7) with the same parameters as in figure 4 and the same values of the shock as panel (a). The parameters of the AR(1) process are $\gamma = 0.9$ and $\sigma = 0.15$, where ϵ_t is i.i.d. $N(0, 1)$.

of the cycle.[13] Panel (b) of figure 5, which presents a full stochastic simulation of this system beginning from a point on the deterministic cycle, illustrates the cumulative effect of these random phase shifts: after ten periods, the simulated path of effort (dark gray) is noticeably out of sync with the deterministic path (light gray), and after around 50 periods the phase has shifted by about half of a cycle, so that effort in the stochastic simulation is at a trough at the same time that the deterministic simulation is at its peak. Instead of observing a smooth, regular cycle as we would in the deterministic model, the stochastic model clearly generates data that look more like that observed in macroeconomic variables; namely, boom-and-bust cycles that have both stochastic amplitudes and durations. We also emphasize that this is the case despite the fact that the shock is relatively small, in that it only has a small impact

---

[13]See section S2 of Marco Pangallo (2025) for an analytical approach to understanding this permanent-phase-shift phenomenon.

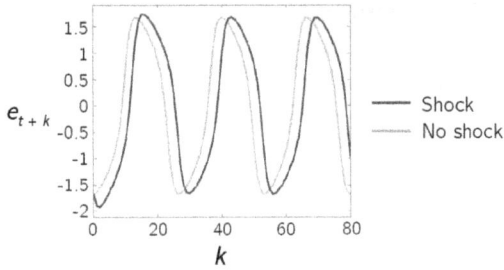

Figure 6. Permanent Phase Shift. Figure shows impact, beginning from a point on the deterministic cycle, of a one-time one-standard-deviation shock to μ at date $t+1$ (i.e., we set $\epsilon_{t+1} = \sigma/\sqrt{1-\gamma^2}$), and $\epsilon_\tau = 0$ for all $\tau \neq t+1$). The light gray line shows the path of e in the unperturbed system, while the dark gray line shows the path of the perturbed system. Parameters are the same as those reported in the notes to figure 5.

on the amplitude of the cycles produced by the model: the amplitude of the fluctuations in the stochastic model in panel (b) of figure 5 (dark gray) are only a little bit different from those in the purely deterministic model (light gray).

PREDICTABILITY IN STOCHASTIC LIMIT CYCLE MODELS

A common criticism of many early models of macroeconomic fluctuations featuring limit cycles (e.g., those of Hicks 1950; Goodwin 1951, etc.), was that they implied an unreasonably high degree of predictability.[14] We wish to emphasize that, once one introduces even small stochastic forces into the environment, this criticism no longer necessarily applies. In particular, as noted above, data simulated from a stochastic limit cycle model can have properties similar to those found in actual economic data, including significantly

---

[14]Another common criticism is that these dynamics would imply the existence of arbitrage forces that would tend to erase any cycles. This proves not to be the case in the equilibrium models with rationally optimizing agents, as illustrated in Beaudry, Galizia, and Portier (2024).

irregular business cycles. A similar property obtains for the *predictability* of the model. In fact, the property is even more stark for predictability than it is for regularity. A deterministic cycle is both perfectly regular and perfectly predictable arbitrarily far into the future. However, while introducing a small amount of stochastic variability in the model would tend to make the cycle only slightly less regular, the degree of unpredictability (as measured, for example, by the forecast-error variance) arbitrarily far into the future will jump up discontinuously.[15] The reason for this discontinuity follows directly from the fact that the "phase" component of the system follows a random walk. As is well known, as long as the variance of the innovations to a random walk process is positive (even if it is arbitrarily small), the forecast-error variance of the process becomes infinite as the forecast horizon increases. In our context, this means that, as you go far enough into the future, no matter how small the variance of the shock process is, as long as it is not actually zero the phase of the cycle becomes completely unpredictable; that is, no matter what the current state is, many periods into the future the system is just as likely to be at the bottom of the cycle as it is at the top.

To illustrate this, figure 7 plots, using the same parameters as in figures 5–6, a forecast of e_{t+k} as of date t (i.e., $\mathbb{E}_t[e_{t+k}]$, dark gray solid line), as well as a 66% conditional confidence interval for e_{t+k} (dark gray dotted lines).[16] For comparison, we also plot the unconditional mean $\mathbb{E}[e_{t+k}]$ (light gray solid line) and an unconditional 66% confidence interval for

---

[15] Formally, let $V_t(\sigma_\mu^2) \equiv \lim_{k \to \infty} Var_t(e_{t+k})$ denote the limit, when $Var(\mu_t) = \sigma_\mu^2$, of the forecast-error variance of e, conditional on information available as of date t, as the forecast horizon k extends infinitely far into the future. Since the deterministic model is perfectly predictable we have $V_t(0) = 0$. However, it can be verified that $\lim_{\sigma_\mu^2 \to 0} V_t(\sigma_\mu^2) > 0$.

[16] The initial state of the system at date t is given by $X_t = -8.00$, $e_{t-1} = -1.68$, and $\mu_t = 0$.

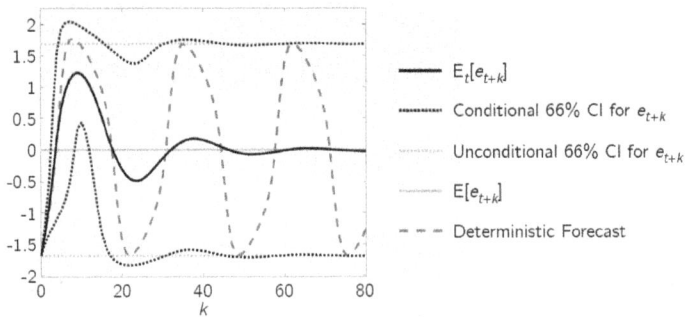

Figure 7. Unpredictability. Parameters are the same as those reported in the notes to figure 5. Unconditional and conditional moments obtained from 100,000 simulations of the model. Unconditional forecasts are computed as time averages, conditional forecasts as ensemble averages.

e_{t+k} (light gray dotted lines), as well as a "deterministic" forecast as of date t (light gray dashed line) obtained by shutting down the stochastic process completely. Several important points emerge from the figure. First, unlike the deterministic forecast, which oscillates indefinitely (reflecting the existence of the limit cycle), the conditional forecast converges to the unconditional forecast as the forecast horizon k increases. Thus, in this stochastic environment even a risk-neutral investor could not earn above-average long-run returns by betting on e. Second, as the conditional confidence interval indicates, even in the shorter run when $\mathbb{E}_t[e_{t+k}]$ differs significantly from $\mathbb{E}[e_t]$, there is quite a bit of forecast uncertainty in this example. Indeed, not only does the mean conditional forecast converge to the unconditional one as k increases, the conditional confidence interval also converges to its unconditional counterpart. To the extent that this pattern holds for all conditional moments, this would imply that the stochastic process is in fact ergodic.

Conclusion

We have presented a reduced form model of accumulation and interactions between agents. When complementarities between agents are increased (but stay weak in the sense that there are no multiple period-t equilibria), nontrivial dynamic properties can emerge, depending on two mechanisms: the impact (positive or negative) of past accumulation on current actions and the strength of sluggishness in actions. Depending on the relative strength of those three forces, hysteresis, two-cycles, or smooth limit cycles equilibria can emerge. Therefore, complementarities in actions and accumulation allow for the emergence of fluctuations in otherwise stable and non-oscillating economies. We view our results as showing that minimal nonlinearity and weak complementarities are enough to generate rich dynamics, and in particular endogenous fluctuations. 🖝

Acknowledgments

The support of the Economic and Social Research Council (ESRC) is gratefully acknowledged, via the Rebuilding Macroeconomics Network (Grant Ref: ES/R00787X/1).

REFERENCES

Anderson, P. W. 1972. "More Is Different." *Science* 177 (4047): 393–396. https://doi.org/10.1126/science.177.4047.393.

Axtell, R. L., and J. D. Farmer. 2022. *Agent-Based Modeling in Economics and Finance: Past, Present, and Future.* INET Oxford Working Papers 2022-10. https://doi.org/10.1257/jel.20221319.

Beaudry, P., D. Galizia, and F. Portier. 2015. *Reviving the Limit Cycle View of Macroeconomic Fluctuations.* Technical report. NBER Working Paper 21241. https://doi.org/10.3386/w21241.

———. 2020. "Putting the Cycle Back into Business Cycle Analysis." *American Economic Review* 110 (1): 1–47. https://doi.org/10.1257/aer.20190789.

———. 2024. *How Do Strategic Complementarity and Substitutability Shape Equilibrium Dynamics?* NBER Working Paper 32661. https://doi.org/10. 3386/w32661.

Borsos, A., A. Carro, A. Glielmo, M. Hinterschweiger, J. Kaszowska-Mojsa, and A. Uluc. 2026. "Agent-Based Modeling at Central Banks: Recent Developments and New Challenges." In *The Economy as an Evolving Complex System IV,* edited by R. M. del Rio-Chanona, M. Pangallo, J. Bednar, E. D. Beinhocker, J. Kaszowska-Mojsa, F. Lafond, P. Mealy, A. Pichler, and J. D. Farmer. Santa Fe, NM: SFI Press.

Dawid, H., D. Delli Gatti, L. E. Fierro, and S. Poledna. 2026. "Implications of Behavioral Rules in Agent-Based Macroeconomics." In *The Economy as an Evolving Complex System IV,* edited by R. M. del Rio-Chanona, M. Pangallo, J. Bednar, E. D. Beinhocker, J. Kaszowska-Mojsa, F. Lafond, P. Mealy, A. Pichler, and J. D. Farmer. Santa Fe, NM: SFI Press.

De Grauwe, P., and Y. Ji. 2019. *Behavioural Macroeconomics: Theory and Policy.* Oxford, UK: Oxford University Press.

Delli Gatti, D., G. Fagiolo, M. Gallegati, M. Richiardi, and A. Russo, eds. 2018. *Agent-Based Models in Economics.* Cambridge Books. Cambridge University Press. https://ideas.repec.org/b/cup/cbooks/9781108414999.html.

Delli Gatti, D., E. Gaffeo, M. Gallegati, G. Giulioni, and A. Palestrini. 2008. *Emergent Macroeconomics: An Agent-Based Approach to Business Fluctuations.* New Economic Windows. Milan, Italy: Springer. https://doi.org/10.1007/978-88-470-0725-3.

Evans, G. W., and S. Honkapohja. 2001. *Learning and Expectations in Macroeconomics.* Princeton, NJ: Princeton University Press.

Goodwin, R. M. 1951. "The Nonlinear Accelerator and the Persistence of Business Cycles." *Econometrica* 19 (1): 1–17. https://www.jstor.org/stable/1907905.

Guckenheimer, J., and P. Holmes. 2002. *Nonlinear Oscillations, Dynamical Systems, and Bifurcations of Vector Fields.* New York, NY: Springer.

Hayek, F. A. 1948. *Individualism and Economic Order*. Chicago, IL: University of Chicago Press.

Hicks, J. R. 1950. *A Contribution to the Theory of the Trade Cycle*. Oxford, UK: Clarendon Press.

Hommes, C., S. Kozicki, S. Poledna, and Y. Zhang. 2026. "How an Agent-Based Model Can Support Monetary Policy in a Complex Evolving Economy." In *The Economy as an Evolving Complex System IV*, edited by R. M. del Rio-Chanona, M. Pangallo, J. Bednar, E. D. Beinhocker, J. Kaszowska-Mojsa, F. Lafond, P. Mealy, A. Pichler, and J. D. Farmer. Santa Fe, NM: SFI Press.

Kydland, F. E., and E. C. Prescott. 1982. "Time to Build and Aggregate Fluctuations." *Econometrica* 50 (6): 1345–1370. https://doi.org/10.2307/1913386.

Pangallo, M. 2025. "Synchronization of Endogenous Business Cycles." *Journal of Economic Behavior & Organization* 229:106827. https://doi.org/10.1016/j.jebo.2024.106827.

Sargent, Thomas J. 1993. *Bounded Rationality in Macroeconomics: The Arne Ryde Memorial Lectures*. Oxford, UK: Oxford University Press.

Schelling, T. C. 1971. "Dynamic Models of Segregation." *Journal of Mathematical Sociology* 1 (2): 143–186. https://doi.org/10.1080/0022250X.1971.9989794.

———. 2006. *Micromotives and Macrobehavior*. New York, NY: W. W. Norton.

Smets, F., and R. Wouters. 2007. "Shocks and Frictions in US Business Cycles: A Bayesian DSGE Approach." *The American Economic Review* 97 (3): 586–606. https://doi.org/10.1257/aer.97.3.586.

Smith, A. 1776. *An Inquiry into the Nature and Causes of the Wealth of Nations*. London, UK: W. Strahan and T. Cadell.

Wan, Y.-H. 1978. "Computations of the Stability Condition for the Hopf Bifurcation of Diffeomorphisms on R^2." *SIAM Journal of Applied Mathematics* 34:167–175. https://doi.org/10.1137/0134013.

Wikan, A. 2013. *Discrete Dynamical Systems with an Introduction to Discrete Optimization Problems*. London, UK: Bookboon.

Appendix

EIGENVALUE ANALYSIS

Many of the proofs of the propositions below involve analyzing the eigenvalues of the linearized system. Specifically, for the general case where we allow for interactions, the relevant eigenvalues are those of the matrix M from equation (6) (the no-interaction matrix M_L in (3) can be obtained by simply setting $\rho = 0$ in M). The eigenvalues of M are the roots of the quadratic equation

$$Q(\lambda) = \lambda^2 - T\lambda + D = 0, \tag{A.1}$$

where T is the trace of M (and also the sum of its eigenvalues) and D is the determinant of M (and also the product of its eigenvalues). The two eigenvalues are

$$\lambda, \bar{\lambda} = \frac{T}{2} \pm \sqrt{\left(\frac{T}{2}\right)^2 - D} \tag{A.2}$$

where

$$T = \frac{\alpha_2 - \alpha_1}{1 - \rho} + (1 - \delta) \tag{A.3}$$

and

$$D = \frac{\alpha_2(1 - \delta)}{1 - \rho}. \tag{A.4}$$

Figure A.1 illustrates the possible eigenvalue configurations in (T, D)-space. There are three lines and a parabola in the figure. The line passing through B and C corresponds to $D = 1$, the line through A and B to the equation $Q(-1) = 0$ ($\Leftrightarrow D = -T - 1$), and the line through A and C to the equation $Q(1) = 0$ ($\Leftrightarrow D = T - 1$). On the perimeter of triangle $\overparen{ABC}$, at least one eigenvalue has a modulus of 1. When (T, D) is inside $\overparen{ABC}$, both eigenvalues of the system will be inside the unit circle, while when (T, D) is outside $\overparen{ABC}$ at least one eigenvalue is outside the unit

Figure A.1. Possible Configurations of the Local Dynamics. This figure shows the plane (T, D), where T is the trace and D the determinant of the matrix M. The point E_0 correspond to the model without demand externalites. It is placed in the zone where the two eigenvalues are real, positive and smaller than 1.

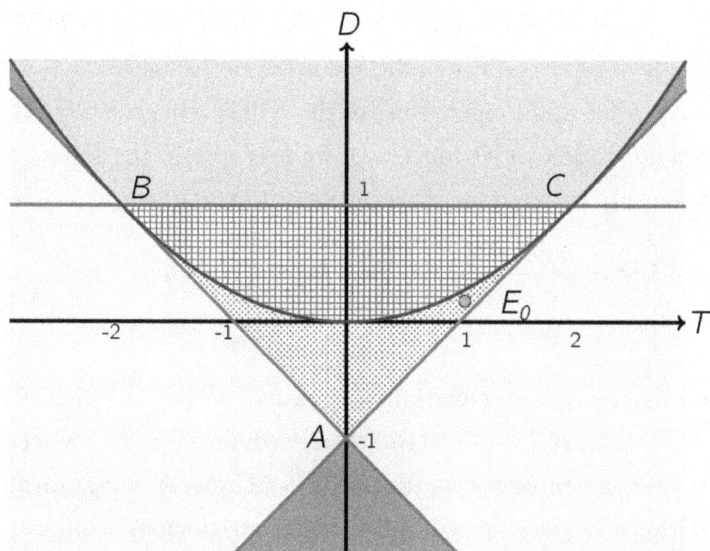

Unstable (saddle),real eigenvalues	Stable, complex eigenvalues
Unstable, complex eigenvalues	Stable, real eigenvalues
Unstable, real eigenvalues	

circle. Whether the eigenvalues are real or complex depends on the sign of the discriminant of the equation $Q(\lambda) = 0$, which is given by $\Delta = T^2/4 - D$. The parabola in figure A.1 corresponds to the equation $\Delta = 0$ (i.e., $D = T^2/4$). The eigenvalues are complex conjugates above the parabola, and real below it. Finally, when the eigenvalues are real, they are both positive in the northeast quadrant of the plane, both negative in the northwest quadrant, and of opposite signs in the lower half of the plane.

Under assumption A1, proposition 1 shows that, when $\rho = 0$, SS_0 corresponds to some point in the region of figure A.1 that is below the parabola (the eigenvalues are real), in the north-east quadrant (the eigenvalues are both positive), and to the left of the line through A and C (both eigenvalues are less than one). An example of such a point is indicated by E_0 in the figure.

As ρ varies, the eigenvalues of the system will generally vary. From equations (A.3) and (A.4), we may obtain the following relationship between the trace and determinant of M:

$$
\begin{cases}
D = \dfrac{\alpha_2(1 - \delta)}{\alpha_2 - \alpha_1}\, [T - (1 - \delta)] & \text{, if } \alpha_1 \neq \alpha_2 \\
T = 1 - \delta & \text{, if } \alpha_1 = \alpha_2
\end{cases}
\qquad (\text{A.5})
$$

Note that (A.5) is the equation of a line in (T, D)-space (where the line is vertical at $T = 1 - \delta$ for the case where $\alpha_1 = \alpha_2$). When ρ varies, (T, D) moves along the line (A.5). Under Assumption 15, equations (A.3) and (A.4) imply that this movement is continuous in ρ, and is such that D strictly increases with ρ, i.e., the vertical component of (T, D) moves in an upward direction as ρ increases.

It is clear from equations (A.3) and (A.4) that

$$
\lim_{\rho \to -\infty} (T, D) = E_{-\infty} \equiv (1 - \delta, 0).
$$

This point is shown in figure A.2. Note that $E_{-\infty}$ corresponds to the case of one null and one positive/stable eigenvalue. Thus, for the limiting case $\rho \to -\infty$, the SS is locally stable and exhibits monotonic dynamics (no cycles).

Depending on the values of α_1 and α_2, there are a number of possible configurations for E_0, some examples of which are illustrated by the gray dots in figure A.2. Starting from such a point, as ρ decreases, (T, D) moves continuously along the line (A.5) (e.g., one of the lines denoted by (a) to (e)), starting from E_0 and moving in a downward direction, approaching $E_{-\infty}$ as $\rho \to -\infty$.

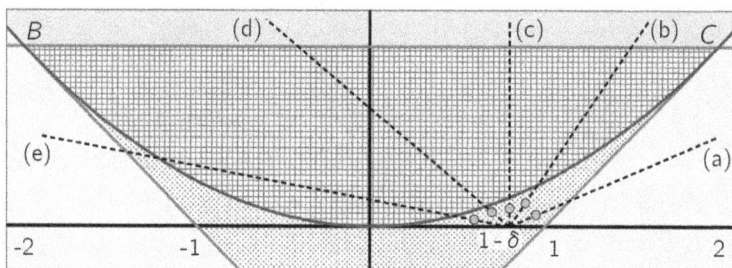

Figure A.2. Varying ρ Between $-\infty$ and 1. This figure shows the plane (T, D), where T is the trace and D the determinant of matrix M. The gray dots correspond to possible values of (T, D) for the model without interactions ($\rho = 0$).

Conversely, as ρ increases, so does D, so that (T, D) moves continuously upward along the line (A.5), and from equation (A.4) we see that $\lim_{\rho \uparrow 1} D = \infty$, so that this movement continues without bound as ρ increases towards 1. Thus, as ρ increases from 0 toward 1, (T, D) will necessarily exit the triangle $\widehat{ABC}$ at some value of $\rho < 1$. Examples of such movements are shown in figure A.2, where starting from one of the gray dots (which indicate possible values of E_0), (T, D) traverses one of the lines (a) to (e) in an upward direction. As (T, D) crosses the perimeter of triangle $\widehat{ABC}$, the SS becomes unstable. In the case of half-line (a), a real eigenvalue increases through 1 as (T, D) crosses the right edge of the triangle. In the case of half-lines (b), (c) and (d), the eigenvalues will first become complex conjugates while still stable, and then subsequently exit the unit circle together as (T, D) crosses the top edge of the triangle. Finally, in the case of half-line (e), a real eigenvalue decreases through -1 as (T, D) crosses the left edge of the triangle.[17]

---

[17]Note that, for cases like (e) where (T, D) departs triangle $\widehat{ABC}$ through the left edge, there will always be an intermediate range of ρ where the eigenvalues are temporarily complex/stable, but transition back to being real/stable before stability is lost. For cases where (T, D) departs triangle

PROOFS OF PROPOSITIONS

Proof of Proposition 1

The trace and determinant of matrix M_L are given respectively by (A.3) and (A.4) for the case where $\rho = 0$, that is, by

$$T = \alpha_2 - \alpha_1 + 1 - \delta, \qquad D = \alpha_2(1 - \delta). \qquad (\text{A.6})$$

We are interested in establishing conditions under which (T, D) is in the zone of Figure ?? corresponding to eigenvalues that are both real, positive, and stable. This zone is defined by the following four conditions:

$$D < T^2/4, \qquad (\text{A.7})$$
$$T > 0, \qquad (\text{A.8})$$
$$D > 0, \qquad (\text{A.9})$$
$$D > T - 1. \qquad (\text{A.10})$$

Condition (A.7) is the condition for real eigenvalues, (A.8) and (A.9) together are conditions for two positive eigenvalues, and lastly, given the other three conditions, (A.10) is the condition for two stable eigenvalues.

Condition (A.7) holds if either $\alpha_1 > (\sqrt{1 - \delta} + \sqrt{\alpha_2})^2$ or $\alpha_1 < (\sqrt{1 - \delta} - \sqrt{\alpha_2})^2$. Note that the latter is precisely assumption A1(ii). Next, condition (A.8) is equivalent to $\alpha_1 < 1 - \delta + \alpha_2$. This is violated under the first possible condition ensuring (A.7), but is implied by the second. Thus, assumption A1(ii) is necessary and sufficient to have both (A.7) and (A.8) hold. Next, from (A.6) and the basic parameter restrictions, condition (A.9) clearly holds. Finally, condition (A.10) is equivalent to assumption A1(i). Thus, assumption A1

$\overline{ABC}$ through the right edge, there may or may not be such an intermediate range of complex/stable eigenvalues.

is necessary and sufficient to ensure that both eigenvalues are real, positive, and stable. □

PROOF OF PROPOSITION 2

The period-t symmetric equilibrium is given by the level of effort e_t that solves

$$e_t = \max\left\{\alpha_t + F(e_t), 0\right\}, \tag{A.11}$$

where $\alpha_t = \alpha_0 - \alpha_1 X_t + \alpha_2 e_{t-1}$ is a given history. The left-hand side of equation (A.11) is strictly increasing, with slope equal to one. Suppose $\alpha_t + F(0) \leq 0$. Then clearly $e_t = 0$ is an equilibrium. Furthermore, since $F'(e) \leq \rho < 1$ by assumptions A2–A3, the expression $\alpha_t + F(e_t)$ remains below the 45-degree line for $e_t > 0$. Thus, $e_t = 0$ is the unique equilibrium for this case.

Suppose instead $\alpha_t + F(0) > 0$. In this case, for e_t sufficiently small $\alpha_t + F(e_t)$ will be above the 45-degree line. Further, since $F'(e) \leq \rho < 1$, $\alpha_t + F(e_t)$ must eventually cross the 45-degree line at some $e_t = e_t^* > 0$, and then remain below the 45-degree line thereafter. Thus, $e_t = e_t^*$ is the unique equilibrium for this case. □

PROOF OF PROPOSITION 3

First, recall that assumption A1 implies $\rho^* > 0$. Further, clearly $\rho^* < 1$ if and only if $\alpha_1 < \delta\alpha_2$. Next, combining the two equations in (5), a SS value of e is any solution to

$$\rho^* e = \alpha_0 + F(e) \equiv G(e). \tag{A.12}$$

If $\rho = 0$, this condition becomes

$$\rho^* e = \alpha_0.$$

which is uniquely satisfied by $e = e^s$.

Thus, suppose $\rho \neq 0$. The left-hand side of (A.12) is a straight line emanating from the origin with slope ρ^*. That slope is positive and smaller than one if $\alpha_1 < \delta\alpha_2$, and larger than one otherwise. Meanwhile, the shape of $G(e)$ (the right-hand side of (A.12)) is determined by the properties of F. Note that the vertical intercept $G(0) = \alpha_0 + F(0)$ is strictly positive by assumption A2(iii), so that G always begins above the ρ^*e line, and by assumptions A2(iv), A2(vi), and A3, we have $G'(e) = F'(e) \leq G'(e^s) = \rho < 1$, where the weak inequality is strict for $e \neq e^s$.

If $\rho < 0$ (see panel (a) in fig. A.3(c)), then G is strictly decreasing, and will therefore intersect the ρ^*e line exactly once, so that the SS is unique. Similarly, if $0 < \rho \leq \rho^*$, then G is increasing but flatter than the ρ^*e line (except possibly exactly at $e = e^s$), so that again the two intersect exactly once, and therefore there is a unique SS. If $\alpha_1 \geq \delta\alpha_2$, then $\rho^* \geq 1 > \rho \geq G'(e)$ (fig. A.3(b)), so that we are necessarily in this case, which completes the proof of part (i) of the proposition.

If $\alpha_1 < \delta\alpha_2$, meanwhile, we have $\rho^* < 1$, so that it is possible to have $G'(e) > \rho^*$. As already established, if $\rho \leq \rho^*$ (fig. A.3(c)) then this does not happen. If, however, $\rho > \rho^*$ (fig. A.3(d)), then since $G(e^s) = e^s$ and $G'(e^s) = \rho$, it follows that G intersects the ρ^*e line at e^s from below. Since, as already noted, G is necessarily above the ρ^*e line for e sufficiently small, this implies the existence of another intersection—and therefore another SS—at a value $e < e^s$.

Meanwhile, by assumption A2(v), G must eventually become flatter than the ρ^*e line, implying the existence of a third intersection/SS at a value $e > e^s$. Finally, by assumption A2(vi), the slope of G monotonically decreases as e moves away from e^s in either direction, and therefore there can only be one additional intersection/SS below e^s and one above e^s, making exactly three SSs total. $\square$

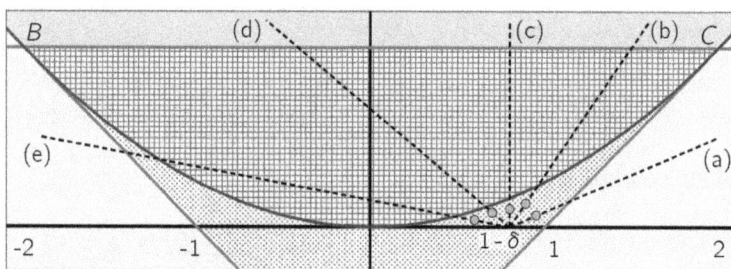

Figure A.3. Steady State(s). This figure depicts the solutions to the equation $\left(1 + \frac{\alpha_1}{\delta} - \alpha_2\right) e = \alpha_0 + F(e)$ that is satisfied at a SS of the model.

PROOF OF PROPOSITION 4

As discussed in the section on "Eigenvalue Analysis," as ρ decreases from o, (T, D) moves continuously along the line (A.5) starting from the no-interaction point E_0 (see fig. A.1) and approaching $E_{-\infty} \equiv (1 - \delta, 0)$ as $\rho \to -\infty$. Since E and $E_{-\infty}$ are both in the intersection of triangle $\widehat{ABC}$ (the region featuring two stable eigenvalues) with the upper-right quadrant of the (T, D) plane (the region where both eigenvalues have positive real parts), and since that intersection is convex, it follows that (T, D) must remain in this "both stable with positive real parts" region for all values of $\rho \leq 0$. This confirms parts (i) and (ii) of the proposition.

Next, in addition to the above properties, E_0 and $E_{-\infty}$ both also lie below the parabola $D = T^2/4$, and therefore for $\rho \leq 0$ sufficiently large or sufficiently small, the eigenvalues must also be real and positive. If $\alpha_1 \geq \alpha_2$, then line (A.5) is vertical or negatively sloped, in which case the entire line segment connecting E and $E_{-\infty}$ must also lie below the parabola, and therefore the eigenvalues are always real and positive for $\rho \leq 0$.

Suppose instead that $\alpha_1 < \alpha_2$. Then there is a part of line (A.5) lying above the $D = T^2/4$ parabola if and only if there are values of T such that $\psi(T) < 0$, where

$$\psi(T) \equiv \frac{1}{4}T^2 - \frac{\alpha_2\,(1-\delta)}{\alpha_2 - \alpha_1}T + \frac{\alpha_2\,(1-\delta)^2}{\alpha_2 - \alpha_1}.$$

$\psi(T)$ is a convex parabola with roots

$$T_1 \equiv 2\frac{(1-\delta)\,\sqrt{\alpha_2}}{\sqrt{\alpha_2}+\sqrt{\alpha_1}} \quad \text{and} \quad T_2 \equiv 2\frac{(1-\delta)\,\sqrt{\alpha_2}}{\sqrt{\alpha_2}-\sqrt{\alpha_1}},$$

so that $\psi(T) < 0$ if and only if T_1, T_2 are real and $T_1 < T < T_2$. If $\alpha_1 < 0$ then the roots are complex, while if $\alpha_1 = 0$ then $T_1 = T_2$ and therefore we cannot have $T_1 < T < T_2$. Thus, if $\alpha_1 \leq 0$, then no part of the line (A.5) can lie above the parabola, and therefore (T, D) must remain real (and positive) as for all $\rho \leq 0$.

Suppose then that $0 < \alpha_1 < \alpha_2$, so that T_1, T_2 are real with $T_1 < T_2$. Using (A.3), the condition $T_1 < T < T_2$ can be equivalently expressed as $\rho \in (\rho_1, \rho_2)$, where

$$\rho_1 \equiv 1 - \frac{\left(\sqrt{\alpha_2}+\sqrt{\alpha_1}\right)^2}{1-\delta}, \qquad \rho_2 \equiv 1 - \frac{\left(\sqrt{\alpha_2}-\sqrt{\alpha_1}\right)^2}{1-\delta}.$$

Suppose $\alpha_2 \leq 1-\delta$. Then by assumption A1(ii), $(\sqrt{\alpha_1}+\sqrt{\alpha_2})^2 < 1-\delta$, in which case $\rho_1 > 0$. Thus, the part of line (A.5) lying above the parabola corresponds to values of $\rho > 0$, and thus for $\rho \leq 0$ the line lies below the parabola, so that again (T, D) remains real for all $\rho \leq 0$. This completes the proof of part (iii).

Finally, suppose again that $0 < \alpha_1 < \alpha_2$, but now $\alpha_2 > 1-\delta$. Then by assumption A1(ii), $(\sqrt{\alpha_1}-\sqrt{\alpha_2})^2 > 1-\delta$, in which case $\rho_2 < 0$. Thus, we have $-\infty < \rho_1 < \rho_2 < 0$, and the part of line (A.5) corresponding to $\rho \in (\rho_1, \rho_2)$ lies above the parabola, so that the eigenvalues are complex for this range of negative values of ρ. This completes the proof of part (iv). □

PROOF OF PROPOSITIONS 5 AND 6

As discussed in "Eigenvalue Analysis," as ρ increases from 0 to 1, (T, D) moves upward along the line (A.5) from point E_0, eventually exiting triangle $\overline{ABC}$ at some point $\rho = \bar{\rho}$. We have a flip bifurcation (eigenvalue equal to -1) if this exit occurs via the left edge of the triangle, a Hopf bifurcation (pair of complex eigenvalues with modulus 1) if it occurs via the top edge, and a $+1$ bifurcation (eigenvalue equal to 1) if it occurs via the right edge.

A flip bifurcation thus occurs if and only if line (A.5) is negatively sloped, with $D < 1$ at the point on the line where $T = -2$ (see fig. A.2). The first condition is met if and only if $\alpha_2 < \alpha_1$. Assuming this is the case, the second condition occurs if and only if

$$\frac{\alpha_2(1-\delta)}{\alpha_2 - \alpha_1}[-2-(1-\delta)] < 1,$$

which can be simplified to the condition given in part *i* of the proposition. Further, since $2 - \delta > 1$, this latter condition on its own actually implies the condition $\alpha_2 < \alpha_1$ for a negative slope, so that the condition given in part (i) is, on its own, necessary and sufficient for a flip bifurcation.

Next, a $+1$ bifurcation occurs if and only if line (A.5) is positively sloped, with $D < 1$ at the point on the line where $T = 2$. The first condition is met if and only if $\alpha_2 > \alpha_1$. Assuming this is the case, the second condition occurs if and only if

$$\frac{\alpha_2(1-\delta)}{\alpha_2 - \alpha_1}[2-(1-\delta)] < 1,$$

which can be simplified to the condition from part (iii) of the proposition. Since $\delta < 1$, that condition on its own implies the condition $\alpha_2 > \alpha_1$ for a positive slope, and therefore again the condition given in part (iii) is, on its own, necessary and sufficient for a $+1$ bifurcation.

Finally, a Hopf bifurcation occurs if and only if either (a) line (A.5) is negatively sloped, with $D > 1$ at the point on the line where $T = -2$; (b) line (A.5) is positively sloped, with $D > 1$

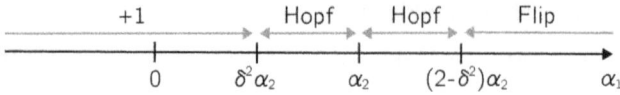

Figure A.4. Bifurcations on the α_1 line. For given δ and α_2, we show on the α_1 line the conditions for $+1$, Hopf and flip bifurcations.

at the point on the line where $T = 2$; or (c) line (A.5) is vertical. Given the analysis for the flip and $+1$ cases, (a) occurs if and only if $\frac{\alpha_1}{(2-\delta)^2} < \alpha_2 < \alpha_1$, (b) occurs if and only if $\alpha_1 < \alpha_2 < \frac{\alpha_1}{\delta^2}$, while (c) occurs if and only if $\alpha_1 = \alpha_2$. It is straightforward to see that one of these cases must occur if and only if the condition given in part (ii) of the proposition holds. □

PROOF OF PROPOSITION 7

First, note that the two-cycle referenced in the flip case and the closed invariant curve referenced in the Hopf case are well established general properties of such bifurcations (see, e.g., Guckenheimer and Holmes 2002) whenever the system is locally non-linear (as assumption A2(vi) ensures in this case). It thus remains only to verify the claims about SSs.

Based on the conditions established in proposition 6, figure A.4 shows, for given α_2 and δ, the values of α_1 corresponding to flip, Hopf, and $+1$ bifurcations.

A necessary condition for the existence of multiple SSs is $\alpha_1 < \delta\alpha_2$ (see proposition 3), a necessary condition for which is $\alpha_1 < \alpha_2$. Let us therefore restrict to the $(-\infty, \alpha_2)$ half-line in figure A.4. In this case, only Hopf and $+1$ bifurcations are possible, which confirms that the SS remains unique for the flip case.

Next, the Hopf bifurcation portion of this half-line corresponds to $\delta^2\alpha_2 < \alpha_1 < \alpha_2$. For the $\delta\alpha_2 \leq \alpha_1 < \alpha_2$ part of this, proposition 3 rules out additional SSs. Thus, consider the $\delta^2\alpha_2 < \alpha_1 < \delta\alpha_2$ portion. Recall from proposition 3 that $\rho^* = 1 - \alpha_2 + \alpha_1/\delta$ is the value of ρ above which we will have multiple SSs. Meanwhile, for a Hopf bifurcation the bifurcation value $\bar\rho$ is such that $D = 1$, and therefore from (A.4) we have $\bar\rho = 1 - \alpha_2(1-\delta)$. Thus, there will not be multiple SSs at the Hopf bifurcation if and only if $\bar\rho < \rho^*$, which is equivalent to the condition $\alpha_1 > \alpha_2\delta^2$, a condition that we have already taken to hold by hypothesis. Thus, we conclude that the SS must remain unique for the flip case as well.

Finally, for the $+1$ case we must have $\alpha_1 < \delta^2\alpha_2$. The bifurcation in this case occurs at the value $\bar\rho$ of ρ at which line (A.5) (which is positively sloped in this case) intersects the $\Delta = T - 1$ line. At this intersection, we have

$$T = \frac{\alpha_2\delta\,(2-\delta) - \alpha_1}{\alpha_2\delta - \alpha_1},$$

and therefore from (A.3) one can solve to obtain $\bar\rho = \rho^*$. Thus, from proposition 3, the bifurcation point is precisely the point at which two additional SSs must appear. □

PROOF OF PROPOSITION 8

For symmetric allocations, and letting hats denote deviations from the no-interaction SS, our non-linear dynamical system can be written

$$H\,(\hat{e}_t) = (\alpha_2 - \alpha_1)\,\hat{e}_{t-1} - \alpha_1\,(1 - \delta)\,\widehat{X}_{t-1}, \qquad (\text{A.13})$$

$$\widehat{X}_t = (1 - \delta)\,\widehat{X}_{t-1} + e_{t-1}, \qquad (\text{A.14})$$

where $H(\hat{e}) = \hat{e} - \widehat{F}(\hat{e})$ and $\widehat{F}(\hat{e}) \equiv F(e^s + \hat{e})$. Note that since $F(e^s) = 0$ by assumption A2(ii), we have $H(0) = \widehat{F}(\hat{e}) = 0$.

Hopf bifurcation case. Since $\widehat{F}'(\hat{e}) = F'(e^s + \hat{e}) \le \rho$ by assumption A2(iv), it follows that, for $\rho < 1$ (which is the relevant case here), H is a strictly increasing function, and is therefore invertible. Letting $G(\cdot) \equiv H^{-1}(\cdot)$ denote its inverse, we can write $(A.13)$ as

$$\hat{e}_t = G\left((\alpha_2 - \alpha_1)\,\hat{e}_{t-1} - \alpha_1\,(1-\delta)\,\widehat{X}_{t-1}\right).$$

Next, note that $G(0) = 0$ and $G'(0) = 1/(1-\rho)$. Thus, we can write this as

$$\hat{e}_t = \frac{\alpha_2 - \alpha_1}{1-\rho}\hat{e}_{t-1} - \frac{\alpha_1\,(1-\delta)}{1-\rho}\widehat{X}_{t-1} + m\left(\hat{e}_{t-1}, \widehat{X}_{t-1}\right),$$

where

$$m\left(\hat{e}, \widehat{X}\right) \equiv G\left((\alpha_2 - \alpha_1)\,\hat{e} - \alpha_1\,(1-\delta)\,\widehat{X}\right)$$
$$- \frac{\alpha_2 - \alpha_1}{1-\rho}\hat{e} + \frac{\alpha_1\,(1-\delta)}{1-\rho}\widehat{X},$$

and by construction m is $O(\|(\hat{e}, \widehat{X})\|^2)$ as $(\hat{e}, \widehat{X}) \to 0$ (i.e., the Taylor expansion of m around 0 contains no constant or linear terms). Thus, the bivariate system $(A.13)$–$(A.14)$ can be written as

$$\begin{pmatrix}\hat{e}_t \\ \widehat{X}_t\end{pmatrix} = M\begin{pmatrix}\hat{e}_{t-1} \\ \widehat{X}_{t-1}\end{pmatrix} + \begin{pmatrix}m\left(\hat{e}_{t-1}, \widehat{X}_{t-1}\right) \\ 0\end{pmatrix}, \qquad (A.15)$$

where M is the first-order matrix defined in (6). Note that M contains all first-order information for this system (with all higher-order information contained in m).

To prove the desired result, we make use of Wan's (1978) theorem, and of the formulation of it given by Wikan (2013). In what follows, we take ρ to be in a sufficiently small neighborhood of $\bar{\rho}$ so that the eigenvalues of the linearized system are complex conjugates. For a given ρ, let $\lambda(\rho)$ denote the eigenvalue with positive imaginary part, and $\lambda \equiv \lambda(\bar{\rho})$ (i.e., the eigenvalue at the bifurcation point).

To study the stability of the limit cycle as the system goes through a Hopf bifurcation, we first need to convert system (A.15) at the bifurcation point (i.e., for $\rho = \bar{\rho}$) into the "standard form"

$$\begin{pmatrix} y_{1t} \\ y_{2t} \end{pmatrix} = \begin{pmatrix} \cos\theta & -\sin\theta \\ \sin\theta & \cos\theta \end{pmatrix} \begin{pmatrix} y_{1t-1} \\ y_{2t-1} \end{pmatrix} + \begin{pmatrix} f(y_{1t-1}, y_{2t-1}) \\ g(y_{1t-1}, y_{2t-1}) \end{pmatrix},$$

(A.16)

where $\theta \in (0, \pi)$, $(y_1, y_2) = \zeta(\hat{e}, \widehat{X})$ for some smooth invertible function ζ whose inverse ζ^{-1} is also smooth and for which $\zeta(0) = 0$, and f and g are $O(\|(y_1, y_2)\|^2)$ as $(y_1, y_2) \to 0$. Given the model written in this standard form, define

$$a = -\mathrm{Re}\left(\frac{(1 - 2\lambda)\bar{\lambda}^2}{1 - \lambda} \xi_{11}\xi_{20} \right) - \frac{1}{2}|\xi_{11}|^2 - |\xi_{02}|^2 + \mathrm{Re}\left(\bar{\lambda}\xi_{21} \right),$$

where $\bar{x}$ and $\mathrm{Re}(x)$ indicate the complex conjugate and real part of x, respectively, and

$$\xi_{20} = \frac{1}{8} \left[(f_{11} - f_{22} + 2g_{12}) + i\,(g_{11} - g_{22} - 2f_{12}) \right],$$

$$\xi_{11} = \frac{1}{4} \left[(f_{11} + f_{22}) + i\,(g_{11} + g_{22}) \right],$$

$$\xi_{02} = \frac{1}{8} \left[(f_{11} - f_{22} - 2g_{12}) + i\,(g_{11} - g_{22} + 2f_{12}) \right],$$

$$\xi_{21} = \frac{1}{16} \left[(f_{111} + f_{122} + g_{112} + g_{222}) \right.$$
$$\left. + i\,(g_{111} + g_{122} - f_{112} - f_{222}) \right].$$

As shown in Wikan (2013, see theorem 2.5.2), the Hopf bifurcation is supercritical if $\partial|\lambda(\rho)|/\partial\rho\big|_{\rho=\bar{\rho}} > 0$ and $a < 0$. We have already verified the first of these conditions (see Proposition ??), so it remains only to check the second.

By definition, the eigenvalues of M at the Hopf bifurcation point are complex and satisfy $|\lambda| = 1$, so that we may write $\lambda = \cos\theta + i\sin\theta$, $\bar{\lambda} = \cos\theta - i\sin\theta$, for some $\theta \in (0, \pi)$. Define

$$C = \begin{pmatrix} \cos\theta & -\sin\theta \\ \sin\theta & \cos\theta \end{pmatrix}.$$

By construction, λ and $\overline{\lambda}$ are the eigenvalues of C. Define also

$$R = \begin{pmatrix} \lambda & 0 \\ 0 & \overline{\lambda} \end{pmatrix}, \qquad V_C = \begin{pmatrix} \sin\theta & \sin\theta \\ -i\sin\theta & i\sin\theta \end{pmatrix},$$

$$V_M = \begin{pmatrix} \lambda + \delta - 1 & \overline{\lambda} + \delta - 1 \\ 1 & 1 \end{pmatrix}.$$

It can be verified that the columns of V_C and V_M are eigenvectors of C and M, respectively, so that $C = V_C R V_C^{-1}$ and $M = V_M R V_M^{-1}$. We therefore have $M = V_M V_C^{-1} C V_C V_M^{-1} = B^{-1} C B$, where

$$B \equiv V_C V_M^{-1} = \begin{pmatrix} 0 & \sin\theta \\ -1 & \cos\theta - (1-\delta) \end{pmatrix}.$$

Thus, making the change of variables $(y_1, y_2)' = B \times (\hat{e}, \widehat{X})'$, we may write (A.15) for the $\rho = \bar{\rho}$ case in the standard form (A.16), where

$$f(y_1, y_2) = 0,$$

$$g(y_1, y_2) = \frac{\gamma_1}{\sin\theta} y_1 - \gamma_2 y_2 - G\left(\frac{\gamma_1}{\sin\theta} y_1 - \gamma_2 y_2\right),$$

and $\gamma_1 \equiv (\alpha_2 - \alpha_1)\cos\theta - \alpha_2(1-\delta)$, $\gamma_2 \equiv \alpha_2 - \alpha_1$.

We can now check the condition $a < 0$ for the Hopf bifurcation to be supercritical. Since $\bar{\rho} \in (0, 1)$, assumptions A2(i) and A2(vi) imply that $F'' = 0$ and $F''' < 0$ (where derivatives without arguments indicate values at the no-interaction SS), and therefore

$$G'' = 0, \qquad G''' = \frac{F'''}{(1-\bar{\rho})^4} < 0.$$

Thus, $g_{11} = g_{12} = g_{22} = 0$, and since $f(y_1, y_2) = 0$, it follows that $\xi_{20} = \xi_{11} = \xi_{02} = 0$ and thus

$$a = \mathrm{Re}\left(\overline{\lambda}\xi_{21}\right) = \frac{\alpha_2(1-\delta)}{16}\left(\frac{\gamma_1^2}{\sin^2\theta} + \gamma_2^2\right) G''' < 0.$$

This confirms that the limit cycle is supercritical.

Flip bifurcation case. A flip bifurcation occurs at the intersection of line (A.5) and the line $D = -T - 1$. It can be verified that this intersection occurs for

$$\rho = \bar{\rho} = 1 + \alpha_2 - \frac{\alpha_1}{2 - \delta} \in (0, 1). \qquad \text{(A.17)}$$

In what follows, we consider only values of ρ in a neighborhood of $\bar{\rho}$.

Let λ be the smallest eigenvalue of M. Since $\rho \approx \bar{\rho}$, $\lambda \approx -1$. An eigenvector of M associated with this λ is $v = (\lambda + \delta - 1, 1)'$, so that the associated eigenspace can be written $\zeta = \{\beta v : \beta \in \mathbb{R}\}$. Tangent to this one-dimensional eigenspace at $(\hat{e}, \widehat{X}) = (0, 0)$ is a (locally) analytic one-dimensional invariant manifold μ. For some open interval $\Xi \subset \mathbb{R}$ containing 0, we can write $\mu = \{(\phi(\widehat{X}), \widehat{X}) : \widehat{X} \in \Xi\}$ for an analytic function ϕ, where tangency to ζ implies $\phi' = \lambda + \delta - 1$ (where derivatives without arguments are evaluated at $\widehat{X} = 0$).

Since μ is invariant, and $(\hat{e}, \widehat{X}) \in \mu$ implies $\hat{e} = \phi(\widehat{X})$, (A.13)–(A.14) together imply that ϕ implicitly solves

$$H\left(\phi\left((1 - \delta)\,\widehat{X} + \phi\left(\widehat{X}\right)\right)\right)$$
$$= (\alpha_2 - \alpha_1)\,\phi\left(\widehat{X}\right) - \alpha_1\,(1 - \delta)\,\widehat{X}. \qquad \text{(A.18)}$$

Differentiating (A.18) twice and evaluating at the SS, we obtain

$$\left(\lambda^2 + \lambda - T\right)\phi'' = 0,$$

where T is as usual the trace of M, and we have used the facts that $\phi' = \lambda + \delta - 1$, $H' = 1 - F' = 1 - \rho$, and $H'' = -F'' = 0$ (where the last equality is implied by assumptions A2(i) and A2(vi)). Since, for ρ near $\bar{\rho}$ we have $T < 0$ and $\lambda \approx -1$, the term in parentheses is non-zero, and therefore $\phi'' = 0$.

Next, differentiating (A.18) three times and evaluating at the SS, we may solve for

$$\phi''' = \frac{(\lambda + \delta - 1)^3 \lambda^3 F'''}{(1 - \rho)(\lambda^3 + \lambda + \delta - 1) - (\alpha_2 - \alpha_1)}, \qquad \text{(A.19)}$$

where we have used the values of ϕ' and ϕ'' obtained above, along with the facts that $H' = 1 - \rho$, $H'' = 0$, and $H''' = -F'''$.

The evolution of the restriction of (A.13)-(A.14) to μ can be written as $\widehat{X}_t = \chi(\widehat{X}_{t-1}) \equiv (1 - \delta)\widehat{X}_{t-1} + \phi(\widehat{X}_{t-1})$, with $\hat{e}_t = \phi(\widehat{X}_t)$. Note that, for $\rho = \bar{\rho}$, we have $\chi' = -1$. From Wikan (2013, see theorem 1.5.1), the flip bifurcation in the system $\widehat{X}_t = \chi(\widehat{X}_{t-1})$ at $\rho = \bar{\rho}$ is supercritical if, for the case where $\rho = \bar{\rho}$, we have $b > 0$, where

$$b \equiv \frac{1}{2}\chi''^2 + \frac{1}{3}\chi'''.$$

Since $\chi'' = \phi'' = 0$, we need only verify that $\chi''' = \phi''' > 0$ when $\rho = \bar{\rho}$. Substituting the value of $\bar{\rho}$ from (A.17) into (A.19) for ρ, and using the fact that $\lambda = -1$ for this case, we may obtain that

$$\phi''' = -\frac{(2 + \delta)^3 (2 - \delta) F'''}{\alpha_1 - \alpha_2 (2 - \delta)^2}.$$

Proposition 6 implies that the denominator of this expression is positive, while since $\rho = \bar{\rho} > 0$, assumption A2(vi) implies that $F''' < 0$. We therefore conclude that $\chi''' = \phi''' > 0$, and thus the flip bifurcation is supercritical.

Pitchfork bifurcation case. Linearizing the system around the SS (e_j, X_j), $j = L, H$, the first-order matrix is identical to M from (6), except with $F'(e_j)$ in place of ρ. Thus, analysis of the stability of the SS (e_j, X_j) is identical to the analysis of the stability of (e^s, X^s), except with $F'(e_j)$ in place of ρ. Using directly analogous arguments to the $+1$ case from the proof of proposition 7, we conclude that both eigenvalues are stable as long as $F'(e_j) < \bar{\rho} = \rho^* \equiv 1 - \alpha_2 + \alpha_1/\delta$. From panel (d) of figure A.3, meanwhile, we observe that indeed this must be the case, and therefore (e_j, X_j) is stable for $j = L, H$. $\qquad\square$

UNDERSTANDING FINANCIAL CONTAGION: A COMPLEXITY-MODELING PERSPECTIVE

Fabio Caccioli, University College London and
Systemic Risk Centre, London School of Economics and Political Science

Abstract

This chapter reviews key contributions of complexity science to the study of systemic risk in financial systems. The focus is on network models of financial contagion, where I explore various mechanisms of shock propagation, such as counterparty default risk and overlapping portfolios. I highlight how the interconnectedness of financial institutions can amplify risk, and I discuss how standard risk-management tools, which neglect these interactions, can increase systemic risk.

Introduction

The global financial crisis of 2008 made it clear that, in addition to managing traditional risks such as market risk (the risk of adverse fluctuations in the value of investments due to market movements), credit risk (the risk of potential default by counterparties), and liquidity risk (the risk of being unable to meet short-term financial obligations), a new kind of risk needs to be managed: systemic risk.

Systemic risk refers to the risk associated with the collapse of the financial system as a whole, rather than the failure of individual institutions. Unlike other categories of risk, systemic risk is a collective property of the system, emerging from the interactions and interconnections between the many heterogeneous players operating in financial markets. Even more concerning, it can arise as an unintended consequence of agents who are attempting to reduce their own individual risks using traditional risk management tools.

The study of systemic risk is of obvious importance to regulators, whose role is to ensure the stability of the financial system and the broader economy. In fact, during his opening address at the 2010 European Central Bank's forum, the then-president of the ECB, Jean-Claude Trichet (2011), famously stated, "I found the available models of limited help. In fact, I would go further: in the face of the crisis, we felt abandoned by conventional tools." However, understanding how systemic risk builds up in the system is equally important for individual institutions. As David Viniar, Goldman Sachs's chief financial officer, reported in 2007: "We were seeing things that were 25-standard deviation moves, several days in a row" (Larsen 2007) Quotes like these, though anecdotal, make it clear that standard risk-management tools, which may work well in normal market conditions, fail under certain market regimes, and that new tools to complement traditional approaches are required.

These new tools need to explicitly account for the interactions and interconnections within the financial system, modeling the buildup of systemic risk from the bottom up. Unlike traditional tools that often view institutions in isolation, effective models of systemic risk must consider how risks propagate through the network of players and how seemingly isolated events can cascade into systemwide crises. This approach requires a fundamental shift in risk management to embrace complexity and interdependence as core features of the financial system.

Given that systemic risk is a collective property that emerges from the interactions between agents, it is particularly well-suited to be studied through the lens of complex-systems science (Haldane and May 2011).

In this chapter, I will review some of the work on modeling systemic risk using tools from complex-systems science. This review will not be exhaustive, as the field of systemic risk is highly interdisciplinary and has been addressed using a variety of approaches. Instead, I will highlight key contributions that complexity science has brought to the field.

Financial Networks

For more than twenty years, networks have been one of the primary tools used in the modeling of complex systems. A network is simply a mathematical object composed of nodes, which represent the components of a system, and links, which connect some of these nodes. As such, networks provide a natural way to encode the structure of interactions between the units that make up a complex system.

Network science's contributions to the study of complex systems are vast, spanning fields such as biological systems (Jeong *et al.* 2000), ecological systems (Dunne 2006), transportation networks (Guimerà *et al.* 2005), technological networks (Faloutsos, Faloutsos, and Faloutsos 1999), and, of course, social networks (Liljeros *et al.* 2001), to name a few.

As mentioned earlier, the financial system is also composed of many interacting units. Some of these interactions, such as lending relationships between financial institutions, can naturally be modeled using networks.

In financial networks, nodes represent financial institutions (such as banks, funds, and insurance companies), and links represent the relationships between them. Financial institutions interact in a variety of ways: lending money to each other, trading derivatives, forming ownership ties, or investing in similar products. Some relationships are direct, such as when one bank lends money to another under a formal contract. Others are indirect, like when two institutions are connected through the market because they invest in the same financial products.

These relationships evolve over time as old contracts come to maturity, new ones are established, and portfolios are rebalanced. A comprehensive description of the financial system would therefore be in terms of a multilayer dynamical network, where each layer represents a specific type of relationships, and where each layer changes over time.

~535~

While dynamical multilayer modeling has become feasible in recent years (Bargigli *et al.* 2015; Poledna *et al.* 2015), historically, most research has focused on individual static layers due to limited data and the need to begin with simpler models.

Due to regulatory boundaries, much of the empirical research on financial networks has focused on national interbank systems (see, e.g., Boss *et al.* 2004; Soramäki *et al.* 2007; De Masi, Iori, and Caldarelli 2006; Iazzetta and Manna 2009; Cont, Moussa, and Santos 2013; Craig and von Peter 2014; Fricke and Lux 2015). Collectively, these studies have revealed several general features common across different networks, much like stylized facts in financial markets. Specifically, interbank networks exhibit the following characteristics:

- **Heavy-tailed degree distributions:** A node's degree is the number of its links, which in interbank networks corresponds to the number of counterparties. Unsurprisingly, these systems are characterized by a few highly connected hubs (which can often be identified as systemically important institutions) and many nodes with low connectivity.

- **High clustering:** In network science, clustering refers to how interconnected a node's neighbors are, often measured by the clustering coefficient. This measures the abundance of triangles (closed loops) around a node. Like many real-world networks, interbank networks have a high level of clustering compared to random networks, meaning that banks tend to have highly interconnected counterparts.

- **Negative assortativity:** Assortativity measures the tendency of nodes to be connected with others of similar degree. In social networks, positive assortativity is common, where high-degree nodes are connected to other high-degree nodes. However, in interbank networks, negative assortativity is observed—highly connected nodes tend to be linked with less-connected ones, similar to technological networks such as the internet.

- **Core–periphery structure:** These networks often feature a core of highly interconnected nodes (typically the systemically important institutions), surrounded by a periphery of less connected institutions.

When studying financial networks, a key goal is to develop models that describe how shocks propagate between institutions. If a bank or group of banks fails, its neighbors may be affected and potentially fail as well, setting off a chain reaction. The central question is: Under what conditions does an initial shock escalate into a full-blown crisis, leading to the collapse of the network? Understanding the structural properties of interbank networks is critical, as the system's topology directly influences how shocks propagate.

Financial contagion refers to the propagation of shocks between financial institutions. The term is borrowed from epidemiology, and the analogy is intuitive: Just as an infected individual can spread a disease to others they interact with, a stressed financial institution can transmit shocks to those it is connected with.

As mentioned earlier, financial institutions interact in various ways, and these interactions give rise to different contagion mechanisms and channels. These mechanisms can be classified along different dimensions, but a key distinction is between direct versus indirect contagion and solvency versus liquidity contagion.

- **Direct contagion** occurs when there is an explicit relationship, such as an interbank loan, between two institutions.

- **Indirect contagion** arises when two institutions are connected through shared investments in the same assets. For example, if one institution is under stress and liquidates its position, the resulting drop in asset prices can cause losses for the other institution.

Similarly, contagion can also be characterized by the type of shock being transmitted:

- **Solvency contagion** occurs when shocks propagate from the borrower to the lender. In the case of interbank loans, if the borrower defaults, the lender suffers losses because it is unable to recover the funds lent out. This is often referred to as *counterparty default risk.*

- **Liquidity contagion** occurs when shocks flow in the opposite direction, from lenders to borrowers. A typical example is a bank run, where depositors withdraw their money, causing the bank to default due to insufficient liquidity to meet all withdrawal demands.

Next I will discuss some results obtained for direct and indirect solvency contagion.

Contagion Due to Counterparty Default Risk

Contagion due to counterparty default risk occurs when an institution incurs a loss because a counterparty it is exposed to defaults or faces an increased risk of default.

For simplicity, I will focus on interbank loans, where the lender is exposed to the borrower and suffers a loss if the borrower defaults. I will also refer primarily to banks throughout the chapter, though most models can be extended to other types of financial institutions.

A stylized representation of a bank can be described by its balance sheet (see top panel of fig. 1), which consists of an asset side (containing investments with positive value) and a liability side (containing debt). On the asset side, I distinguish between external assets and interbank assets. Interbank assets represent exposures to other banks in the network, such as interbank lending, while external assets are investments outside the network, such as stocks, bonds, and other securities.

Similarly, on the liability side, I differentiate between interbank liabilities and external liabilities. A bank's interbank assets are the interbank liabilities of other banks, meaning the network consists of interconnected balance sheets (see the bottom panel of fig. 1 for a pictorial representation). The liability side also includes equity, which is the difference between assets and liabilities and represents the value owned by the bank's shareholders. Equity is the portion that remains after a bank liquidates all its assets and pays off its debt.

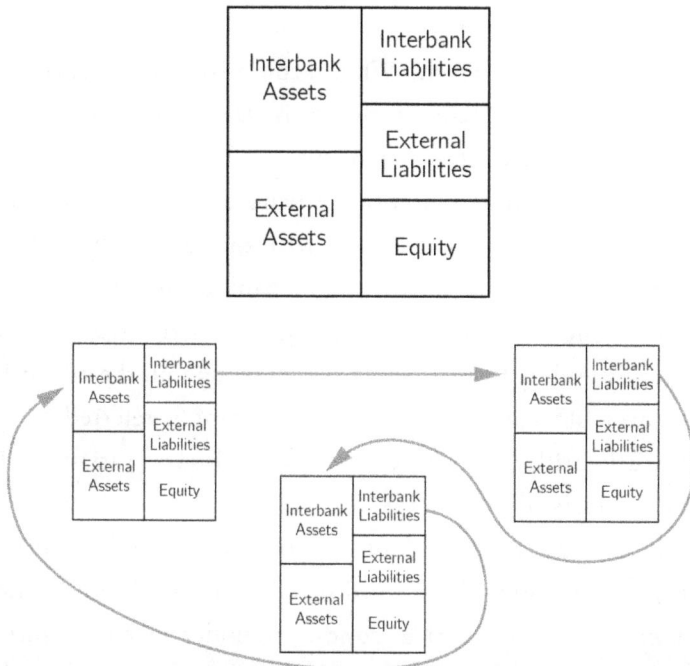

Figure 1. Top panel: Stylized representation of balance sheet. Assets and liabilities are split into interbank and external. The value of assets equals that of liability plus equity. Bottom panel: Pictorial representation of a network of interconnected balance sheets. Shocks can propagate from borrowers to lenders through the links.

Mathematically, this relationship is expressed by the balance sheet identity:

$$A_i^{\text{int}} + A_i^{\text{ext}} = L_i^{\text{int}} + L_i^{\text{ext}} + E_i, \qquad (1)$$

where A_i^{int} and A_i^{ext} are the interbank and external assets of bank i, L_i^{int} and L_i^{ext} are its interbank and external liabilities, and E_i is the equity of bank i.

A shock to a bank corresponds to a devaluation of its assets, and, according to the balance sheet identity, this must result in a reduction on the liability side. Since debt obligations must be honored, the equity absorbs losses first. If a bank's losses exceed its equity, even selling its entire assets would not cover its debt, rendering the bank insolvent. While technical default occurs when a debtor misses a payment, insolvency is often used as a proxy for default in financial contagion modeling.

Now, let's assume that when a bank defaults, its creditors lose the interbank assets associated with that bank. This is how contagion occurs: The default of one bank causes losses for its creditors, who may then default if the losses are large enough, potentially leading to a cascading effect throughout the network. This threshold contagion model is commonly referred to as Furfine's algorithm (Furfine 2003), and, with some variations, it has been widely adopted in the literature.

I focus here on the work of Prasanna Gai and Sujit Kapadia (2010), who theoretically studied this dynamic on Erdős–Rényi random networks and derived conditions under which the initial default of a randomly selected bank could lead to a global cascade of defaults.

I highlight this paper because the model's behavior can be understood in terms of a percolation problem, making it an excellent example of how complex systems contribute to the understanding of systemic risk.

Gai and Kapadia consider a system of N banks with the same amount of interbank assets, total assets, and equity. They model the system using directed Erdős–Rényi networks, where each potential interbank loan between any two directed pairs of banks is present with a probability p.

The probability p is related to the average degree $\langle k \rangle$ of the network. The degree k_i of node i is the number of its links, and the average degree is the mean number of connections per node across the network $\langle k \rangle = \sum_i k_i / N$. Since the network is directed, we can further distinguish between in-degree (k_i^{in}) and out-degree (k_i^{out}), which are the number of incoming or outgoing links of node i, respectively.

Gai and Kapadia assume in their model that the interbank assets of a bank i are uniformly spread across its counterparties, so that each borrows an amount $A_i^{\text{int}}/k_i^{\text{in}}$. Here a link from j to i means that j borrowed from i, that is, the direction of the links indicates the propagation of losses.

If we denote by σ_i the state of bank i—with $\sigma_i = 1$ meaning default and $\sigma_i = 0$ meaning that the bank is running—the loss ℓ_i experienced by bank i because of the failure of its counterparties is equal to

$$\ell_i = \sum_j \frac{A_i^{\text{int}}}{k_i^{\text{in}}} \sigma_j, \qquad (2)$$

from which we can write the condition for the default as i as

$$\sigma_i = \begin{cases} 1 \text{ if } \frac{1}{k_i^{\text{in}}} \sum_j \sigma_j > \frac{E_i}{A_i} \\ 0 \text{ otherwise} \end{cases}. \qquad (3)$$

This dynamic mirrors Duncan Watts's model of global cascades in random networks (Watts 2002), here generalized to directed networks. In fact, many financial contagion dynamics can be seen as variations of linear threshold models, where a node's state depends on the states of its neighbors (counterparties in interbank networks) and whether a certain threshold (such as equity) is crossed.

The model's goal is to understand the conditions under which the initial default of a bank can trigger a global cascade of defaults, meaning that a significant fraction of banks in the network default. I will refer to this in the following as contagion probability.

A characteristic of sufficiently sparse random networks is that they are locally treelike, meaning there are no short loops in the

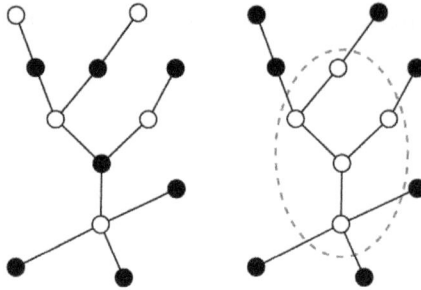

Figure 2. Pictorial representation of the network. White nodes represent vulnerable banks. Black nodes represent non-vulnerable ones. In the left panel, vulnerable banks are not connected between them, which prevents the occurrence of global cascades. In the right panel, global cascades can occur because vulnerable banks form a connected cluster in the network— denoted by the red dashed line. If one vulnerable bank defaults, the entire cluster will go down.

network. Since the process starts with only one bankrupted bank, a chain reaction requires the network to contain banks that can default because of the failure of a counterparty—these are referred to as vulnerable banks. Furthermore, these vulnerable banks need to be connected within the network for the cascade to spread, that is, global cascades of bankruptcies are possible when a percolating cluster of vulnerable banks exists in the network (see fig. 2).

Figure 3 shows the probability of observing global cascades as a function of the average degree. Whether these cascades actually occur depends on whether vulnerable banks are connected to each other through paths composed exclusively of other vulnerable banks. When such a percolating cluster of vulnerable banks exists, the initial failure of one bank within this cluster can trigger the default of a large number of banks, thereby resulting in high systemic risk (see right panel of fig. 2 for a pictorial representation).

If, instead, vulnerable banks are scattered throughout the network and separated by non-vulnerable banks, the non-

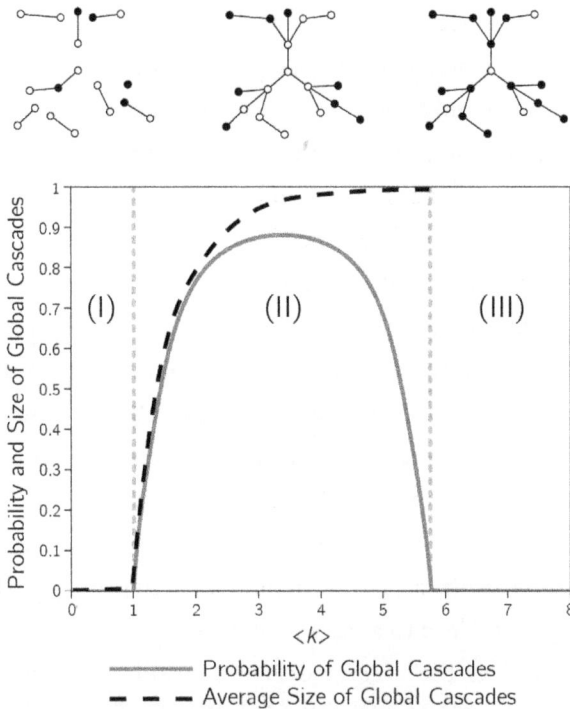

Figure 3. Probability of observing a global cascade on an Erdős–Rényi network as a function of its average degree. The activation threshold is the same for all nodes. Vertical lines mark the two transitions that separate regions where no global cascades occur from those where cascades occur with nonzero probability. In region (I), the network contains many vulnerable nodes, but it is not well-connected, so the initial activation can only propagate locally, preventing global cascades. In region (II), a percolating cluster of vulnerable nodes emerges, allowing cascades to spread through the network. In region (III), vulnerable nodes are sparse and do not form a percolating cluster, preventing cascades. The network structure is pictorially represented at the top, with three diagrams corresponding to each region. White nodes indicate vulnerable nodes.

vulnerable banks effectively shield vulnerable ones from one another (see left panel of fig. 2 for a pictorial representation). In this configuration, cascades are unlikely to occur, and systemic risk remains low.

When the average degree is below one, the network consists

of small, disconnected components without a giant component. In this case, when a bank goes bankrupt, its effect is local because it can only reach a limited number of other banks. In the Gai–Kapadia model, the size of each interbank loan is inversely proportional to the number of loans a bank issues (the in-degree of the node). As the average degree increases above one, a giant component emerges in the network, making banks more connected. At the same time, the number of vulnerable banks decreases as their exposures become more spread out across multiple connections. The combination of these two effects initially increases systemic vulnerability, because the presence of both a giant component and a sufficient number of vulnerable banks means that vulnerable banks can be reached from one another. As a result, the failure of one bank can propagate through the network and lead to a cascade of defaults.

However, as the average degree continues to increase, the size of each interbank exposure decreases further, reducing the number of vulnerable banks overall. Once the average degree exceeds a critical value (the value that separates regions (II) and (III) in the example of fig. 3), vulnerable banks become too dispersed within the network, making it difficult for a percolating cluster to emerge. As a result, the probability of global cascades drops, and systemic risk is significantly reduced.

From a risk-management perspective, increasing diversification is beneficial because it reduces exposure to individual counterparties. Traditional risk-management tools, however, tend to treat investors as isolated entities. When accounting for the interconnectedness of banks, the situation becomes more complex. Greater diversification also implies more potential pathways for shocks to propagate through the network. It is the combination of these two effects that gives rise to the nonmonotonic behavior illustrated in figure 3.

In addition to the probability of observing global cascades of defaults, Gai and Kapadia study the behavior of the average size of global cascades, which appears to monotonically increase with the average degree. In particular, Gai and Kapadia identify a region—this is the region close to the second transition in figure 3—where the system is characterized by a robust yet fragile behavior: On the one hand, the system is robust because the probability of observing a cascade is small. On the other hand, if a cascade is triggered, the whole system is affected. From a statistical perspective, there is nothing that would distinguish between a stable and unstable system in this regime, and small fluctuations in the structure of the system can determine whether a shock is absorbed or propagated throughout the entire network. This behavior is analogous to that discussed by Watts (2002), and it is a good example of what Andrew Haldane referred to as the knife-edge nature of the financial system (Haldane 2013).

Gai and Kapadia's paper considers a relatively homogeneous system, where banks are of the same size and the network follows a Poisson degree distribution. Additionally, the network is neutral in terms of assortativity. However, as discussed earlier, real interbank networks typically do not exhibit these characteristics. Instead, they are characterized by heavy tails in the degree distribution, negative assortativity, and high clustering.

Fabio Caccioli, Thomas Catanach, and J. Doyne Farmer (2012) study the Gai–Kapadia model on scale-free networks as well as Poisson networks with degree–degree correlations. Through numerical simulations, it is shown that heavy tails in the degree distribution make the system more stable against the failure of a randomly selected bank, but more fragile if the failure involves the most connected institutions. Negative assortativity, on the other hand, is found to enhance systemic stability, a result consistent with the findings in Joshua Payne, Peter Sheridan Dodds, and Margaret Eppstein (2009) for the linear threshold model on undirected networks.

Rigorous analytical results for the existence of global cascades in large networks with arbitrary in- and out-degree sequences are provided in Hamed Amini, Rama Cont, and Andreea Minca (2012) and (2016). Additionally, Thomas Hurd, James Gleeson, and Sergey Melnik (2017) present an analytical derivation of the conditions for global cascades in assortative interbank networks.

Finally, Y. Ikeda, T. Hasegawa, and K. Nemoto (2010) study the effect of clustering for the linear threshold model, where it is shown that increasing clustering makes global cascades more likely to occur.

As mentioned at the start of this section, the dynamic underlying the Gai–Kapadia model is the Furfine contagion algorithm. Beyond its theoretical exploration of financial contagion, it has also been applied to study the stability of real interbank systems (Furfine 2003; Wells 2004; Degryse and Nguyen 2004; Upper and Worms 2004; Amundsen and Arnt 2005; Lublóy 2005; van Lelyveld and Liedorp 2006; Mistrulli 2011). A common finding in the literature is that, typically, not much contagion occurs once the network is calibrated to a real-world scenario. This raises the question: Are interbank networks truly relevant? If financial networks are not as critical as initially thought, why do regulators and practitioners react with concern whenever a financial institution is at risk of going under?

In the next sections, I will explore two potential solutions to this apparent paradox. The first relates to the idea that stress can propagate within the system before actual defaults occur. The second focuses on the coupling of the interbank network with another contagion channel: contagion due to overlapping portfolios.

Contagion Due to Credit Quality Deterioration

In the threshold models considered above, shocks propagate from the borrower to the lender only after the default of the borrower. However, one could argue that the value of an interbank asset should be reduced even before the default of the counterparty. This is because the more a bank is under stress, the more likely it is that it will default, and the less its creditors expect to receive when the loan comes to maturity.

Proceeding from this intuition, and with the aim of providing a tool to estimate the buildup of systemic risk prior to the occurrence of defaults, Stefano Battiston and collaborators introduced a new algorithm, which they dubbed DebtRank (Battiston *et al.* 2012) and which was somehow inspired by network centrality measures such as Google PageRank.

The main assumption of the algorithm is that losses propagate from borrowers to lenders linearly: If a borrower loses $x\%$ of its equity after a loan is issued, the lender writes down the value of the corresponding interbank asset by $x\%$. The original algorithm further assumes that losses are propagated only once by a borrower to its creditors, meaning no further devaluation of a given interbank asset occurs after the initial write-down. This second assumption was relaxed in Marco Bardoscia *et al.* (2015), whose formulation we will use for ease of exposition.

Consider a given interbank network at time $t = 0$, with $W_{ij}(0)$ denoting the amount lent by bank i to bank j and $E_i(0)$ its equity. If the interbank network is hit by a shock at time $t = 1$, we can write a recursive map for the relative loss $h_i(t)$ experienced by bank i at time t: $h_i(t) = (E_i(0)-E_i(t))/E_i(0)$.

By iterating the balance sheet identity over discrete time steps and considering the linear propagation of shocks:

$$h_i(t) = \min\left\{1, \sum_j \Lambda_{ij} h_j(t) + h_i(1)\right\},\qquad(4)$$

where $\Lambda_{ij} = W_{ij}(0)/E_i(0)$, and the upper bound of 1 ensures that lenders do not lose more than the amount they were owed.

Equation (4) expresses the losses of bank i as the sum of two contributions: $h_i(1)$, corresponding to exogenous losses, and $\sum_j \Lambda_{ij} h_j(t)$, coming from the network. A given counterparty j contributes to the loss of i as soon as it displays an equity loss, that is, $h_j > 0$, even if the bank remains solvent (insolvency occurs when $h_j = 1$). Furthermore, in equation 4 the contribution to i's loss coming from j is weighted by Λ_{ij}, which is called matrix of interbank leverage (Battiston, D'Errico, and Gurciullo 2016).

Leverage λ refers to the practice of borrowing money to invest, and it is typically measured as the ratio of assets to equity. Leverage is related to risk because it amplifies losses: If an investor has a leverage ratio of λ, a 1% devaluation of its assets translates into a λ% equity loss. The matrix Λ, composed of elements Λ_{ij}, captures the contribution of interbank exposures to the leverage of banks within the network.

By linearizing equation (4), we can compute an iterative map for the quantity $\Delta h_i(t) = h_i(t) - h_i(t-1)$, which represents the marginal losses experienced by i between iterations $t-1$ and t as

$$\Delta h_i(t) = \sum_j \Lambda_{ij} \Delta h_j(t).\qquad(5)$$

From this, we see that $\Delta h_j = 0$ for all j is a fixed point of the dynamic, whose stability depends on the largest eigenvalue $\lambda_{\max}$ of the matrix Λ. If $\lambda_{\max} > 1$, the fixed point is unstable, meaning that even the tiniest perturbation would lead to the default of a bank—after which the map (eq. (5)) would no longer hold, as the upper bound in equation (4) cannot be neglected. Taking some latitude, we could say that λ_{max} generalizes the notion of leverage to the network context: The larger λ_{max} is, the greater the endogenous amplification of losses becomes.

Bardoscia *et al.* (2017) study the stability of interbank networks under the DebtRank framework as these networks become more interconnected. Starting from a sparse directed acyclic graph, they randomly add links to the network while keeping the total interbank assets and liabilities of each bank constant. They observe that, on average, the system tends to become more unstable as the network's connectivity increases, a phenomenon they attribute to the emergence of particular cyclic structures within the network. The result is reminiscent of Lord May's findings that increasing the complexity of ecosystems leads to their stability (May 1972), and it provides a clear example of how an action that, according to standard models, should reduce individual risk— by increasing diversification—can, in fact, inadvertently lead to greater systemic risk.

A generalization of DebtRank that accounts for nonlinear propagation of shocks and interpolates between DebtRank and the Furfine algorithm was introduced in Bardoscia *et al.* (2016). The behavior of this nonlinear model is qualitatively similar to that of DebtRank, with the system characterized by a transition between a regime where small shocks can be amplified and a regime where shocks do not propagate. However, accounting for nonlinear propagation introduces a soft threshold to the propagation of shocks, offering a more realistic representation of how financial distress spreads through the system. This approach allows shocks not to be amplified for lenders when borrowers experience small losses, while still retaining the idea that losses can be transmitted before defaults occur.

Additionally, Paolo Barucca *et al.* (2020) present a framework for valuing interbank assets that incorporates both the interconnectedness of financial institutions and uncertainty in external asset values. This framework extends the DebtRank approach by focusing on how shocks and credit quality deterioration propagate through the system via mark-to-market revaluations of interbank assets.

DebtRank has been extensively applied to assess the stability and systemic risk of financial systems across various regions using real-world empirical data (Poledna *et al.* 2015; Silva, Tabak, and Guerra 2017; Caceres-Santos *et al.* 2020; Cuba *et al.* 2021; Landaberry *et al.* 2021; León, Martínez, and Cepeda 2019), and it has also been used as a tool to study policy measures aimed at containing systemic risk. In particular, Sebastian Poledna and Stefan Thurner (2013; 2016) consider an agent-based model of the interbank system coupled with the economic system. In the model, firms approach banks for capital to fund investments, and banks establish interbank connections to ensure the flow of such funds. Depending on their production outcomes, some firms may go bankrupt, passing their losses to the financial system.

Poledna and Thurner measure the frequency and size of cascades in the interbank market as the model runs, and they propose introducing a tax based on the contribution to systemic risk that a new interbank loan imposes on the system, measured using DebtRank. By comparing the benchmark of no tax with the scenario where the tax is implemented, they show that it is possible to reduce the probability of large cascades occurring in the interbank market while maintaining the same level of credit for the economy. In contrast, a tax based purely on the

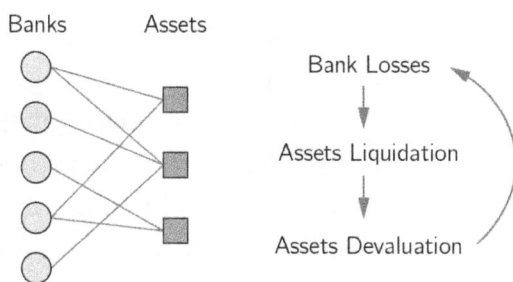

Figure 4. Left panel: A network of overlapping portfolios. Banks interact indirectly through their common assets. Right panel: Feedback mechanism of contagion due to overlapping portfolios.

size of the interbank loan (without considering its contribution to systemic risk) leads to a reduction in available credit.

Contagion Due to Overlapping Portfolios

I have so far discussed the case of contagion through direct linkages between financial institutions. As mentioned earlier, contagion can also occur when financial institutions invest in common assets. In this case, while there is no direct contract between the institutions, shocks can still propagate from one to another indirectly through asset prices. If one institution is under stress and needs to sell part of its assets, those assets may be devalued due to market impact—the tendency of prices to respond to trading activity (Bouchaud, Farmer, and Lillo 2009). This devaluation leads to mark-to-market losses for other institutions holding the same assets. In turn, these institutions may be forced to liquidate their own assets in response to their losses, further depressing prices and potentially triggering a fire sale.

Network models can be adapted to represent this contagion mechanism (see fig. 4). Instead of focusing on direct interbank connections, the network becomes a bipartite network of banks and assets that represents banks' portfolios. In this structure, banks interact indirectly through their overlapping portfolios, that is, the common assets they hold.

Network models of contagion due to overlapping portfolios were first studied in Fabio Caccioli $et\ al.$ (2014), where the authors introduced two approaches: a linear threshold model, in which banks remain passive until they become insolvent and then liquidate their entire portfolios, and a model where banks actively manage risk, responding to losses by selling off part of their portfolios to reduce exposure. In both cases, asset prices respond to liquidation through a linear market impact on log-prices

$$\log(p_a(t)/p_a(0)) = -\eta_a q_a(t), \tag{6}$$

where $p_a(t)$ is the price at time t, $q_a(t)$ is the volume of asset a that has been liquidated up to time t, and η_a is a market impact parameter that is associated with the asset's liquidity, that is, its sensitivity to trading.

Similarly to Gai and Kapadia's analysis, the key question is understanding the conditions under which a global cascade of defaults can be triggered by the initial default of a bank or the devaluation of an asset. As in the previous section, the threshold model can be understood as a network percolation problem, where a vulnerable bank is one that can default due to the devaluation of a single asset. An equivalent description can be provided using a branching process, which is the one given in Caccioli $et\ al.$ (2014). In a branching process, a progenitor generates a number of offspring n drawn from a distribution $P(n)$, and each offspring in turn produces more offspring according to the same distribution $P(n)$. The question is whether the process will eventually die out or if there is a nonzero probability it will continue indefinitely. If the expected number of offspring per individual is $\bar{n} = \sum_n nP(n)$ and $\bar{n} < 1$, the population will go extinct with probability one. If, however, $\bar{n} > 1$, there is a chance the population will survive. This is analogous to epidemiology, where n is equivalent to the basic reproduction number.

The analogy with financial contagion is clear. Given a generation 0 defaulted bank, it can trigger a certain number of generation 1 defaults among banks that share common investments, which can in turn trigger generation 2 defaults. If the expected number of next-generation defaults per current bankrupted bank is larger than one, the contagion can continue to spread through the network with a nonzero probability. In reality, the situation is more complex because banks differ in size, investments, leverage, and other factors. However, the result can be generalized by introducing a matrix B, where each element B_{ij} represents the probability that the default of bank j will trigger the default of bank i due to their common investments. The matrix B can be estimated from data about portfolio holding and the market impact parameters, and its largest eigenvalue plays the role of the basic reproduction number in this case, further strengthening the analogy between financial contagion and epidemiology.

Caccioli *et al.* (2014) study the behavior of the model on the bipartite equivalent of Erdős–Rényi random graphs using the same balance sheet setting as Gai and Kapadia. Similar behaviors to those observed for counterparty default risk are found, particularly regarding both the probability of global cascades and the average size of those cascades. The case of heavy-tailed degree distributions is studied via numerical simulations in Opeoluwa Banwo *et al.* (2016), while an analytical solution of a simplified model is given in Tomokatsu Onaga, Fabio Caccioli, and Teruyoshi Kobayashi (2023) using techniques developed in Peter Fennell and James Gleeson (2019) for multistate dynamical processes on networks.

A bipartite network threshold model for contagion due to overlapping portfolios is also introduced in Xuqing Huang *et al.* (2013), where it is calibrated using balance sheet data from US commercial banks in 2007. The paper shows that abrupt transitions in the number of defaults occur as a function of asset liquidity and the size of the external shock (devaluation of an asset class). Furthermore, they validate their model through an analysis of true positives and false positives for defaults observed in the United States from 2008 to 2011, demonstrating that the network model has some explanatory power.

While the threshold model is a useful benchmark for understanding how network structure affects stability, in practice, banks actively manage their own risk by reducing investments before becoming insolvent in an attempt to contain their losses.

Caccioli *et al.* (2014) consider a simple protocol for such preventative liquidation, based on leverage targeting. When a bank suffers a loss, its leverage—and therefore its risk—increases. In response, the bank may liquidate part of its assets to bring its leverage back to the desired level.

However, Caccioli *et al.* (2014) demonstrate that preemptive liquidation actually makes the system more unstable compared to the benchmark of a passive investor. Moreover, the more aggressively a bank tries to reach its leverage target, the more unstable the system becomes. The reason is intuitive: When banks liquidate assets to reduce leverage, they push prices down, leading to further losses. In the absence of activation thresholds for liquidation, this makes it easier for a downward spiral to take hold. In fact, as shown in Caccioli, Jean-Philippe Bouchaud, and J. Doyne Farmer (2012), it is even possible for an individual investor using leverage to drive themselves into bankruptcy by attempting to exit their position. The presence of a network further exacerbates this tendency.

Stylized models of contagion, such as those discussed above, are useful for understanding the mechanics of the system and how its response to shocks is influenced by structural parameters (such as connectivity) and dynamical ones (such as the liquidity of assets). Beyond theoretical analyses, more detailed frameworks have been developed and calibrated to real systems. A leverage targeting fire-sale model was introduced and applied to European banks by Robin Greenwood, Augustin Landier, and David Thesmar (2015), and further extended by Fernando Duarte and Thomas Eisenbach (2021), who present a study of the US banking system. Rama Cont and Eric Schaanning (2017) introduced a hybrid model, where banks remain passive below a certain threshold but switch to leverage targeting once the threshold is crossed. They calibrated this model to bond holdings of European Union banks. Notably, they also introduced the concept of indirect exposure, accounting for additional risk arising from the network of overlapping portfolios. In related work, Cont and Schaanning introduced an index to monitor indirect contagion, derived from the spectrum of the matrix of liquidity-weighted overlapping portfolios (2019).

Historically, research on systemic risk focused primarily on banks, as they were regulated by central banks, where part of the early literature emerged, and because data, albeit sparse, were available for them. When considered collectively, empirical studies on banking systems suggest that the contagion effect of overlapping portfolios is larger than that of interbank lending. More recently, research has expanded to model other types of financial institutions. For instance, Paolo Barucca, Tahir Mahmood, and Laura Silvestri (2021) provide an empirical analysis of the network of overlapping portfolios between European investment funds and UK regulated banks and insurance companies, while Christoph Fricke and Daniel Fricke

(2021) and Daniel Fricke and Hannes Wilke (2023) adapt Greenwood, Landier, and Thesmar's model (2015) to study the case of funds.

While from a technical perspective, the dynamics in models of funds, such as leverage targeting, may resemble those observed for banks, the underlying mechanisms behind these rules differ. It is crucial that these distinctions are modeled as realistically as possible to ensure accurate calibration to real data.

Additionally, new system-wide stress-testing frameworks have been proposed in the literature to model various institutions, such as banks, funds, and insurers (Farmer *et al.* 2020; Caccioli, Ferrara, and Ramadiah 2024). These studies demonstrate that neglecting intersectoral overlapping portfolios can result in a severe underestimation of systemic risk.

Multiple Contagion Channels

In the previous sections, we considered two channels of financial contagion: contagion due to counterparty default risk and contagion due to overlapping portfolios. However, as discussed earlier, banks simultaneously interact through different kinds of relationships. The multilayer nature of financial networks has been explored in Bargigli *et al.* (2015) and Poledna *et al.* (2015). In practice, multiple contagion channels are often active simultaneously, making it essential to study their combined effects. This has been explored to some extent in works such as Gai and Kapadia (2010), Rodrigo Cifuentes, Gianluigi Ferrucci, and Hyun Song Shin (2005), and Erlend Nier *et al.* (2007), which examine scenarios where all banks interact in an interbank network and invest in the same asset. The asset dynamics follow a similar pattern to what was discussed in the previous section. Their findings indicate that including these additional interactions clearly leads to additional losses compared to the benchmark scenario of only counterparty default risk. Robert May and Nimalan Arinaminpathy (2010) take this further by considering interactions between asset classes in a stylized setting. Caccioli *et al.* (2015) attempt to address the question raised at the end section 3, namely, why regulators are concerned about bank defaults if contagion in interbank lending networks is not typically significant. The authors study the Austrian interbank network, assuming a certain degree of correlation between banks' external assets. They perform stress tests using the Furfine algorithm, accounting for the devaluation of external assets upon liquidation under three different scenarios: (i) contagion occurs only due to counterparty default risk, (ii) contagion occurs only due to overlapping portfolios, and (iii) contagion occurs due to both channels simultaneously.

They find that, in some regimes, while each contagion channel may not result in large cascades when considered individually, the combination of the two can lead to systemic instability. This suggests that while the interbank lending network may not be a significant risk factor on its own, it can play a crucial role in amplifying contagion arising from overlapping portfolios.

Using data on interbank exposures and portfolio holdings of Mexican banks, Poledna *et al.* (2021) generalize DebtRank to multilayer networks to study their combined effect, and they find that neglecting the interaction between the two contagion channels significantly underestimates systemic losses. Building on the idea of interacting contagion channels, Garbrand Wiersema *et al.* (2023) develop an eigenvalue-based approach, demonstrating that the instability caused by interacting channels can far exceed the sum of the instabilities from individual channels acting in isolation.

All in all, this line of research makes it clear that to properly estimate and manage systemic risk, one must account for as many contagion channels as possible simultaneously. Neglecting their interaction can lead to significant underestimation of systemic losses, as the combined effect of these channels can be much greater than the sum of their individual impacts.

Dynamical Models

The models discussed so far are essentially static, where a network is given, a shock is applied, and the system evolves according to a deterministic dynamic. These models are highly useful for understanding how the structural constraints of the network and simple behavioral rules of financial institutions can lead to the endogenous amplification of shocks. Moreover, they have practical applications in stress testing, where the focus is on analyzing scenarios that unfold over a relatively short time period, during which the network structure can be considered effectively fixed.

However, to fully understand how a system can endogenously transition from a normal state to a crisis regime, we need dynamic models in which investments and connections evolve over time. This is where agent-based modeling becomes particularly valuable. In agent-based models, individual financial institutions (agents) interact with one another based on a set of behavioral rules, allowing for the emergence of complex system-wide phenomena. By simulating the evolution of investments, network structures, and decision-making processes, agent-based models can offer a more realistic representation of how systemic risk develops and propagates over time. I report here a few results obtained using simple agent-based models, which clearly demonstrate how standard financial risk-management tools can, under certain conditions, lead to systemic instabilities.

Fulvio Corsi, Stefano Marmi, and Fabrizio Lillo (2016) introduce and study a dynamical model of overlapping portfolios that includes stochasticity. In this model, banks actively manage their risk through leverage targeting, and asset prices fluctuate over time due to both the trading activity of banks and random fluctuations. The model shows that financial innovation, through a reduction in the cost of diversification, can trigger a transition from a stable market regime to one characterized by bubbles and bursts.

Thurner, Farmer, and Geanakoplos (2012) consider a dynamical model of leveraged investment funds. In this model, funds borrow money from banks and use leverage to purchase undervalued assets, providing a stabilizing mechanism for the market, which is also populated by noise traders. However, the bank requires them to keep their leverage below a specified cap, and they are subject to margin calls when this limit is breached. A margin call forces the fund to repay part of the borrowed money. If a fund is fully leveraged, even a small market fluctuation can trigger a margin call, forcing the fund to sell some of its assets to meet its creditors' demands. This, in turn, depresses the asset price further, potentially triggering more margin calls and creating a downward spiral. Thurner and coauthors demonstrate that this mechanism can explain the emergence of heavy tails and clustered volatility in financial markets.

~559~

Using Thurner, Farmer, and Geanakoplos's model as a benchmark, Poledna *et al.* (2014) study the effect of two different regulatory schemes for banks. The first is the one mandated by Basel II, where exposures are mitigated solely by collateral, and the second is a perfect-hedging scheme, where banks are additionally required to hold options to hedge their exposures to funds. Interestingly, they find that both policies increase systemic stability when leverage is low, but reduce it when leverage is high. This is because these policies lead to increased synchronization among funds that need to deleverage, inadvertently increasing systemic risk. This is another example of how measures intended to reduce risk can sometimes exacerbate systemic risk.

The impact of Basel II regulations is also central to the work of Tobias Adrian and Hyun Song Shin (2010), who explore how a Value-at-Risk (VaR) constraint, such as that required by Basel II, leads to leverage procyclicality. This phenomenon causes banks to increase their leverage during favorable market conditions (low volatility and rising prices) and reduce it during downturns (high volatility and falling prices).

From an interacting systems perspective, this procyclicality creates a feedback loop that can fuel market bubbles and crashes. In periods of high prices and low volatility, banks increase their leverage and expand their investments, driving prices even higher and reducing volatility further. This positive feedback loop promotes continued balance sheet expansion. Conversely, during market downturns, banks sell off assets, pushing prices down and increasing volatility, which prompts even more asset liquidations, reinforcing the negative spiral.

Christoph Aymanns and J. Doyne Farmer (2015) present a simple agent-based model illustrating leverage procyclicality. Their model demonstrates how endogenous market oscillations— characterized by gradual increases in leverage and asset prices followed by sudden crashes—emerge when banks manage risk based on historical volatility estimates to determine their target leverage. These oscillations occur within a chaotic regime of the system. In an extended version of the model, which includes a leverage-targeting bank and an unleveraged fundamentalist investor, Aymanns *et al.* (2016) show that, when roughly calibrated with realistic parameters, the model can replicate leverage cycles lasting ten to fifteen years, akin to those observed in real markets.

Aymanns *et al.* (2016) also investigate the effects of different capital buffer policies, ranging from procyclical to countercyclical measures, and find that the optimal policy depends on the size of the banking system. Basel II's procyclical rules are found to be effective when the banking system is small and leverage is low, as in such a case the bank's impact on the market is limited, making it reasonable to approximate it as an isolated agent, an assumption implicit in standard risk-management tools. However, when the banking system is large and leverage is high, maintaining constant leverage becomes the optimal strategy, as Basel II's procyclical rules tend to destabilize the system. Finally, the authors show that slowing down the response of the bank, that is, giving the leveraged investor more time to pay the loan back, is in fact the most effective strategy to improve systemic stability.

Conclusions

This chapter explored complex-systems approaches to the study of financial systemic risk, an area that has seen significant research activity from complexity scientists since 2008. I discussed how domino effects in financial networks can be understood through the emergence of a percolating cluster of vulnerable banks, or equivalently, as a supercritical branching process. Additionally, I examined simple agent-based models whose dynamics can endogenously lead to market instability. I also highlighted several examples where actions intended to reduce individual risk, according to standard risk-management tools, can inadvertently increase systemic risk.

Network models of financial contagion began as simple, stylized models designed to gain intuition about how shocks spread across banks and how network structure might influence this propagation. Over time, these models have evolved to account for more complex contagion mechanisms, incorporate institutions beyond banks, and introduce more realistic dynamics. This evolution has paved the way for the development of network-based macroprudential stress-testing frameworks, which can be calibrated with real-world data and complement traditional microprudential stress testing, where the interconnections between financial institutions and feedback loops are often overlooked.

Clearly, the literature on systemic risk is much broader than can be covered here. Key topics that have emerged include network reconstruction (Squartini *et al.* 2018; Anand *et al.* 2018), liquidity contagion (Anand, Gai, and Marsili 2012; Cimini and Serri 2016), the estimation of systemic risk and connectivity from market data (Brownlees and Engle 2017; Billio *et al.* 2012; Aste 2021), and the development of stress-testing frameworks that incorporate climate risks and estimate transition risks (Battiston *et al.* 2017; Monasterolo 2020). For a more comprehensive understanding of financial networks and contagion, I also recommend Paul Glasserman and H. Peyton Young (2016) and Bardoscia *et al.* (2021).

In the broader field of complexity applied to finance, it is also worth mentioning the literature on minority games in physics (Challet and Zhang 1997; Challet *et al.* 2001) and the literature on heterogeneous agents in economics (Brock and Hommes 1998), which have contributed to understanding potential sources of market instability (see, e.g., Brock, Hommes, and Wagener 2009; Marsili 2014; Bardoscia, Livan, and Marsili 2012; Caccioli, Marsili, and Vivo 2009).

Beyond its scientific value, the primary motivation for managing systemic risk is to mitigate its negative consequences on the real economy. The financial system's role should be to ensure that funding is properly allocated to firms and households. When a financial crisis occurs, it can spill over to the real economy, causing rising unemployment, lower production, and gross domestic product loss. Therefore, models of financial systemic risk must eventually be coupled with macroeconomic models, such as those recently developed in Cars Hommes *et al.* (2022) and Poledna *et al.* (2023), where agent-based modeling approaches have already proven fruitful (Delli Gatti *et al.* 2011; Tedeschi *et al.* 2012; Gualdi *et al.* 2015b; Gualdi *et al.* 2015a; Geanakoplos *et al.* 2012; Baptista *et al.* 2016). However, these models still lack a direct connection to financial models, making their integration an important area for future research.

In conclusion, complexity science has significantly advanced our understanding of systemic risk by providing tools to model the intricate interconnections and feedback loops that shape financial systems. As the field progresses, future research must continue to integrate these approaches with macroeconomic modeling and policy frameworks to create actionable insights for regulators and policymakers. By doing so, we can better prepare for the risks posed by financial crises and help safeguard the stability of the real economy. ✒

REFERENCES

Adrian, T., and H. S. Shin. 2010. "Liquidity and Leverage." *Journal of Financial Intermediation* 19 (3): 418–437. https://doi.org/10.1016/j.jfi.2008.12.002.

Amini, H., R. Cont, and A. Minca. 2012. "Stress Testing the Resilience of Financial Networks." *International Journal of Theoretical and Applied Finance* 15 (01): 1250006. https://doi.org/10.1142/S0219024911006504.

———. 2016. "Resilience to Contagion in Financial Networks." *Mathematical Finance* 26 (2): 329–365. https://doi.org/10.1111/mafi.12051.

Amundsen, E., and H. Arnt. 2005. *Contagion Risk in the Danish Interbank Market.* Working Paper. Danmarks Nationalbank. https://www.nationalbanken.dk/en/news-and-knowledge/publications-and-speeches/archive-publications/2005/contagion-risk-in-the-danish-interbank-market.

Anand, K., P. Gai, and M. Marsili. 2012. "Rollover Risk, Network Structure and Systemic Financial Crises." *Journal of Economic Dynamics and Control* 36 (8): 1088–1100. https://doi.org/10.1016/j.jedc.2012.03.005.

Anand, K., I. van Lelyveld, A. Banai, S. Friedrich, R. Garratt, G. Hałaj, J. Fique, *et al.* 2018. "The Missing Links: A Global Study on Uncovering Financial Network Structures from Partial Data." *Journal of Financial Stability* 35:107–119. https://doi.org/10.1016/j.jfs.2017.05.012.

Aste, T. 2021. "Stress Testing and Systemic Risk Measures Using Elliptical Conditional Multivariate Probabilities." *Journal of Risk and Financial Management* 14 (5): 213. https://doi.org/10.48550/arXiv.2004.06420.

Aymanns, C., F. Caccioli, J. D. Farmer, and V. W. C. Tan. 2016. "Taming the Basel Leverage Cycle." *Journal of Financial Stability* 27:263–277. https://doi.org/10.1016/j.jfs.2016.02.004.

Aymanns, C., and J. D. Farmer. 2015. "The Dynamics of the Leverage Cycle." *Journal of Economic Dynamics and Control* 50:155–179. https://doi.org/10.1016/j.jedc.2014.09.015.

Banwo, O., F. Caccioli, P. Harrald, and F. Medda. 2016. "The Effect of Heterogeneity on Financial Contagion Due to Overlapping Portfolios." *Advances in Complex Systems* 19 (08): 1650016. https://doi.org/10.1142/S0219525916500168.

Baptista, R., J. D. Farmer, M. Hinterschweiger, K. Low, D. Tang, and A. Uluc. 2016. *Macroprudential Policy in an Agent-Based Model of the UK Housing Market.* https://doi.org/10.2139/ssrn.2850414.

Bardoscia, M., P. Barucca, S. Battiston, F. Caccioli, G. Cimini, D. Garlaschelli, F. Saracco, T. Squartini, and G. Caldarelli. 2021. "The Physics of Financial Networks." *Nature Reviews Physics* 3 (7): 490–507. https://doi.org/10.1038/s42254-021-00322-5.

Bardoscia, M., S. Battiston, F. Caccioli, and G. Caldarelli. 2015. "DebtRank: A Microscopic Foundation for Shock Propagation." *PloS One* 10 (6): e0130406. https://doi.org/10.1371/journal.pone.0130406.

———. 2017. "Pathways Towards Instability in Financial Networks." *Nature Communications* 8 (1): 14416. https://doi.org/10.1038/ncomms14416.

Bardoscia, M., F. Caccioli, J. I. Perotti, G. Vivaldo, and G. Caldarelli. 2016. "Distress Propagation in Complex Networks: The Case of Nonlinear DebtRank." *PloS One* 11 (10): e0163825. https://doi.org/10.1371/journal.pone.0163825.

Bardoscia, M., G. Livan, and M. Marsili. 2012. "Financial Instability from Local Market Measures." *Journal of Statistical Mechanics: Theory and Experiment* 2012 (08): P08017. https://doi.org/10.1088/1742-5468/2012/08/P08017.

Bargigli, L., di Iasio G., L. Infante, F. Lillo, and F. Pierobon. 2015. "The Multiplex Structure of Interbank Networks." *Quantitative Finance* 15 (4): 673–691. https://doi.org/10.1080/14697688.2014.968356.

Barucca, P., M. Bardoscia, F. Caccioli, M. D'Errico, G. Visentin, G. Caldarelli, and S. Battiston. 2020. "Network Valuation in Financial Systems." *Mathematical Finance* 30 (4): 1181–1204. https://doi.org/10.1111/mafi.12272.

Barucca, P., T. Mahmood, and L. Silvestri. 2021. "Common Asset Holdings and Systemic Vulnerability across Multiple Types of Financial Institution." *Journal of Financial Stability* 52:100810. https://doi.org/10.1016/j.jfs.2020.100810.

Battiston, S., M. D'Errico, and S. Gurciullo. 2016. "DebtRank and the Network of Leverage." *The Journal of Private Equity*, 58–71. https://www.jstor.org/stable/44396818.

Battiston, S., A. Mandel, I. Monasterolo, F. Schütze, and G. Visentin. 2017. "A Climate Stress-Test of the Financial System." *Nature Climate Change* 7 (4): 283–288. https://doi.org/10.1038/nclimate3255.

Battiston, S., M. Puliga, R. Kaushik, P. Tasca, and G. Caldarelli. 2012. "DebtRank: Too Central to Fail? Financial Networks, the FED and Systemic Risk." *Scientific Reports* 2 (1): 541. https://doi.org/10.1038/srep00541.

Billio, M., M. Getmansky, A. W. Lo, and L. Pelizzon. 2012. "Econometric Measures of Connectedness and Systemic Risk in the Finance and Insurance Sectors." *Journal of Financial Economics* 104 (3): 535–559. https://doi.org/10.1016/j.jfineco.2011.12.010.

Boss, M., H. Elsinger, M. Summer, and S. Thurner. 2004. "Network Topology of the Interbank Market." *Quantitative Finance* 4 (6): 677–684. https://doi.org/10.1080/14697680400020325.

Bouchaud, J.-P., J. D. Farmer, and F. Lillo. 2009. "How Markets Slowly Digest Changes in Supply and Demand." In *Handbook of Financial Markets: Dynamics and Evolution,* 57–160. Amsterdam, Netherlands: Elsevier. https://doi.org/10.48550/arXiv.0809.0822.

Brock, W. A., and C. H. Hommes. 1998. "Heterogeneous Beliefs and Routes to Chaos in a Simple Asset Pricing Model." *Journal of Economic Dynamics and Control* 22 (8-9): 1235–1274. https://doi.org/10.1016/S0165-1889(98)00011-6.

Brock, W. A., C. H. Hommes, and F. O. O. Wagener. 2009. "More Hedging Instruments May Destabilize Markets." *Journal of Economic Dynamics and Control* 33 (11): 1912–1928. https://doi.org/10.1016/j.jedc.2009.05.004.

Brownlees, C., and R. F. Engle. 2017. "SRISK: A Conditional Capital Shortfall Measure of Systemic Risk." *The Review of Financial Studies* 30 (1): 48–79. https://doi.org/10.1093/rfs/hhw060.

Caccioli, F., J.-P. Bouchaud, and J. D. Farmer. 2012. "Impact-Adjusted Valuation and the Criticality of Leverage." *Risk* 25 (12): 74–77.

Caccioli, F., T. A. Catanach, and J. D. Farmer. 2012. "Heterogeneity, Correlations and Financial Contagion." *Advances in Complex Systems* 15 (supp02): 1250058. https://doi.org/10.48550/arXiv.1109.1213.

Caccioli, F., J. D. Farmer, N. Foti, and D. Rockmore. 2015. "Overlapping Portfolios, Contagion, and Financial Stability." *Journal of Economic Dynamics and Control* 51:50–63. https://doi.org/10.1016/j.jedc.2014.09.041.

Caccioli, F., G. Ferrara, and A. Ramadiah. 2024. "Modelling Fire Sale Contagion across Banks and Non-Banks." *Journal of Financial Stability* 71:101231. https://doi.org/10.1016/j.jfs.2024.101231.

Caccioli, F., M. Marsili, and P. Vivo. 2009. "Eroding Market Stability by Proliferation of Financial Instruments." *The European Physical Journal B* 71:467–479. https://doi.org/10.48550/arXiv.0910.0064.

Caccioli, F., M. Shrestha, C. Moore, and J. D. Farmer. 2014. "Stability Analysis of Financial Contagion Due to Overlapping Portfolios." *Journal of Banking & Finance* 46:233–245. https://doi.org/10.1016/j.jbankfin.2014.05.021.

Caceres-Santos, J., A. Rodriguez-Martinez, F. Caccioli, and S. Martinez-Jaramillo. 2020. "Systemic Risk and Other Interdependencies Among Banks in Bolivia." *Latin American Journal of Central Banking* 1 (1-4): 100015. https://doi.org/10.1016/j.latcb.2020.100015.

Challet, D., A. Chessa, M. Marsili, and Y.-C. Zhang. 2001. "From Minority Games to Real Markets." *Quantitative Finance* 1 (1): 168–176. https://doi.org/10.1080/713665543.

Challet, D., and Y.-C. Zhang. 1997. "Emergence of Cooperation and Organization in an Evolutionary Game." *Physica A: Statistical Mechanics and its Applications* 246 (3-4): 407–418. https://doi.org/10.1016/S0378-4371(97)00419-6.

Cifuentes, R., G. Ferrucci, and H. S. Shin. 2005. "Liquidity Risk and Contagion." *Journal of the European Economic Association* 3 (2-3): 556–566. https://www.jstor.org/stable/40004998.

Cimini, G., and M. Serri. 2016. "Entangling Credit and Funding Shocks in Interbank Markets." *PLoS One* 11 (8): e0161642. https://doi.org/10.1371/journal.pone.0161642.

Cont, R., A. Moussa, and E. B. Santos. 2013. "Network Structure and Systemic Risk in Banking Systems." Chap. 5 in *Handbook on Systemic Risk.* Cambridge, UK: Cambridge University Press. https://doi.org/10.1017/CBO9781139151184.018.

Cont, R., and E. Schaanning. 2017. *Fire Sales, Indirect Contagion and Systemic Stress Testing.* https://doi.org/10.2139/ssrn.2541114.

———. 2019. "Monitoring Indirect Contagion." *Journal of Banking & Finance* 104:85–102. https://doi.org/10.1016/j.jbankfin.2019.04.007.

Corsi, F., S. Marmi, and F. Lillo. 2016. "When Micro Prudence Increases Macro Risk: The Destabilizing Effects of Financial Innovation, Leverage, and Diversification." *Operations Research* 64 (5): 1073–1088. https://doi.org/10.1287/opre.2015.1464.

Craig, B., and G. von Peter. 2014. "Interbank Tiering and Money Center Banks." *Journal of Financial Intermediation* 23 (3): 322–347. https://doi.org/10.1016/j.jfi.2014.02.003.

Cuba, W., A. Rodriguez-Martinez, D. A. Chavez, F. Caccioli, and S. Martinez-Jaramillo. 2021. "A Network Characterization of the Interbank Exposures in Peru." *Latin American Journal of Central Banking* 2 (3): 100035. https://doi.org/10.1016/j.latcb.2021.100035.

De Masi, G., G. Iori, and G. Caldarelli. 2006. "Fitness Model for the Italian Interbank Money Market." *Physical Review E* 74 (6): 066112. https://doi.org/10.1103/PhysRevE.74.066112.

Degryse, H., and G. Nguyen. 2004. *Interbank Exposures: An Empirical Examination of Systemic Risk in Belgium Banking System.* Working Paper 43. National Bank of Belgium. https://doi.org/10.2139/ssrn.1691645.

Delli Gatti, D., S. Desiderio, E. Gaffeo, P. Cirillo, and M. Gallegati. 2011. *Macroeconomics from the Bottom-Up.* Milan, Italy: Springer Science & Business Media.

Duarte, F., and T. M. Eisenbach. 2021. "Fire-Sale Spillovers and Systemic Risk." *The Journal of Finance* 76 (3): 1251–1294. https://doi.org/10.1111/jofi.13010.

Dunne, J. A. 2006. "The Network Structure of Food Webs." In *Ecological Networks: Linking Structure to Dynamics in Food Webs,* 27–86. Oxford, UK: Oxford University Press.

Faloutsos, M. M., P. Faloutsos, and C. Faloutsos. 1999. "On Power-law Relationships of the Internet Topology." *ACM SIGCOMM Computer Communication Review* 29 (4): 251–262. https://doi.org/10.1145/316194.316229.

Farmer, J. D., A. M. Kleinnijenhuis, P. Nahai-Williamson, and T. Wetzer. 2020. *Foundations of System-Wide Financial Stress Testing with Heterogeneous Institutions.* Technical report 861. Bank of England.

Fennell, P. G., and J. P. Gleeson. 2019. "Multistate Dynamical Processes on Networks: Analysis Through Degree-Based Approximation Frameworks." *SIAM Review* 61 (1): 92–118. https://doi.org/10.1137/16M1109345.

Fricke, C., and D. Fricke. 2021. "Vulnerable Asset Management? The Case of Mutual Funds." *Journal of Financial Stability* 52:100800. https://doi.org/10.1016/j.jfs.2020.100800.

Fricke, D., and T. Lux. 2015. "Core–Periphery Structure in the Overnight Money Market: Evidence from the e-MID Trading Platform." *Computational Economics* 45:359–395. https://doi.org/10.1007/s10614-014-9427-x.

Fricke, D., and H. Wilke. 2023. "Connected Funds." *The Review of Financial Studies* 36 (11): 4546–4587. https://doi.org/10.1093/rfs/hhad030.

Furfine, C. H. 2003. "Interbank Exposures: Quantifying the Risk of Contagion." *Journal of Money, Credit and Banking,* 111–128. https://www.jstor.org/stable/3649847.

Gai, P, and S. Kapadia. 2010. "Contagion in Financial Networks." *Proceedings of the Royal Society A: Mathematical, Physical and Engineering Sciences* 466 (2120): 2401–2423. https://doi.org/10.1098/rspa.2009.0410.

Geanakoplos, J., R. Axtell, D. J. Farmer, P. Howitt, B. Conlee, J. Goldstein, M. Hendrey, N. M. Palmer, and Yang C.-Y. 2012. "Getting at Systemic Risk via an Agent-Based Model of the Housing Market." *American Economic Review* 102 (3): 53–58. https://doi.org/10.2139/ssrn.2018375.

Glasserman, P., and H. P. Young. 2016. "Contagion in Financial Networks." *Journal of Economic Literature* 54 (3): 779–831. https://doi.org/10.1257/jel.20151228.

Greenwood, R., A. Landier, and D. Thesmar. 2015. "Vulnerable Banks." *Journal of Financial Economics* 115 (3): 471–485. https://doi.org/10.1016/j.jfineco.2014.11.006.

Gualdi, S., J.-P. Bouchaud, G. Cencetti, M. Tarzia, and F. Zamponi. 2015a. "Endogenous Crisis Waves: Stochastic Model with Synchronized Collective Behavior." *Physical Review Letters* 114 (8): 088701. https://doi.org/10.1103/PhysRevLett.114.088701.

Gualdi, S., M. Tarzia, F. Zamponi, and J.-P. Bouchaud. 2015b. "Tipping Points in Macroeconomic Agent-based Models." *Journal of Economic Dynamics and Control* 50:29–61. https://doi.org/10.1016/j.jedc.2014.08.003.

Guimerà, R., S. Mossa, A. Turtschi, and L. A. N. Amaral. 2005. "The Worldwide Air Transportation Network: Anomalous Centrality, Community Structure, and Cities' Global Roles." *Proceedings of the National Academy of Sciences* 102 (22): 7794–7799. https://doi.org/10.1073/pnas.0407994102.

Haldane, A. G. 2013. "Rethinking the Financial Network." In *Fragile Stabilität - - stabile Fragilität,* edited by Stephan A. Jansen, Eckhard Schröter, and Nico Stehr, 243–278. Wiesbaden, Germany: Springer Fachmedien Wiesbaden.

Haldane, A. G., and R. M. May. 2011. "Systemic Risk in Banking Ecosystems." *Nature* 469 (7330): 351–355. https://doi.org/10.1038/nature09659.

Hommes, C., Mario He, S. Poledna, M. Siqueira, and Y. Zhang. 2022. *CANVAS: A Canadian Behavioral Agent-based Model.* Technical report. Bank of Canada. https://doi.org/10.1016/j.jedc.2024.104986.

Huang, X., I. Vodenska, S. Havlin, and H. E. Stanley. 2013. "Cascading Failures in Bi-Partite Graphs: Model for Systemic Risk Propagation." *Scientific Reports* 3 (1): 1219. https://doi.org/10.1038/srep01219.

Hurd, T. R., J. P. Gleeson, and S. Melnik. 2017. "A Framework for Analyzing Contagion in Assortative Banking Networks." *PloS One* 12 (2): e0170579. https://doi.org/10.1371/journal.pone.0170579.

Iazzetta, C., and M. Manna. 2009. *The Topology of the Interbank Market: Developments in Italy Since 1990.* Temi di Discussione (Working Paper) 711. Bank of Italy. https://doi.org/10.2139/ssrn.1478472.

Ikeda, Y., T. Hasegawa, and K. Nemoto. 2010. "Cascade Dynamics on Clustered Network." *Journal of Physics: Conference Series* 221:012005. https://doi.org/10.1088/1742-6596/221/1/012005.

Jeong, H., B. Tombor, R. Albert, Z. N. Oltvai, and A.-L. Barabási. 2000. "The Large-Scale Organization of Metabolic Networks." *Nature* 407 (6804): 651–654. https://doi.org/10.1038/35036627.

Landaberry, V., F. Caccioli, A. Rodriguez-Martinez, A. Baron, S. Martinez-Jaramillo, and R. Lluberas. 2021. "The Contribution of the Intra-Firm Exposures Network to Systemic Risk." *Latin American Journal of Central Banking* 2 (2). https://doi.org/10.1016/j.latcb.2021.100032.

Larsen, P. T. 2007. "Goldman Pays the Price of Being Big." Published on August 13, 2007, *Financial Times,* https://www.ft.com/content/d2121cb6-49cb-11dc-9ffe-0000779fd2ac.

León, C., C. Martínez, and F. Cepeda. 2019. "Short-Term Liquidity Contagion in the Interbank Market." *Cuadernos de Economía* 38 (76): 51–79.

Liljeros, F., C. R. Edling, L. A. N. Amaral, H. E. Stanley, and Y. Åberg. 2001. "The Web of Human Sexual Contacts." *Nature* 411 (6840): 907–908. https://doi.org/10.1038/35082140.

Lublóy, Á. 2005. "Domino Effect in the Hungarian Interbank Market." *Hungarian Economic Review* 52 (4): 377–401.

Marsili, M. 2014. "Complexity and Financial Stability in a Large Random Economy." *Quantitative Finance* 14 (9): 1663–1675. https://doi.org/10.1080/14697688.2013.765061.

May, R. M. 1972. "Will a Large Complex System be Stable?" *Nature* 238 (5364): 413–414. https://doi.org/10.1038/238413a0.

May, R. M., and N. Arinaminpathy. 2010. "Systemic Risk: The Dynamics of Model Banking Systems." *Journal of the Royal Society Interface* 7 (46): 823–838. https://doi.org/10.1098/rsif.2009.0359.

Mistrulli, P. E. 2011. "Assessing Financial Contagion in the Interbank Market: Maximum Entropy Versus Observed Interbank Lending Patterns." *Journal of Banking & Finance* 35 (5): 1114–1127. https://doi.org/10.1016/j.jbankfin.2010.09.018.

Monasterolo, I. 2020. "Climate Change and the Financial System." *Annual Review of Resource Economics* 12 (1): 299–320. https://doi.org/10.1146/annurev-resource-110119-031134.

Nier, E., J. Yang, T. Yorulmazer, and A. Alentorn. 2007. "Network Models and Financial Stability." *Journal of Economic Dynamics and Control* 31 (6): 2033–2060. https://doi.org/10.1016/j.jedc.2007.01.014.

Onaga, T., F. Caccioli, and T. Kobayashi. 2023. "Financial Fire Sales as Continuous-State Complex Contagion." *Physical Review Research* 5 (4): 043123. https://doi.org/10.1103/PhysRevResearch.5.043123.

Payne, J. L., P. S. Dodds, and M. J. Eppstein. 2009. "Information Cascades on Degree-Correlated Random Networks." *Physical Review E* 80 (2): 026125. https://doi.org/10.1103/PhysRevE.80.026125.

Poledna, S., S. Martínez-Jaramillo, F. Caccioli, and S. Thurner. 2021. "Quantification of Systemic Risk from Overlapping Portfolios in the Financial System." *Journal of Financial Stability* 52:100808. https://doi.org/10.1016/j.jfs.2020.100808.

Poledna, S., M. G. Miess, C. Hommes, and K. Rabitsch. 2023. "Economic Forecasting with an Agent-based Model." *European Economic Review* 151:104306. https://doi.org/10.1016/j.euroecorev.2022.104306.

Poledna, S., J. L. Molina-Borboa, S. Martínez-Jaramillo, M. van der Leij, and S. Thurner. 2015. "The Multi-Layer Network Nature of Systemic Risk and Its Implications for the Costs of Financial Crises." *Journal of Financial Stability* 20:70–81. https://doi.org/10.1016/j.jfs.2015.08.001.

Poledna, S., and S. Thurner. 2016. "Elimination of Systemic Risk in Financial Networks by Means of a Systemic Risk Transaction Tax." *Quantitative Finance* 16 (10): 1599–1613. https://doi.org/10.1080/14697688.2016.1156146.

Poledna, S., S. Thurner, J. D. Farmer, and J. Geanakoplos. 2014. "Leverage-Induced Systemic Risk Under Basle II and Other Credit Risk Policies." *Journal of Banking & Finance* 42:199–212. https://doi.org/10.1016/j.jbankfin.2014.01.038.

Silva, T. C., B. M. Tabak, and Solange Maria Guerra. 2017. "Why Do Vulnerability Cycles Matter in Financial Networks?" *Physica A: Statistical Mechanics and its Applications* 471:592–606. https://doi.org/10.1016/j.physa.2016.12.063.

Soramäki, K., M. L. Bech, J. Arnold, R. J. Glass, and W. E. Beyeler. 2007. "The Topology of Interbank Payment Flows." *Physica A: Statistical Mechanics and its Applications* 379 (1): 317–333. https://doi.org/10.1016/j.physa.2006.11.093.

Squartini, T., G. Caldarelli, G. Cimini, A. Gabrielli, and D. Garlaschelli. 2018. "Reconstruction Methods for Networks: The Case of Economic and Financial Systems." *Physics Reports* 757:1–47. https://doi.org/10.1016/j.physrep.2018.06.008.

Tedeschi, G., A. Mazloumian, M. Gallegati, and D. Helbing. 2012. "Bankruptcy Cascades in Interbank Markets." *PLoS One* 7 (12): e52749. https://doi.org/10.1371/journal.pone.0052749.

Thurner, S., J. D. Farmer, and J. Geanakoplos. 2012. "Leverage Causes Fat Tails and Clustered Volatility." *Quantitative Finance* 12 (5): 695–707. https://doi.org/10.48550/arXiv.0908.1555.

Thurner, S., and S. Poledna. 2013. "DebtRank-Transparency: Controlling Systemic Risk in Financial Networks." *Scientific Reports* 3 (1): 1888. https://doi.org/10.1038/srep01888.

Trichet, J.-C. 2011. "Reflections on the Nature of Monetary Policy Non-Standard Measures and Finance Theory." In *Approaches to Monetary Policy Revisited: Lessons from the Crisis; Sixth ECB Central Banking Conference, 18–19 November 2010.* Frankfurt, Germany: European Central Bank.

Upper, C., and A. Worms. 2004. "Estimating Bilateral Exposures in the German Interbank Market: Is There a Danger of Contagion?" *European Economic Review* 48 (4): 827–849. https://doi.org/10.1016/j.euroecorev.2003.12.009.

van Lelyveld, I., and F. Liedorp. 2006. "Interbank Contagion in the Dutch Banking Sector: A Sensitivity Analysis." *International Journal of Central Banking*, https://EconPapers.repec.org/RePEc:ijc:ijcjou:y:2006:q:2:a:4.

Watts, D. J. 2002. "A Simple Model of Global Cascades on Random Networks." *Proceedings of the National Academy of Sciences* 99 (9): 5766–5771. https://doi.org/10.1073/pnas.082090499.

Wells, S. J. 2004. *Financial Interlinkages in the United Kingdom's Interbank Market and the Risk of Contagion*. Working Paper No. 230. Bank of England. https://www.bankofengland.co.uk/working-paper/2004/financial-interlinkages-in-the-uks-interbank-market.

Wiersema, G., A. M. Kleinnijenhuis, T. Wetzer, and J. D. Farmer. 2023. "Scenario-Free Analysis of Financial Stability with Interacting Contagion Channels." *Journal of Banking & Finance* 146:106684. https://doi.org/10.1016/j.jbankfin.2022.106684.

Protektorat Böhmen und Mähren
Protektorat Čechy a Morava
THE JAPANESE GOVERNMENT
TEN DOLLARS
MP
REICHSBANKNOTE
2 MILLIONEN MARK
Eine Krone
JEDNA KORUNA
10 sen in silver.
ISLA DE CUBA

THE ART IN THESE VOLUMES

Both volumes of *The Economy as an Evolving Complex System IV* feature public-domain imagery of numerous forms of currency spanning time and space, an acknowledgment of how technological developments and innovation impact economies and societal norms. All images are out of copyright and were sourced via The New York Public Library, the Internet Archive, and Wikimedia Commons.

VOL. II
TABLE OF CONTENTS

Part V: Inequality, Labor & Structural Resilience

Part VI: Innovation and Technological Disruption

Part VII: Political Economy and Public Policy

CONTRIBUTORS TO THESE VOLUMES

Joos Akkerman, *Delft University of Technology*

Pia Andres, *Durham University and Centre for Economic Performance*

W. Brian Arthur, *Santa Fe Institute and SRI International*

Robert Axtell, *George Mason University and Santa Fe Institute*

Stefano Battiston, *University of Zurich, Ca' Foscari, University of Venice, and CEPR*

Paul Beaudry, *University of British Columbia and NBER*

Jenna Bednar, *University of Michigan and Santa Fe Institute*

Eric D. Beinhocker, *University of Oxford and Santa Fe Institute*

András Borsos, *Magyar Nemzeti Bank, Complexity Science Hub, and University of Oxford*

Jean-Philippe Bouchaud, *Capital Fund Management and Académie des Sciences*

William Brock, *University of Wisconsin, Madison, and University of Missouri, Columbia*

Fabio Caccioli, *University College London and Systemic Risk Centre, London School of Economics and Political Science*

Adrian Carro, *University of Oxford and Banco de España*

Diane Coyle, *University of Cambridge*

Herbert Dawid, *Bielefeld University*

Domenico Delli Gatti, *Università Cattolica del Sacro Cuore*

R. Maria del Rio-Chanona, *University College London*

Giovanni Dosi, *Sant'Anna School of Advanced Studies*

Marion Dumas, *London School of Economics and Political Science*

Salva Duran-Nebreda, *Institute of Evolutionary Biology (CSIC-UPF)*

Steven N. Durlauf, *University of Chicago*

J. Doyne Farmer, *University of Oxford and Santa Fe Institute*

Luca Eduardo Fierro, *International Institute for Applied Systems Analysis*

Tatiana Filatova, *Delft University of Technology*

Morgan R. Frank, *University of Pittsburgh*

Koen Frenken, *Utrecht University*

Dana Galizia, *Carleton University*

John Geanakoplos, *Yale University and Santa Fe Institute*

Aldo Glielmo, *Banca d'Italia*

Omar Guerrero, *University of Helsinki*

Marc Hinterschweiger, *Bank of England*

Cars Hommes, *Bank of Canada, University of Amsterdam, and Tinbergen Institute*

Jagoda Kaszowska-Mojsa, *University of Oxford, Narodowy Bank Polski, and Institute of Economics, Polish Academy of Sciences*

Sharon Kozicki, *Bank of Canada*

François Lafond, *University of Oxford*

Francesco Lamperti, *Sant'Anna School of Advanced Studies, RFF-CMCC European Institute on Economics and the Environment, and Euro-Mediterranean Center on Climate Change (CMCC)*

Rosario N. Mantegna, *Università degli Studi di Palermo and Complexity Science Hub*

David McMillon, *Emory University*

Penny Mealy, *University of Oxford, Santa Fe Institute, and Monash University*

Irene Monasterolo, *Utrecht University, CEPR, and Wirtschaftsuniversität Wien*

José Moran, *Macrocosm Inc., University of Oxford, and Complexity Science Hub*

Esteban Moro, *Northeastern University and Massachusetts Institute of Technology*

Ljubica Nedelkoska, *Complexity Science Hub and Central European University*

Frank Neffke, *Complexity Science Hub*

Scott E. Page, *University of Michigan–Ann Arbor and Santa Fe Institute*

Marco Pangallo, *CENTAI Institute*

Marcelo C. Pereira, *Universidade Estadual de Campinas and Scuola Superiore Sant'Anna*

Anton Pichler, *Vienna University of Economics and Business and Complexity Science Hub*

Sebastian Poledna, *Austrian Institute of Economic Research*

Franck Portier, *University College London and CEPR*

Massimo Riccaboni, *IMT School for Advanced Studies Lucca and Scuola Superiore IUSS*

Matteo Richiardi, *University of Essex*

Andrea Roventini, *Sant'Anna School of Advanced Studies and OFCE Sciences Po*

Angelica Sbardella, *Enrico Fermi Research Center*

Ulrich Schetter, *University of Pavia*

Andrea Tacchella, *Enrico Fermi Research Center*

Arthur Turrell, *Bank of England*

Arzu Uluc, *Bank of England*

Sergi Valverde, *Institute of Evolutionary Biology (CSIC-UPF)*

Justin van de Ven, *University of Essex*

Blai Vidiella, *Theoretical and Experimental Ecology Station (CNRS)*

Maria Enrica Virgillito, *Scuola Superiore Sant'Anna and Universitá Cattolica del Sacro Cuore*

Yang Zhang, *Bank of Canada*

COORDINATING EDITORS

R. MARIA DEL RIO-CHANONA is a lecturer in computer science at University College London. Her research draws from large language models, network science, and agent-based modeling and focuses on the net-zero transition and the impact of new technologies in the economy, with a particular focus on labor markets.

Del Rio-Chanona completed her BSc in physics at UNAM, Mexico and her PhD in mathematics at the University of Oxford, where she was part of the complexity economics group at the Oxford Martin School's Institute for New Economic Thinking. She was a JSMF research fellow at the Complexity Science Hub in Vienna and a visiting scholar at the Harvard Kennedy School. Maria has worked alongside international policy organizations, including the International Monetary Fund, the World Bank, and the International Labour Organization. She is currently a member of the CEPR Artificial Intelligence Research Policy Network.

MARCO PANGALLO is a senior researcher at the Center for Artificial Intelligence (CENTAI), where he leads the complexity economics team. Previously, he was a James S. McDonnell Foundation postdoctoral fellow at the Sant'Anna School of Advanced Studies, Italy. Pangallo obtained his PhD in mathematics at the University of Oxford and was part of the complexity economics group at the Oxford Martin School's Institute for New Economic Thinking.

Pangallo is generally interested in understanding the economy quantitatively through a combination of data-driven and theoretical approaches. He believes that traditional economic models—based on optimization and equilibrium—are not well suited to quantitatively account for the complexity of the economy. Instead, agent-based models are the best tool to assimilate increasingly available granular data and produce more reliable economic forecasts.

EDITORS

JENNA BEDNAR is a professor of political science and public policy at the University of Michigan, a member of the external faculty at the Santa Fe Institute, and serves in the provost's office as the inaugural faculty director of UMICH Votes and Democratic Engagement. She leads a campus-wide collaborative effort to elevate democracy-related research, curriculum, and engagement. Her research focuses on how collective action builds social goods and the role that institutions play in that collaboration. Bednar's current work includes robust system design, especially of federal democracies; the interdependence of norms, culture, and institutions; and place-based public policy to support human social flourishing. Her book *The Robust Federation: Principles of Design* (2009) was awarded the APSA Martha Derthick Best Book Award in recognition of its enduring contribution to the study of federalism. In 2020, she was named the APSA Daniel Elazar Distinguished Federalism Scholar Award. She earned her BA from the University of Michigan and MA and PhD from Stanford University.

ERIC D. BEINHOCKER is a professor of public policy practice at the Blavatnik School of Government, University of Oxford. He is also the founder and executive director of the Institute for New Economic Thinking at the University's Oxford Martin School. INET Oxford is an interdisciplinary research centre dedicated to the goals of creating a more just, sustainable, and prosperous economy. Beinhocker is also a Supernumerary Fellow in economics at Oriel College, Oxford, and an external professor and chairman of the Science Board at the Santa Fe Institute.

Prior to joining Oxford, Beinhocker had an eighteen-year career as a partner at McKinsey & Company, where he held leadership roles in McKinsey's Strategy Practice, its Climate Change and Sustainability Practice, and the McKinsey Global Institute. Beinhocker writes frequently on economic and public-policy issues; his work has appeared in the *Financial Times, The Wall Street Journal, Bloomberg, The Times, The Guardian, The Atlantic,* and *The Washington Post.* He is the author *The Origin of Wealth: The Radical Remaking of Economics and What It Means for Business and Society* (2007). Originally from Boston, Massachusetts, Beinhocker is a graduate of Dartmouth College and the MIT Sloan School.

JAGODA KASZOWSKA-MOJSA is an economic expert at the National Bank of Poland and a Research Associate at the Institute for New Economic Thinking and the Oxford Smith School of Enterprise and the Environment. She also works as a research fellow at the Institute of Economics, Polish Academy of Sciences, and as a lecturer at Cracow University of Economics in Poland. Her research focuses on systemic risk, financial (in)stability, macroprudential policies and regulations, and agent-based modeling.

Kaszowska-Mojsa has degrees in economics from the University of Alcalá and mathematics from Jagiellonian University, as well as a PhD in economics and finance from Cracow University of Economics. Her dissertation was nominated for the Prime Minister's Award in Economics for the best doctoral dissertation, and also won the Central Statistical Office's competition for outstanding doctoral dissertation. As a Fulbright Scholar, she conducted her PhD research in the United States. Her research was funded by the National Bank of Poland, Ministry of Education in Poland, National Science Centre, European Commission (H2020) and Santander Bank.

FRANÇOIS LAFOND is deputy director of the complexity economics group at the Institute for New Economic Thinking, University of Oxford; senior researcher at the Smith School for Enterprise and the Environment; an Oxford Martin fellow; an associate member of Nuffield College, Oxford; and external faculty at the Complexity Science Hub in Vienna.

Lafond's main areas of research are in the economics of innovation and productivity, environmental economics, complex systems, and forecasting. Currently, his research interests lie in the macroeconomics of the net-zero transition, the structure and evolution of production networks, and the future of technology. His research has appeared in economics and interdisciplinary journals, and has been featured in books, news articles, and public-sector reports.

PENNY MEALY is a senior economist at the World Bank, a research associate at the Institute for New Economic Thinking (INET) and the Oxford Smith School of Enterprise and the Environment, an adjunct senior research fellow at SoDa Labs at the Monash Business School, and an external Applied Complexity Fellow at the Santa Fe Institute. Her work applies various methods from complex systems and data science to analyze the interrelated challenges of climate change and economic development. Her research has developed novel, data-driven approaches for analyzing structural change, occupational mobility and the future of work, and the transition to the green economy.

Mealy completed a PhD at INET, University of Oxford. She has held various research fellow roles at the Oxford Martin School, the Oxford Smith School of Enterprise and the Environment, the Bennett Institute for Public Policy at Cambridge University, and SoDa Labs, Monash University. Penny has also frequently advised international organizations, governments, and businesses on green growth and development strategies.

ANTON PICHLER is an assistant professor in supply-chain analytics at the Vienna University of Economics and Business. Previously, he was a James S. McDonnell Foundation postdoctoral fellow at the Complexity Science Hub in Vienna. He holds a PhD in mathematics from the University of Oxford, as well as degrees in quantitative finance, economics, and political science.

His current research focuses on the economics of the energy transition and modeling the impacts from climate-induced disasters. In his research, Pichler builds on and contributes to various quantitative methods including agent-based simulations, mathematical optimization, time-series analysis, machine learning and complex network theory. He has published research papers on topics spanning the propagation of economic shocks in production networks, forecasting technological change, systemic risk in financial networks, and energy supply security.

J. DOYNE FARMER is director of the complexity economics program at the Institute for New Economic Thinking and Baillie Gifford Professor of Complex-Systems Science at the Smith School of Enterprise and the Environment, University of Oxford. He is also an external professor at the Santa Fe Institute and chief scientist at Macrocosm.

Farmer's current research is in economics, including agent-based modeling, financial instability, and technological progress. He was a founder of Prediction Company, a quantitative automated trading firm that was sold to UBS in 2006. His past research includes complex systems, dynamical systems theory, time-series analysis, and theoretical biology. His book, *Making Sense of Chaos: A Better Economics for a Better World*, was published in 2024.

ABOUT THE SANTA FE INSTITUTE

THE SANTA FE INSTITUTE is the world headquarters for complexity science, operated as an independent, nonprofit research and education center located in Santa Fe, New Mexico. Our researchers endeavor to understand and unify the underlying, shared patterns in complex physical, biological, social, cultural, technological, and even possible astrobiological worlds. Our global research network of scholars spans borders, departments, and disciplines, bringing together curious minds steeped in rigorous logical, mathematical, and computational reasoning. As we reveal the unseen mechanisms and processes that shape these evolving worlds, we seek to use this understanding to promote the well-being of humankind and of life on Earth.

SFI PRESS BOARD OF ADVISORS

Sam Bowles
Professor, University of Siena;
SFI Resident Faculty

Sean Carroll
Homewood Professor of
Natural Philosophy, Johns Hopkins
University; SFI Fractal Faculty &
External Faculty

Katherine Collins
Head of Sustainable Investing,
Putnam Investments; SFI Trustee
& Board Chair Emerita

Jennifer Dunne
SFI Resident Faculty &
SFI Vice President for Science

Chris Kempes
SFI Resident Faculty & Chair,
SFI Science Steering Committee

Simon Levin
James S. McDonnell Distinguished
University Professor in Ecology and
Evolutionary Biology at Princeton
University; SFI External Faculty

Ian McKinnon
Founding Partner, Sandia
Holdings LLC; SFI Trustee
& Chair of the Board

John H. Miller
Professor, Carnegie Mellon
University; SFI External Faculty

William H. Miller
Chairman & CEO, Miller
Value Partners; SFI Trustee
& Board Chair Emeritus

Melanie Mitchell
SFI Davis Professor of Complexity

Cristopher Moore
SFI Resident Faculty

Samuel Peters
Manager, ClearBridge Value Trust;
SFI Trustee, Vice-Chair of the Board

Sidney Redner
SFI Resident Faculty

Geoffrey West
SFI Resident Faculty,
Toby Shannan Professor of
Complex Systems & Past President

David Wolpert
SFI Resident Faculty

Andrea Wulf
SFI Miller Scholar

EDITORIAL

David C. Krakauer
Publisher/Editor-in-Chief

Zato Hebbert
Production Coordinator

Sienna Latham
Managing Editor

Ellis B. Wylie
Post-Production Coordinator

*With appreciation to Rachel Fudge for copyediting these volumes,
and to Nicholas Graham for design assistance*

SFI PRESS
THE SANTA FE INSTITUTE PRESS

The SFI Press endeavors to communicate the best of complexity science and to capture a sense of the diversity, range, breadth, excitement, and ambition of research at the Santa Fe Institute.

To provide a distillation of work at the frontiers of complex-systems science across a range of influential and nascent topics.

To change the way we think.

SEMINAR SERIES
New findings emerging from the Institute's ongoing working groups and research projects, for an audience of interdisciplinary scholars and practitioners.

ARCHIVE SERIES
Fresh editions of classic texts from the complexity canon, spanning the Institute's four decades of advancing the field.

COMPASS SERIES
Provocative, exploratory volumes aiming to build complexity literacy in the humanities, industry, and the curious public.

SCHOLARS SERIES
Affordable and accessible textbooks and monographs disseminating the latest findings in the complex-systems-science world.

— ALSO FROM SFI PRESS —

Complexity Economics:
Proceedings of the Santa Fe Institute's 2019 Fall Symposium
W. Brian Arthur, Eric D. Beinhocker, and Alison Stanger, eds.

Foundational Papers in Complexity Science
David C. Krakauer, ed.

For additional titles, inquiries, or news about the Press, visit us at
www.sfipress.org

COLOPHON

The body copy for this book was set in EB Garamond, a typeface designed by Georg Duffner after the Ebenolff-Berner type specimen of 1592. Headings are in Kurier, created by Janusz M. Nowacki, based on typefaces by the Polish typographer Małgorzata Budyta. For footnotes and captions, we have used CMU Bright, a sans serif variant of Computer Modern, created by Donald Knuth for use in TeX, the typesetting program he developed in 1978. Additional type is set in Cochin, a typeface based on the engravings of Nicolas Cochin, for whom the typeface is named

The SFI Press complexity glyphs used throughout this book were designed by Brian Crandall Williams.

SANTA FE INSTITUTE
COMPLEXITY
GLYPHS

ZERO

ONE

TWO

THREE

FOUR

FIVE

SIX

SEVEN

EIGHT

NINE

-A-

-B-

-C-

-D-

-E-

-F-

-G-

-H-

-I-

-J-

-K-

-L-

-M-

-N-

-O-

-P-

-Q-

-R-

-S-

-T-

-U-

-V-

-W-

-X-

-Y-

-Z-

SFI PRESS

SEMINAR SERIES

www.ingramcontent.com/pod-product-compliance
Lightning Source LLC
Chambersburg PA
CBHW031427180326
41458CB00002B/482